ENCYCLOPEDIA OF GENOMICS

ENCYCLOPEDIA OF GENOMICS

Vol. 1

GENOMIC DISEASES

By
Dr. Shiv Kant Prasad
School of Studies of Zoology & Biotechnology
Vikram University
Ujjain
(India)

DPH

DISCOVERY PUBLISHING HOUSE PVT. LTD.
NEW DELHI-110 002

Published by:
Tilak Wasan
DISCOVERY PUBLISHING HOUSE PVT. LTD.
4383/4A, Ansari Road, Darya Ganj
New Delhi-110 002 (India)
Phone : +91-11-23279245, 43596064-65
Fax : +91-11-23253475
E-mail : parul.wasan@gmail.com
discoverypublishinghouse@gmail.com
web : www.discoverypublishinggroup.com

***First Edition:* 2012**

ISBN: 978-93-5056-032-7 (Set)
978-93-5056-033-4 (Vol. 1)
978-93-5056-034-1 (Vol. 2)
978-93-5056-035-8 (Vol. 3)
978-93-5056-036-5 (Vol. 4)
978-93-5056-037-2 (Vol. 5)

Encyclopedia of Genomics

Printed at:
Shree Balaji Art Press
Delhi

Preface

Progress in any scientific discipline is dependent on the availability of techniques and methods that extend the range and sophistication of experiments which may be performed. Over the last 30 years or so this has been demonstrated in a spectacular way by the emergence of genetic engineering. This field has grown rapidly to the point where, in many laboratories around the world, it is now routine practice to isolate a specific DNA fragment from the genome of an organism, determine its base sequence, and assess its function. The technology is also now used in many other applications, including pharmaceutical analysis of samples, paternity disputes, medical diagnosis, genomic and the biotechnology industry.

The present title **"Encyclopedia of Genomics"** is a fast growing area of research for undergraduate and post-graduate students in life sciences, chemistry, engineering and medicine. The overall objective of this publication is to provide a professional level reference work with comprehensive coverage of the molecular basis of genome and the application of that knowledge in genetics, evolution, medicine, and agriculture. It deals with the life processes at a molecular level genetic disease diagnosis and genetic therapy; the theory and techniques for understanding manipulating, and synthesizing biological molecules and their aggregates; and the application of biological processes to make or modify products to improve plants or animals or to develop microorganisms for specific uses.

The development of techniques for genomic science is based on an understanding of the molecular aspect of genomic processes occurring in nature. These techniques have, in turn, enabled expansion of our knowledge and understanding of how genes are organized and expressed in living cells and has provided major impetus to the development of biotechnological enterprises.

The present title has been written to provide readers with a thorough understanding of the techniques which underpin the developments in the field of genomic science. By gaining this understanding, readers will be able to apply the techniques of genomic manipulation to tackle specific problems or to achieve desired objectives in a wide variety of fields.

There can be no claim to originality except in the manner or treatment and much of the information has been obtained from the books and scientific journals available in the different libraries.

The author express his thanks to his friends and colleagues whose continue inspirations have initiated him to bring out this book.

The author expresses her gratitude to Mr. Wasan and staff of M/s Discovery Publishing House Pvt. Ltd. for their whole hearted co-operation in the publication of this book.

S.K. Prasad

Contents

Chapter 1

Finding Genes

National Center for Biotechnology Information (NCBI) provides several focused *bioinformatics* mini-courses (http://www.ncbi.nlm.nih.gov/Class/ minicourses/). The mini-courses are either problem-based such as "Identification of Disease Genes" or NCBI resource-based such as "BLAST Quick Start."

The courses are 2.5 h in length with the first 90 min devoted to an overview and an online demonstration of a problem or problem set by an *instructor*. This is followed by a 1-h hands-on session where students practice a similar problem or problem set to the one demonstrated at their own computers. This chapter describes the mini-course that *focuses* on the identification of a disease gene using NCBI's human genome assembly.

The reference human genome assembly along with *integrated* maps, literature, and *expression* information comprises a powerful discovery system for *exploring* candidate human disease genes.

MINIMAL INFORMATION REQUIRED

Within a Cell

The pathway of genetic information transfer in a cell begins with the transcription of genes within a genome to produce mRNAs and ends with the translation of mRNAs to produce proteins.

The sequence databases contain genomic, mRNA, and protein sequences representing all three stages in the pathway.

DNA
Transcription ↓
mRNA
Translation ↓
Protein

Data for Analysis

One way to identify genes in a genome is to generate a cDNA library from the pool of RNA messages. To generate a cDNA library, the RNA messages from a tissue or from cells representing a developmental stage are copied into more stable cDNA molecules, which are then placed into an appropriate vector to generate a collection of cDNA clones (vector and the individual cDNA insert).

The single pass, short 300-500 nucleotide sequences obtained from sequencing either end of the cDNA insert are called expressed sequence tags (ESTs). One way to solve the problem of identifying genes responsible for a particular *phenotype* is to generate a cDNA library from patient tissues/samples and obtain a number of ESTs.

Then, use the ESTs to determine the genes expressing them and to determine whether they contain any *nucleotide* variations or single *nucleotide polymorphisms* (SNPs) when compared to normal individuals. Sites of DNA sequences where individuals differ at a single nucleotide are called SNPs. We will obtain more information about the SNP database in the latter part of this chapter.

METHODOLOGY

Essential Steps

1. Compare the sequences of ESTs from a patient to the sequences of the human genome (using Basic Local Alignment Search Tool [BLAST]).
2. Identify the genes aligning to the ESTs and download their sequences (using Map Viewer).
3. Identify whether the EST sequences contain any known SNPs (using dbSNP).
4. Determine whether a gene variant is known to cause a *phenotype* (using Online Mendelian Inheritance in Man [OMIM]).

Thus, starting from the transcribed sequences derived from patients, we will obtain information about expressed genes and determine whether these genes contain known variations that lead to the disease *phenotype*.

Resources Used

NCBI assembles component sequences from the human genome sequencing project into longer sequences called contigs whose accession numbers begin with prefix "NT ". NCBI also performs a number of *annotations* on the assembly to identify *genes, transcripts, clones, repeats, markers,* and SNPs. NCBI releases the updated human genome assembly or the new "Build" periodically.

BLAST

BLAST provides a method for rapid searching of nucleotide and protein databases for similarities with a query nucleotide or protein sequence. The human genome BLAST page at provides centralized access to the NCBI human genome assembly and annotated transcript and protein sequences.

The BLAST output links directly to the Human Genome Map Viewer, where database hits can.be analyzed in their genomic context to see the relationship with other annotated features.

Map Viewer

The Map Viewer allows us to view and search an organism's complete genome. It shows integrated views of a collection of genetic, physical, and sequence maps for annotated genes, expressed sequences, SNPs, and other features, and, thus, is a valuable tool for the identification and localization of genes that contribute to human disease (as demonstrated in this mini-course).

dbSNP

NCBI's SNP database contains both single nucleotide substitutions, and short deletion and insertions. The data in dbSNP are integrated with other NCBI genomic data. SNPs are aligned to the human genome and the locations of SNPs with respect to the annotated genes and mRNAs are identified.

OMIM

OMIM is the database of human genes and genetic disorders developed and edited by Dr. Victor A. McKusick and his colleagues at Johns Hopkins and elsewhere, and adapted for the Internet by NCBI.

PROCEDURE

In this problem, we will use as an example the hemochromatosis disease, which is characterized by an iron overload. Consider that a researcher is working on the *hemochromatosis* disease and needs to obtain information about the gene(s) causing the phenotype.

The following steps will describe the analysis of EST sequences that might have been obtained from a *hemochromatosis* patient. It is recommended to follow the link to the "Identification of Disease Genes" through the mini-course webpage. This page contains a link to a file containing the up-to-date screen images of each of the steps described below.

Referring to the file is strongly recommended to follow the mini-course steps. However, a number of screen images are provided in this chapter as well for a reader to follow along. These screen images are from human Build 35.1.

Stage I

One way to identify the genes expressing the ESTs is to compare their sequences using BLAST with the human genome assembly and the genes annotated on it. The specialized BLAST page for searching against the annotated human genome assembly is at.

The user may directly access the human genome BLAST page through the "Identification of Disease Genes" mini-course web page by clicking on the "BLAST (human genome)" link. We can concatenate a number of EST sequences to run the search as a batch. However, we will use only one EST sequence as a query for this *analysis*.

Paste the EST sequence provided on the mini-course page in the query box of the BLAST page and select the "genome (reference only)" database from the pull down menu and use the default

```
>ref NT _007592.14 Hs6_7749     Home sapiens chromosome 6 genomic contig
          Lenght=48945890

 Features in this part of subject sequence:
   hemochromatosis protein isoform 11 precursor
   hemochromatosis protein isoform 10 precursor

 Score = 525 bites  (273) , Expect = 4e-147
 Identitles = 275/276  (99) , Gaps = 0/276 ()
 Strand=Plus/Plus

Query   1         TGCCTCCTTTGGTGAAGGTGACACATCATGTGACCTCTTCAGTGACCACTCTACGGTGTC  60
                  ||||||||||||||||||||||||||||||||||||||||||||||||||||||||||||
Subjct  16951164  TGCCTCCTTTGGTGAAGGTGACACATCATGTGACCTCTTCAGTGACCACTCTACGGTGTC  16951223

Query   61        GGGCCTTGAACTACTACCCCCAGAACATCACCATGAAGTGGCTGAAGGATAAGCAGCCAA  120
                  ||||||||||||||||||||||||||||||||||||||||||||||||||||||||||||
Subjct  16951224  GGGCCTTGAACTACTACCCCCAGAACATCACCATGAAGTGGCTGAAGGATAAGCAGCCAA  16951283

Query   121       TGGATGCCAAGGAGTTCGAACCTAAAGACGTATTGCCCAATGGGGATGGGACCTACCAGG  180
                  ||||||||||||||||||||||||||||||||||||||||||||||||||||||||||||
Subjct  16951284  TGGATGCCAAGGAGTTCGAACCTAAAGACGTATTGCCCAATGGGGATGGGACCTACCAGG  16951343

Query   181       GCTGGATAACCTTGGCTGTACCCCCTGGGAAGAGCAGAGAATACGTACCAGGTGGAGC  240
                  ||||||||||||||||||||||||||||||||||||||||||||||||||||||||||||
Subjct  16951344  GCTGGATAACCTTGGCTGTACCCCCTGGGAAGAGCAGAGAATACGTACCAGGTGGAGC  16951400

Query   241       ACCCAGGCCTGGATCAGCCCCTCATTGTGATCTGGG  276
                  ||||||||||||||||||||||||||||||||||||||||||||||||||||||||||||
Subjct  16951404  ACCCAGGCCTGGATCAGCCCCTCATTGTGATCTGGG  16951439
```

Figure 1.1: MegaBlast of the query EST against the human genome: alignment overview. The difference between the EST and genomic sequence (a G to A variation at the nucleotide 16951392 of the contig NT 007592.14) is highlighted by a rectangle.

program MegaBlast. Start the search by clicking on the "*Begin Search*" button and obtain the results by clicking on the "*Format*" button.

The BLAST results page shows only one match to the contig sequence NT 007592.14 on chromosome 6 in the human genome Build 35.1. In certain cases, there may be multiple matches to the human genome assembly. The alignment of the query EST sequence (indicated by "query") and the matched sequence from chromosome 6 (indicated by "sbjct") shows that the EST sequence is only 99 % identical to the genomic sequence.

Note the location of the nucleotide that is different between the two sequences (a G to A variation at the nucleotide 16951392 of the contig NT 007592.14). The difference may be due to a sequencing error in the low quality EST sequence or it may represent a real SNP in the human genome.

For future reference, paste your results, such as the alignment and the nucleotide difference (sequence difference at the nucleotide 16951392 on NT 007592.14; G in the genomic and A in the query EST sequence), in the window provided in the mini-course webpage.

Stage II

We will now take advantage of the NCBI annotation of the human genome assembly to identify

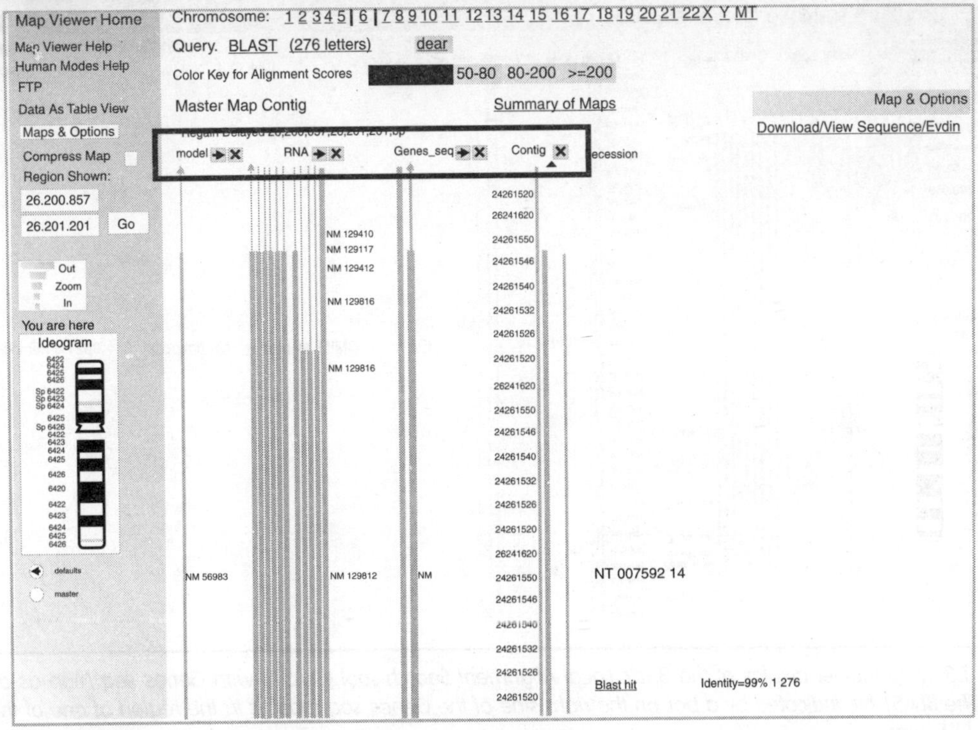

Figure 1.2: Map Viewer display of the Basic Local Alignment Search Tool. The four maps displayed in this view, Model, RNA, Gene-seq, Contig, are highlighted by a rectangle.

the gene corresponding to the EST by using the Map Viewer. To visualize the BLAST hit on the genome using Map Viewer, click the "Genome View" button at the top of the BLAST results page, then on the Map element "NT 007592." Currently, four maps should be displayed.

The Genes_seq map shows the "known" genes annotated by alignment of EST and/or mRNA sequences to the assembly. The Contig map shows the assembled genome contig sequence in the region, the Model map shows the Ab *initio* model genes predicted by the NCBI's program Gnomon and the RNA map shows the alignments of the known alternatively spliced transcripts. Make the Genes_seq map the master map by clicking on the arrow at its top.

The BLAST hit, indicated by the red bar, is in the region of one of the exons of the *HFE* gene annotated on the human genome. The thick bars in the Genes seq map indicate the exons and the thin lines joining them indicate introns of the gene. Zoom out several times until the user sees the entire *HFE* gene structure by clicking on the gray line and selecting option "Zoom out 2 times" from the menu that appears.

The query EST represents a known gene, HFE. The orientation of the arrow next to the gene link indicates the orientation of the gene on the forward or the reverse strand. A gene *annotated* on the forward strand is indicated by an arrow pointing downward whereas a gene annotated on the reverse strand is indicated by an arrow pointing upward.

The HFE gene is annotated on the forward strand of chromosome 6. The right-most map is called the master map and links that give more information about map elements are provided

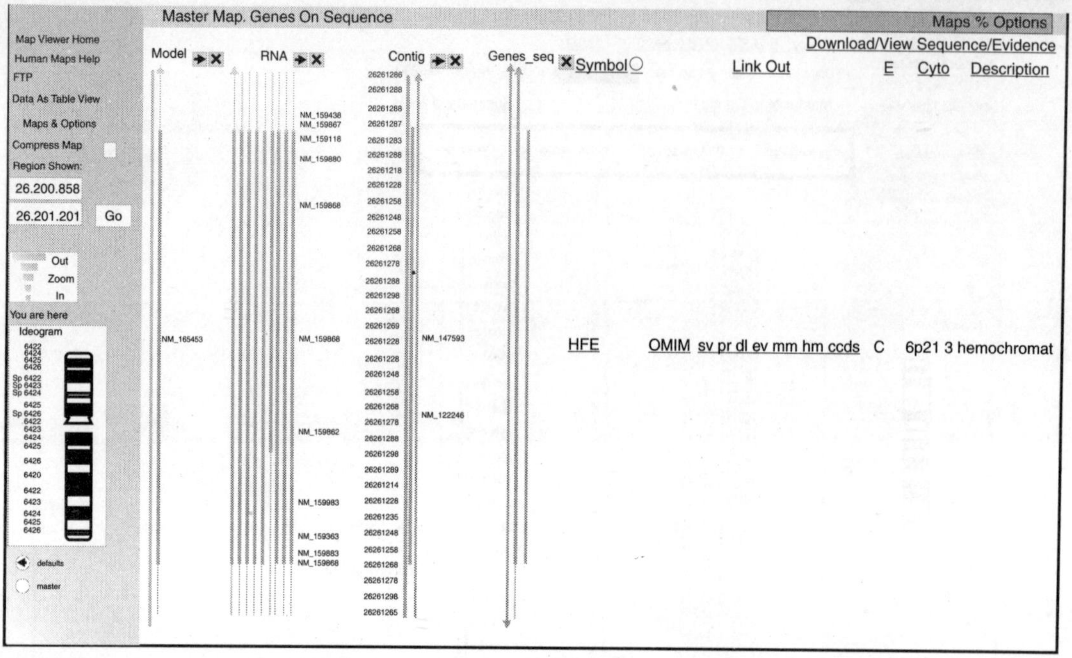

Figure 1.3: Map Viewer display of the Basic Local Alignment Search Tool (BLAST). with Genes seq Map as a master map. The BLAST hit, indicated by a bar on the right side of the Genes seq map, is in the region of one of the exons of the HFE *gene.*

next to it. For example, the current master map, Genes seq map, has links to resources that provide more information about the *HFE* gene such as OMIM, sv (Sequence Viewer), pr (Reference Proteins), dl (Download Sequence), ev (Evidence Viewer), mm (Model Maker), and hm (Homologene).

For more information, please refer to the Human Maps Help document http://www.ncbi.nlm.nih.gov/mapview/ static/humansearch.html. We will use some of these links in this mini-course. Display the entire HFE gene sequence by clicking on the download "dl" link and then on "Display" on the next page. Copy the sequence and paste it in the area provided on the mini-course page. Note the *accession* number of the longest transcript, NM 000410. We will use this information in the next step.

Stage III

Go back to the Map Viewer report by clicking on the back button of the browser twice. Click the Maps and Options link. Remove all the maps except the Genes_seq map by selecting the map under the "Maps Displayed" menu and clicking on "Remove." Now add the Variation map from the "Available maps" menu (by selecting the map and clicking Add).

Make the Variation map the master map by selecting it and clicking the "Make Master/Move to Bottom" option. Then click "Apply." Now two maps are displayed, Variation (it is the rightmost and the master map) and Genes _seq. The master map provides detailed information for the map features, in this case SNPs. Zoom in on the blast hit area (bar on the right side of each map) by clicking on the map line next to it and choosing the appropriate zoom level.

There are two SNPs in the area; rs 1800562 and rs4986950. Click any of the links and obtain

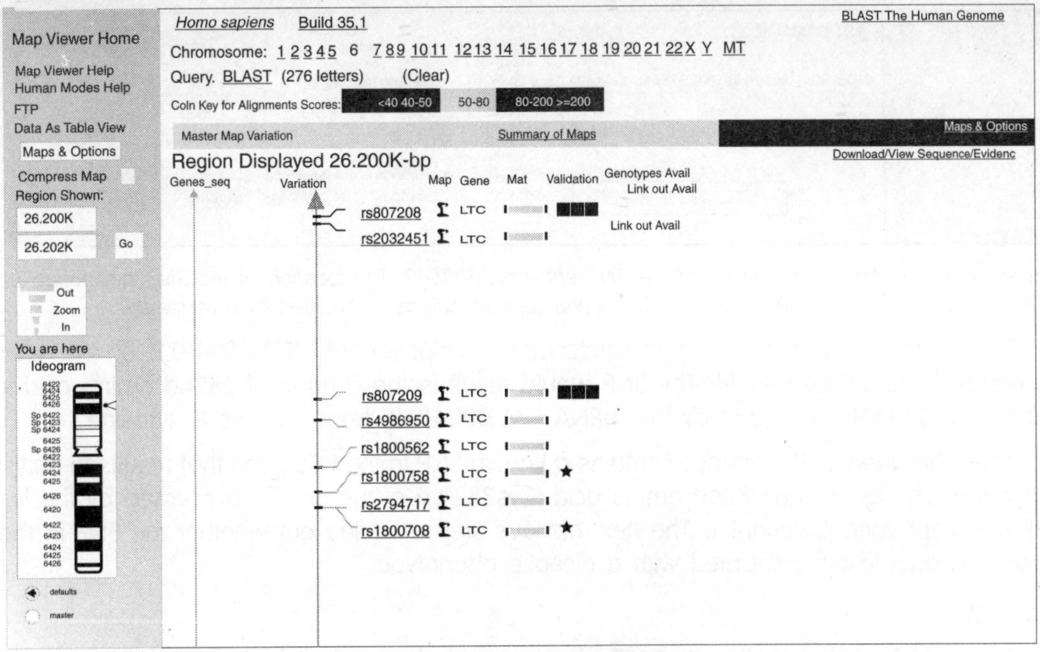

Figure 1.4: Map Viewer display, containing the variation and Genes_seq maps, zoomed in the region of the Basic Local Alignment Search Tool (BLAST). There are two SNPs, rs 1800562 and rs4986950, in the BLAST hit area indicated by a bar on the right side of each map.

information about the location and the nucleotide variation from the "Fasta sequence" and "Integrated maps" panels. The SNP, rs 1800562, represents an A/G SNP at the nucleotide position 16951392 on the contig NT 007592.14 of the reference assembly. This is the same nucleotide variation on the contig NT 007592.14 found in the BLAST. (16951392 G to A).

To identify whether this change represents a change in an encoded amino acid, we will refer to the GeneView panel. This view shows the location of the SNP in the alternatively spliced products annotated on all the assemblies. It also provides information at the protein level; the amino acid number and the change in the sequence, if any.

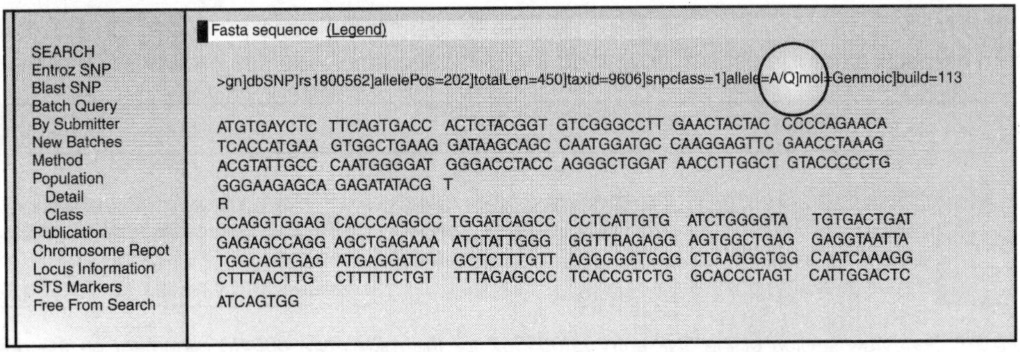

Figure 1.5: Fasta sequence section of the SNP entry rs 1800562. The A/G allele in the SNP, indicated in the definition line on the record, is highlighted by an oval.

Integrated Maps:

NCBI Map Viewer : rs1800562 maps exactly once on NCBI human chromosome 6

Chromosome	Contig accession	Contig position	Chromosome position	Hit orientation	Group term	Group label	Contig label
6	NT 086686 1	25684223	25967140	plus strand	alt_assembly_2	Celera	Celera
6	NT 007592 14	16951392	26201120	plus strand	ref_haplotype	reference	reference

Figure 1.6: Integrated maps section of the SNP entry rs 1800562. The location of the SNP, nucleotide position 16951392 on the contig NT 007592.14 of the reference assembly, is highlighted by a rectangle.

Refer to the panel for the longest transcript, transcript variant 1 NM 000410.2, on the reference assembly contig NT 007592.14. The SNP would result in the change of 282nd amino acid in the protein NP 000401.1, encoded by the mRNA NM 000410.2, from cysteine to tyrosine.

Thus, the query EST sequence contains a known SNP in the HFE gene that results in a cysteine to tyrosine change in the 282nd amino acid (Cys282Tyr) of the protein expressed by the longest HFE transcript variant, variant 1. The next obvious step is to find out whether the SNP in the *HFE* gene is known to be associated with a disease phenotype.

Stage IV

To determine whether the Cys282Tyr amino acid change is linked to a phenotype, we will access the OMIM database. Go back to the Map Viewer report by clicking the back button of the web browser. Make the Genes_seq map a master map by clicking the arrow at the top of the Genes_seq map. Click on the OMIM link next to the *HFE* gene.

This takes us to the OMIM report for the *HFE* gene. It describes the relationship between the mutations in the *HFE* gene and the *hemochromatosis phenotype.* Click the Allelic Variants "View list" in the side blue bar to get information about the mutant proteins from patients. One variant, Cys282Tyr, is reported to cause the hemochromatosis phenotype.

The query EST contains a known variation that would lead to the expression of the Cys282Tyr variant protein associated with the hemochromatosis phenotype.

Results

This Mini-Course describes the steps needed to identify the gene producing an EST obtained from a hemochromatosis patient, download the gene sequence, identify known SNPs in the gene,

GeneView

Group lable	Contig→mRNA→Protein	Contig position	mRNA orientation	Function	dbSNP allele	Protein recidue	Codon pasition	Amino acid pogition
reference	NT 007592→NM 000410→NP 000401	16951392	forward	nonsynonymous	A	Tyr [Y]	2	282
				contigreference	G	Cys [C]	2	282

Figure 1.7: GeneView section of the SNP entry rs1800562 for the mRNA NM 000410 alignment on the reference assembly contig NT 007592. The resulting amino acid change, 282nd amino acid in the protein NP 000401.1, from cysteine to tyrosine, is highlighted by a rectangle.

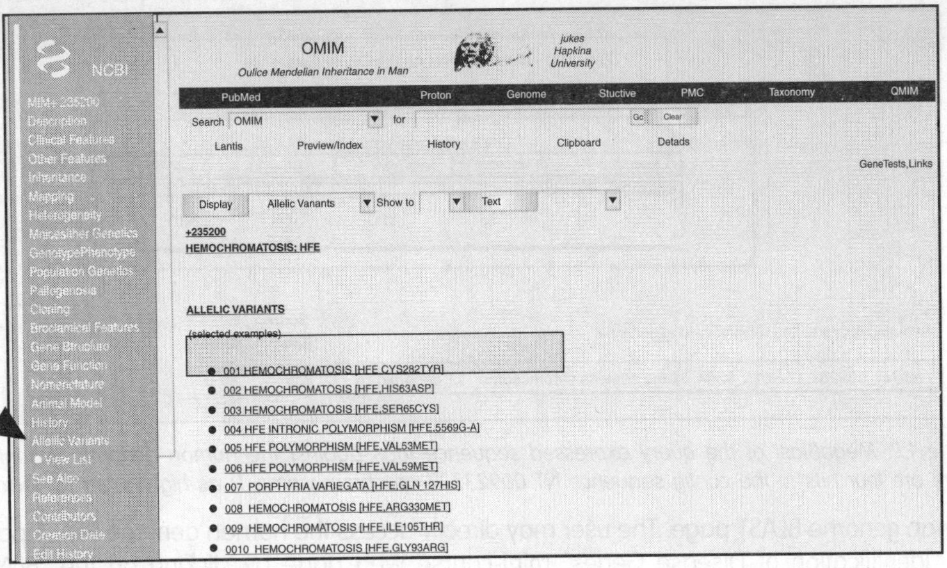

Figure 1.8: Allelic variants list section from the Online Mendelian Inheritance in Man report for the HFE *gene. The Cys282Tyr variant, highlighted by a rectangle, is reported to be associated with hemochromatosis.*

and find SNP-associated phenotypes.

Result of Stage I: The query EST sequence was found to align to contig NT 007592.14 on chromosome 6 with one nucleotide difference (G to A with respect to the nucleotide 16951392 on the contig).

Result of Stage II: The query EST was found to align to the *HFE* gene.

Result of Stage III: The query EST sequence contains a known SNP (G/A with respect to the nucleotide 16951392 on contig NT 007592.14) that results in the Cys282Tyr change in the hemochromatosis protein expressed by the longest HFE mRNA variant

Result of Stage IV: The Cys282Tyr change in the *HFE* protein is associated with hemochromatosis.

DETAILED PROTOCOL

For more practice, we will now perform a similar analysis using another EST sequence from a sickle anemia patient. Sickle cell *anemia* is a disease in which the red blood cells are curved in shape and have difficulty passing through small blood vessels.

It is recommended to follow along from the webpage (http://www.ncbi.nlm.nih.gov/Class/minicourses/diseasegene2.html). This page contains a link to a file containing the screen images of each of the steps described next. Referring to the file is strongly recommended to follow the mini-course steps. However, a number of screen images are provided in this chapter as well for a reader to follow along. These screen images are from human Build 35.1.6.1.

Step 1

Paste the EST sequence that is provided in the mini-course page into the query box of the

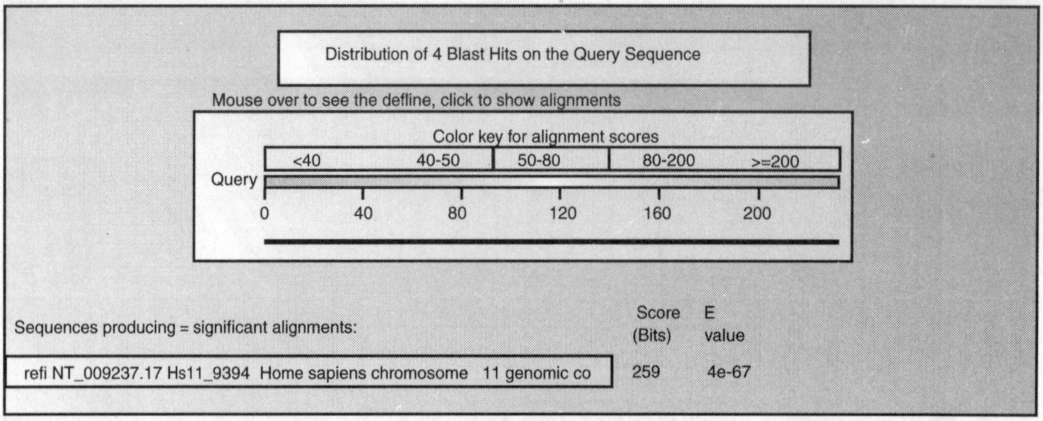

Figure 1.9: MegaBlast of the query expressed sequence tags against the human genome: graphical overview. There are four hits to the contig sequence NT 009237.17 on chromosome 11 as highlighted by rectangles.

human genome BLAST page. The user may directly access the human genome BLAST page through the "Identification of Disease Genes" mini-course web page by clicking on the "BLAST (*human genome*)" link.

Select the "genome" database from the pull down menu, start the search by clicking on the "Begin Search" button and obtain the results by clicking on the "Format" button. The BLAST results page shows four hits to the contig sequence NT *009237.17* on *chromosome* 11, in the human genome build 35.1, with varying percent identity. These multiple hits could arise from similarity to multiple gene family members and/or the query EST sequence originating from multiple exons.

Step 2

To determine the gene expressing the EST in this case, it is much easier to view the BLAST hits in the Map Viewer. Click the "Genome View" button at the top of the BLAST results page, then on the Map element "NT *009237.*" Currently, four maps are displayed; Model, RNA, Genes_seq, and Contig. Refer to stage I for more description of these maps.

The four BLAST hits are indicated by the shaded areas on the right side of each map. Two of these align to the two exons of the *HBB* gene and two align to the two exons of the *HBD* gene (highlighted by the ovals). Note the percent identity of the BLAST hits (highlighted by the rectangles).

The EST sequence is more similar to the HBB exons than to the HBD exons (100 and 99 % compared to 98 and 92 %, respectively). Thus, the query EST is probably expressed by the *HBB* gene but also aligns to the *HBD* gene because of its sequence similarity to the *HBB* gene sequence. One of the exons of the *HBB* gene is only 99 % identical to the query EST.

To note the location of the nucleotide difference between the two sequences, click on the corresponding "Blast hit" link to go back to the BLAST results page. Note that the alignment is on the minus (reverse) strand of the contig NT 009237.17. One nucleotide sequence is different between the query EST and the genomic sequence "Sbjct." Identify the *nucleotide* number at the site of difference with respect to the closest nucleotide number 4035482 of "Sbjct" contig NT 009237.17 and count *downward* by 10 (because the site of difference is 10 nucleotides away from 4035482 and the *alignment* is on the *minus strand*).

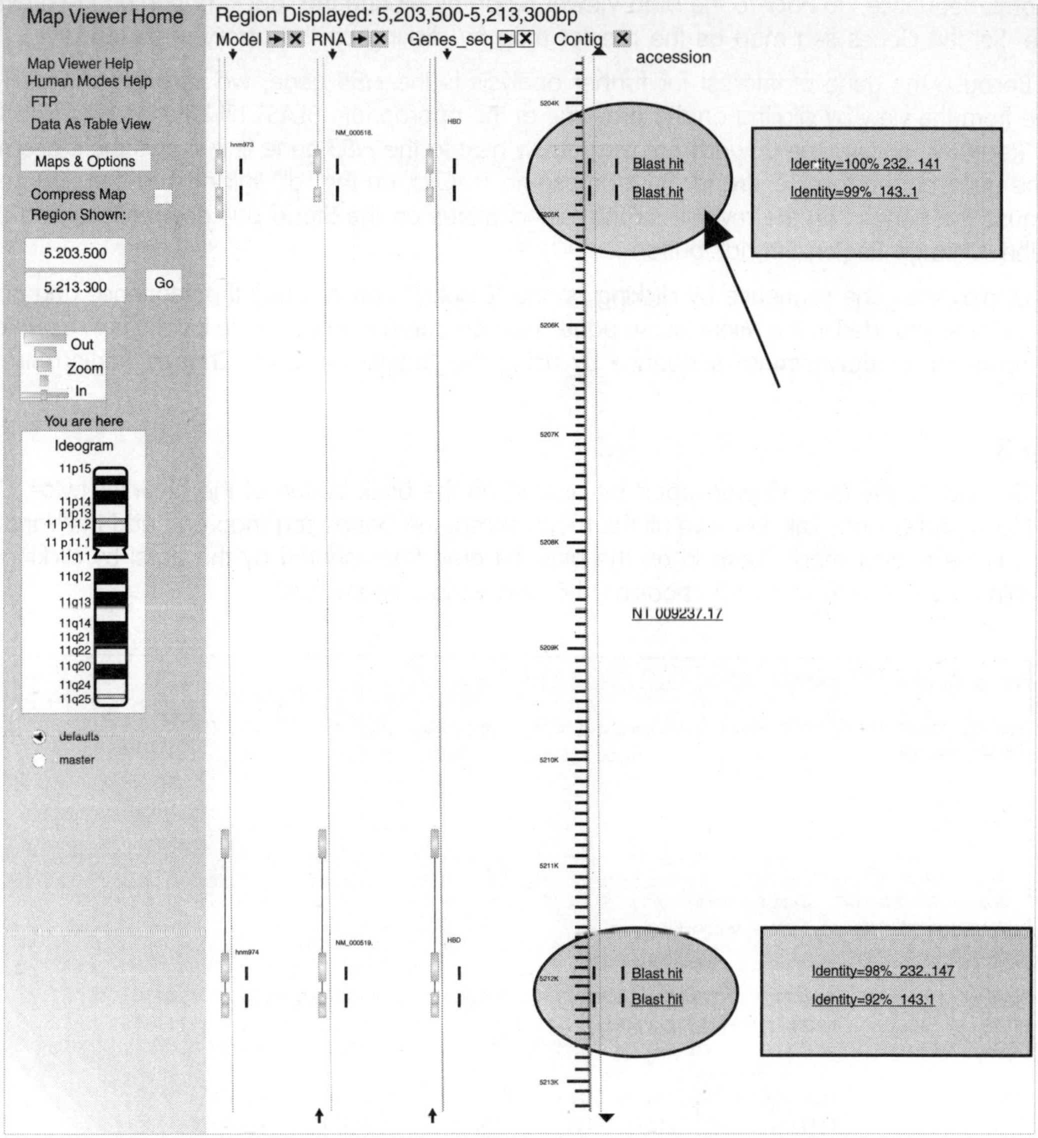

Figure 1.10: Map Viewer display obtained from the Genome View button link on the BLAST results. The four BLAST hits are indicated by the shaded areas on the right side of each map. Two of these align to the two exons of the HBB *gene and two align to the two exons of the* HBD *gene (highlighted by the ovals). Note the percent identity of the BLAST hits.*

Thus the query EST has a T to A variation with respect to the nucleotide 4035473 of the contig NT 009237.17. The difference could be due to a sequencing error in the low quality EST sequence or it may represent a real SNP in the human genome.

For future reference, paste your results, such as the alignment and the nucleotide difference (sequence difference at the nucleotide 4035473 on NT 009237.17; A in the genomic and T in the query EST sequence), in the window provided in the mini-course webpage. To download the HBB genomic sequence, go back to the Map Viewer report by clicking the "back" button on the browser once. Set the Genes seq map as the master map by clicking on the arrow at its top.

Because the gene of interest for further analysis is the *HBB* gene, we can remove the *HBD* gene from the view by clicking on the gray line at the appropriate BLAST hit location and selecting the "Recenter" option. The upward pointing arrow next to the *HBB* gene link shows the placement of the gene on the reverse strand of chromosome 11. Click on the "dl" link next to the HBB gene. Because the gene is on the reverse strand, select minus on the Stand pull down menu and click on the "Change Region/Strand" button.

Display the gene sequence by clicking on the "Display" option. Copy the sequence and paste it in the area provided in the Mini-Course page. You can adjust the nucleotide locations to download the upstream or downstream sequence by using the "adjust by" and "Change Region/Strand" option.

Step 3

Go back to the Map Viewer report by clicking on the back button of the browser twice. Click the Maps and Options link. Remove all the maps except the Genes seq map and add the Variation map as the master map. Zoom in on the blast hit area (represented by the bars) by clicking on the thin gray line next to it and choosing the appropriate zoom level.

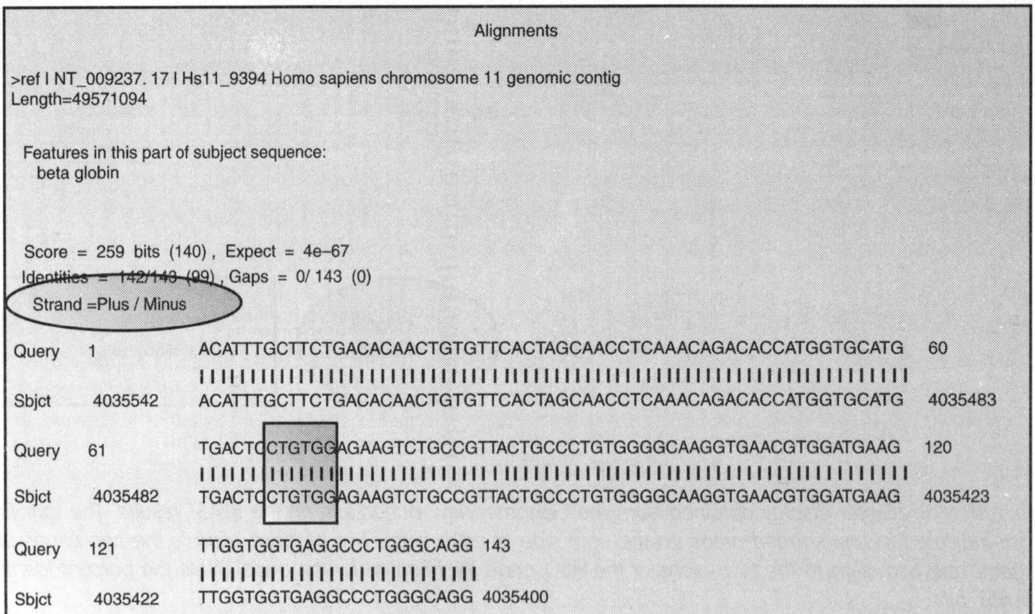

Figure 1.11: Alignment of the 99 % identical Basic Local Alignment Search Tool (BLAST) hit in stage step 1. The BLAST hit is on the minus (reverse) strand (highlighted by an oval) of the contig NT 009237.17. There is one nucleotide difference (highlighted by a rectangle) between the query EST and genomic sequences at nucleotide 4035473 of the contig NT 009237.17.

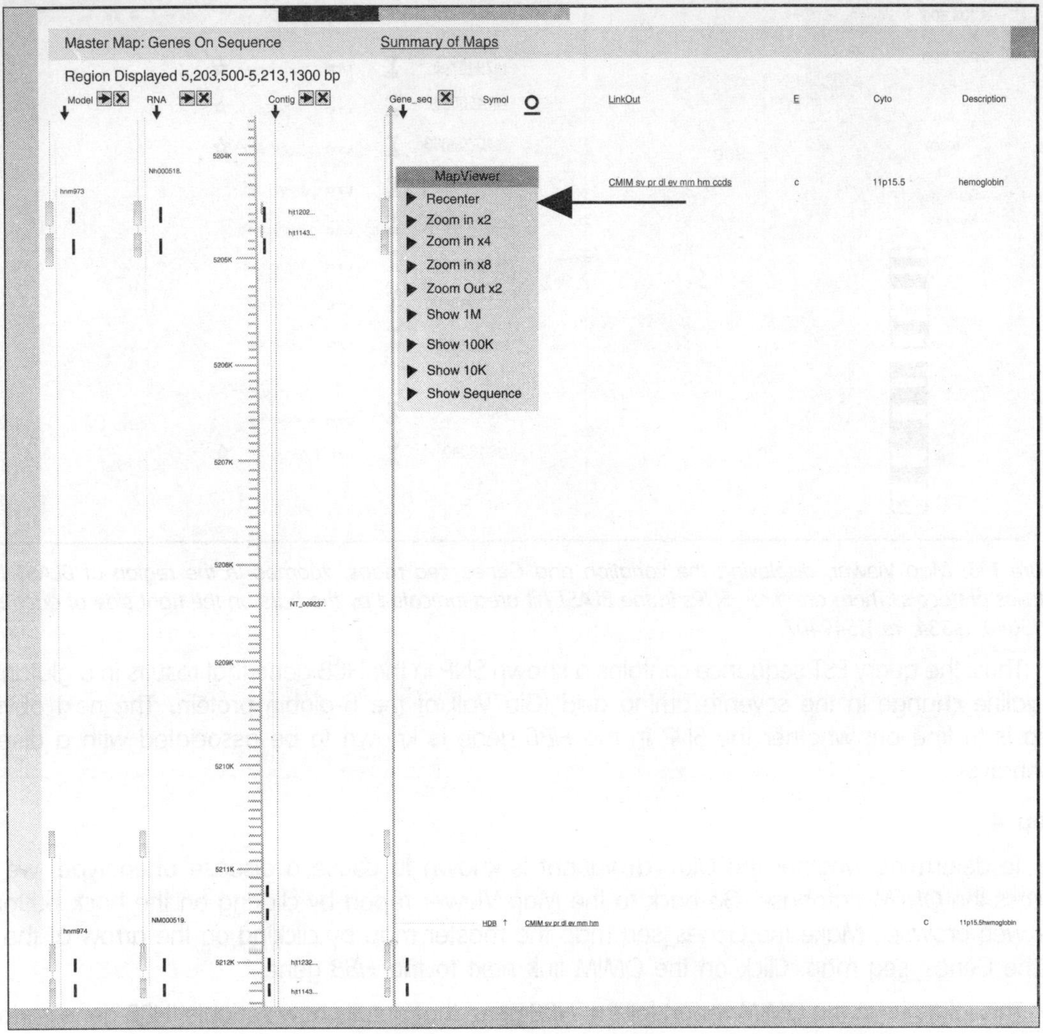

Figure 1.12: Map Viewer display showing the recenter option by clicking on the gray line indicating the contig Map.

There are three SNPs in the area; rs713040, rs334, rs 11549407. Click on any of the links and obtain information about the location and the nucleotide variation from the "Fasta sequence" and "Integrated maps" panels. The SNP, rs334, represents an A/T SNP at the nucleotide position 4035473 on the contig NT 009237.17. This is the same nucleotide variation on the contig NT 009237.17 found in the BLAST result in step 1. (4035473 T to A). Next, to identify whether this change represents a change in an encoded amino acid, we will refer to the GeneView panel.

This view shows the location of the SNP in the alternatively spliced products annotated on all the assemblies. It also provides information at the protein level; the amino acid number and the change in the sequence, if any. Refer to the panel for the transcript, NM_000518, on the reference assembly contig, NT 009237.17. The SNP would result in the change at the seventh amino acid in the protein NP 000509.1, encoded by the mRNA NM 000518.4, from glutamate to valine.

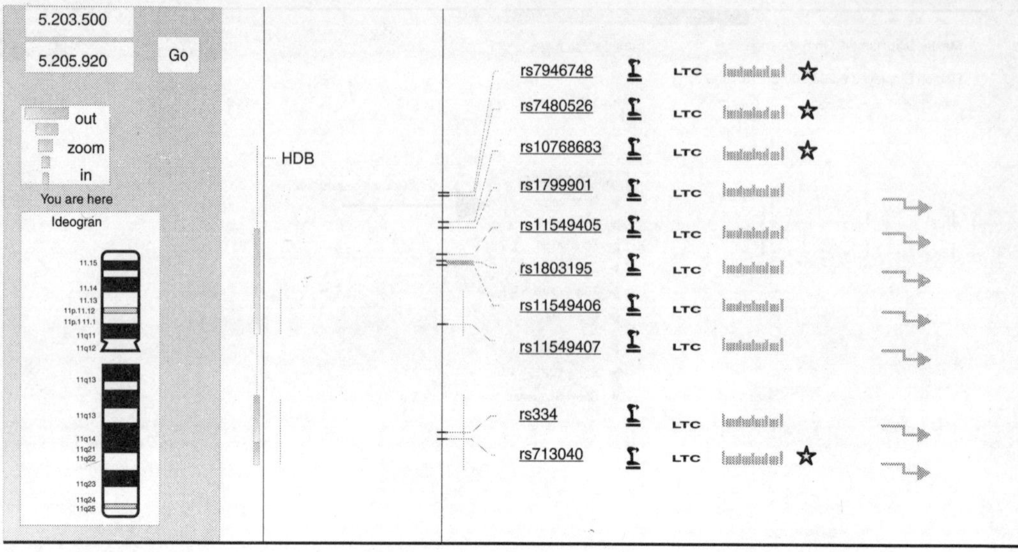

Figure 1.13: Map Viewer, displaying the variation and Genes_seq maps, zoomed in the region of BLAST hit in detailes protocols. There are three SNPs in the BLAST hit area indicated by the bars on the right side of each map; rs713040, rs334, rs 11549407.

Thus, the query EST sequence contains a known SNP in the HBB gene that results in a glutamate to valine change in the seventh amino acid (Glu7Val) of the β-globin protein. The next obvious step is to find out whether the SNP in the *HBB* gene is known to be associated with a disease phenotype.

Step 4

To determine whether the Glu7Va1 variant is known to cause a disease phenotype, we will access the OMIM database. Go back to the Map Viewer report by clicking on the back button of the web browser. Make the Genes_seq map the master map by clicking on the arrow at the top of the Genes_seq map. Click on the OMIM link next to the *HBB* gene.

This takes us to the OMIM report for the *HBB* gene that details how variants (HBB gene variants)

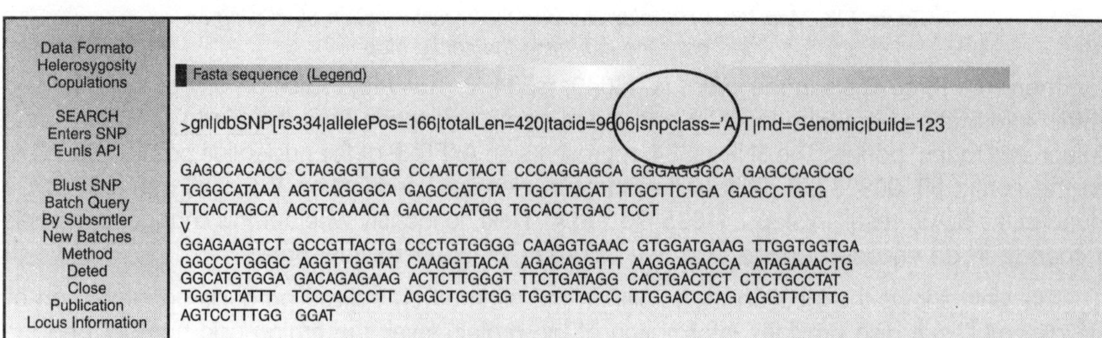

Figure 1.14: Fasta sequence section of the SNP entry rs334. The SNP contains an A/T SNP, indicated in the definition line of the record, highlighted by an oval.

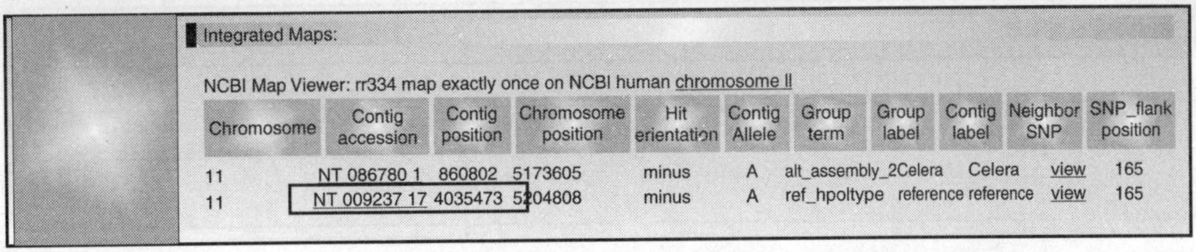

Figure 1.15: Integrated maps section of the SNP entry rs334. The location of the SNP, nucleotide position 4035473 on the contig NT 009237.17 of the reference assembly, is highlighted by a rectangle.

the *HBB* gene are associated with various phenotypes. As mentioned in the OMIM report under the "Psuedogenes" section, the allelic variants are listed for the mature HBB (β-globin) protein which lacks the initiator methionine.

The SNP database reports them for the precursor protein. Hence, the allelic variants in the OMIM report are off by one amino acid compared to the variants in the SNP report.

Thus, the Glu7Val variant in the SNP report corresponds to the Glu6Val variant in the OMIM report. Access the allelic variants list by clicking on the "*View list*" in the blue side bar.

The Glu6Val variant, called hemoglobin S, is reported to cause the sickle cell anemia phenotype. The query EST contains a known variation that leads to the expression of the Glu7Val variant protein associated with the sickle cell anemia *phenotype*.

Results

This mini-course describes steps to identify the gene expressing the ESTs obtained from a sickle cell anemia patient, download the gene sequence, identify known SNPs in the gene and find SNP-associated phenotypes.

Result of Step 1: The query EST sequence was found to align to the contig NT 009237.17 on chromosome 11 with one nucleotide difference (T to A with respect to the nucleotide 4035473 on the contig).

Result of Step 2: The query EST was found to be expressed by the *HBB* gene.

Result of Step 3: The query EST sequence contains a known SNP (T/A with respect to the nucleotide 4035473 on contig NT 009237.17).

Result of Step 4: The Glu7Val change in the *HBB* protein is associated with sickle cell anemia.

APPLICATIONS

The mini-course describes a procedure to identify a known gene and a SNP from the NCBI databases starting from one EST sequence. The same procedure can be used with a batch of EST sequences in the initial human genome BLAST search followed by a similar analysis to identify genes corresponding to them. Some ESTs may be produced by known genes (as described in the mini-course) and some may be produced by novel genes not yet annotated on the Genes_seq map.

The Model map may be useful to identify the novel genes. Also, some ESTs may contain new SNPs, which can be deposited in dbSNP. By comparing the DNA sequence from patients and

Gene View

Group lable	Contig→mRNA→Protein	Contig position	mRNA orientation	Function	dbSNP allele	Protein tesidne	Codon nosition	Amino acid nosition
reference	NT 009237→NM 000518→NP 000509	4035473	reverse	nonsynonymous	T	Val [V]	2	7
				contig reference	A	Glu [B]	2	7

Figure 1.16: GeneView section of the SNP entry rs334 for the mRNA NM 000518 alignment on the reference assembly contig NT 009237. The SNP results in the change at the seventh amino acid in the protein NP 000509.1 from glutamate to valine.

normal individuals, it can be discerned whether the novel SNP and/or the novel gene aré associated with the disease.

FURTHER ANALYSIS

The mini-course "Correlating Disease Gene and Phenotype" elucidates the biochemical and structural basis for the function of the mutant proteins and their relationship to the particular phenotype.

POINTS TO REMEBER

1. Online Mendelian Inheritance in Animals is a database of genes, inherited disorders and traits in animal species (other than human and mouse) authored by Professor Frank Nicholas of the University of Sydney, Australia, with help from many *collaborators* over the years.
2. In addition to the human genome, a number of other genomes are available as BLAST databases. A complete list is available under the Genomes panel on the BLAST page (http://www.ncbi.nlm.nih.gov/BLAST/).
3. MegaBlast also accepts a batch of query sequences. Each query sequence must have a unique identifier written on a separate line before the sequence and the identifier line should begin with a greater than (">") sign. For example,

   >identifier 1

   atgcggctta...

   >identifier2

   ttggcatactg...

   > identifier3

   ggatcgatcag...
4. Since the human Build 36, NCBI also provides access to the previous assembly release (Build) as a BLAST database. More information about each build is provided in the release notes at http://www.ncbi.nlm.nih.gov/genome/guide/ human/release notes.html. You may choose to run the BLAST search against the previous build by using the appropriate option in the database field.

5. In stage 1, there is only one hit to the database sequence. Multiple hits are possible for several reasons such as finding similarity to other gene family members. The query EST sequence shows similarity to two members of the globin family, HBB (encoding β-globin), and HBD (encoding delta globin). The gene encoding the EST may be the one with high similarity, HBB in this case.
6. BLAST may align a single EST in two or more segments if the EST sequence spans two or more exons. The query EST sequence aligns to two exons of both *the HBB* and *HBD* genes.
7. The orientation of the gene on the chromosome can also be discerned from the placement of the blue bars representing the gene structure with respect to the gray line. If the gene is placed on the forward strand then the blue bars representing the gene structure are drawn to the right of the gray line. If the gene is placed on the reverse strand, then the blue bars are drawn to the left of the gray line.
8. If the gene of interest is on the reverse strand, then change the Strand pull down menu to minus and click on the "Change Region/Strand" option before displaying the sequence.
9. The user can also adjust the nucleotide locations to download the *upstream* or *downstream* sequence by using the "adjust by" and "Change Region/Strand" options.
10. When a single nucleotide polymorphism is submitted to dbSNP, an identifier with prefix "ss" is assigned to the entry. It is possible that multiple laboratories may submit information on the same SNP as new techniques are developed to assay variation, or new populations are typed for frequency information. Each of these SNP entries is assigned a unique identifier with prefix "ss". When two or more submitted SNP records refer to the same location in the genome, a Reference SNP record is created, with an "rs" prefix on the identifier, by NCBI during periodic "builds" of the SNP database. This reference record provides a summary list of submitted "ss" records in dbSNP.
11. When the "variation" map is selected as the master map in the Map Viewer, the user is presented with a graphical summary of several properties of the SNP, such as the quality of the SNP computed from mapping, location in the gene region, marker heterozygosity and validation information. For more information, refer to http://www.ncbi.nlm.nih.gov/SNP/get html.cgi?whichHtml=verbose.

    For example, the green triangle next to rs 1800562 indicates that this marker is mapped to a unique position in the genome. The highlighted L, T, and C symbols indicate that the SNP is in the Locus, Transcript, and the coding region of the gene. For the SNP rs807209, only the L symbol is highlighted indicating that the SNP is not in the transcript or the coding region but is in the locus region of the gene. A locus in this report is defined as any part of the marker position on sequence map within a 2000-base interval 5' of the most 5' feature of the gene (CDS, mRNA, gene), or the marker position within a 500-base interval 3' of the most 3' feature of the gene.
12. NCBI has additional human genome assemblies such as the assembly submitted by Celera and chromosome 7 assembly from the Center for Applied Genomics, TCAG. The assembly information is provided under the contig label heading in the Integrated Maps panel. On the Celera assembly, the SNP in the *HFE* gene, rs18000562, is at the nucleotide 25684223

on the contig NT 086686.1. On the Celera assembly, the SNP in the *HBB* gene, rs334, is at the nucleotide 860802 on the contig NT 086780.1.

13. The GeneView panel shows the locations of the SNPs on the genomic assemblies with respect to the genes, their alternatively spliced mRNAs and encoded proteins. The view is colour coded for quick identification of the location and to show whether the change is synonymous (not altering the amino acid translation) or nonsynonymous (altering the amino acid translation). For example, nonsynonymous SNPs are represented in red, synonymous in green and those in introns are in yellow. A link to the "Colour Legend" is provided next to the Gene Model under the GeneView panel.

14. Some OMIM entries report the allelic variants for the mature protein, whereas dbSNP reports variants for the precursor protein. Thus, for the same SNP, amino acid numbering for the allelic variant may be different in these databases. OMIM reports allelic variants for the beta globin mature protein (after removal of the initiator methionine). Thus, the Glu7Val change reported in dbSNP is the same as Glu6Val allelic variant reported in the OMIM database.

15. The OMIM report and thus its allelic variants list are manually derived from publications. dbSNP contains the SNPs reported by the submitters. Currently, a link is provided from dbSNP to OMIM if the amino acid number of the *allelic variant* in the OMIM report matches the number of the changed amino acid due to a SNP in dbSNP. Since the sources of the two databases, OMIM and dbSNP, are different, each may contain information not found in the other.

Chapter 2

Homologous Doubling

Until recent times, the identification and characterization of *homologous* doubling in the form of *segmental* duplications has often been based purely on *anecdotal* reports. However, the completion of the Human Genome Project has now made possible the *systematic* analysis of the extent and distribution of duplicated sequences in humans.

Both *in situ* hybridization and *in silico* approaches have shown that approx 5% of our genome is composed of highly *homologous* duplicated sequence, with enrichments of six- to sevenfold in *pericentromeric*, and two- to threefold in *subtelomeric* regions, respectively.

Not only does the presence of these paralogous segments represent a significant challenge to the correct assembly of the human genome, but there is also an increasing awareness of their role in human evolution, *variation*, and disease. We present a review of homologous doubling or segmental duplications in the *mammalian genome*. We describe their basic characteristics, *distribution*, and *dynamic* nature during recent *evolutionary* history.

Based on these features, we discuss models to account for the proliferation of these sequences in the mammalian lineage, and also their contribution towards *karyotypic* evolution and phenotypic differences between primates. Finally, we highlight the role of segmental duplications as *mediators* of human variation at the genomic level.

FEATURES OF HOMOLOGOUS DOUBLING

Human segmental duplications (also termed low-copy repeats) are blocks of DNA ranging from 1 to 200 kb in length that occur at more than one site within the genome and share a high level (often >90%) of sequence identity. They can include both genic sequences and high-copy repeats, such as long interspersed elements and short interspersed elements, and, unlike *tandem* duplications, are *interspersed* throughout the genome.

This unique distribution pattern suggests that segmental duplications arise through a process termed duplicative transposition, whereby whole blocks of sequence are first duplicated and then transposed from one genomic location to another. Although they have been identified on every human chromosome, their distribution is nonuniform with some chromosomes and chromosomal regions showing peculiar enrichment.

Segmental duplications tend to cluster within pericentromeric and, to a lesser extent, subtelomeric regions. Many human *pericentromeric* regions in particular contain high concentrations of duplicated sequence, often arranged in large blocks composed of complex *modular* structures of individual duplications.

Homologous or segmental duplications may be further classified based on their chromosomal distribution. Although some blocks of sequence may be duplicated to multiple locations within a single chromosome (termed intrachromosomal duplication), others may be located on *non-homologous* chromosomes (inter- or *transchromosomal* duplication).

Intriguingly, the distribution of these two types of duplication appears to be largely exclusive of one another, suggesting mechanistic- differences in their mode of propagation.

Chromosome-Specific Duplications

Initial identification of chromosome-specific duplications often came about through the study of common microdeletion/micoduplication syndromes, such as Prader-Willi/Angelmail syndromes, Williams-Beuren syndrome, Smith-Magenis syndrome, Charcot-Marie-Tooth disease type 1 A, and DiGeorge/Velocardiofacial syndrome.

As more of these syndromes were analyzed, it became apparent that the presence of large and highly homologous segmental duplications flanking these sites of rearrangement was a recurring theme. Available evidence suggests that homology between these duplicated sequences acts as a substrate for unequal meiotic recombination, leading to the deletion, duplication, or inversion of the intervening sequence.

Review of known genomic disorders caused by chromosome-specific duplications shows that these usually involve duplications that are more than 95% similar and 10-500 kb in length, separated by 50 kb to 4 Mb of DNA. Although *intrachromosomal* duplications are found throughout the *euchromatic* portions of virtually every human chromosome, some chromosomes, such as 1, 9, 16, 17, and 22, show particularly high *concentrations* within their most proximal regions.

In these cases the density of duplication can be such that little or no unique sequence occurs across relatively large stretches (300 kb to 4 Mb) of DNA. The organization of these regions can be complex, with large duplication blocks often composed of smaller modules, which have been derived from different genomic locations.

This feature has often led to difficulties in characterizing these regions, necessitating the development of specialized methods to allow these regions to be successfully mapped and sequenced. Many intrachromosomal duplications also share very high levels of sequence identity, with the majority having more than 95% identity between *paralogous* copies. In extreme cases, this level of nucleotide identity approaches the frequency of allelic variation found in the genome as a whole (approx 1 base per kilobase).

This property further confounds the mapping and identification of individual duplicated

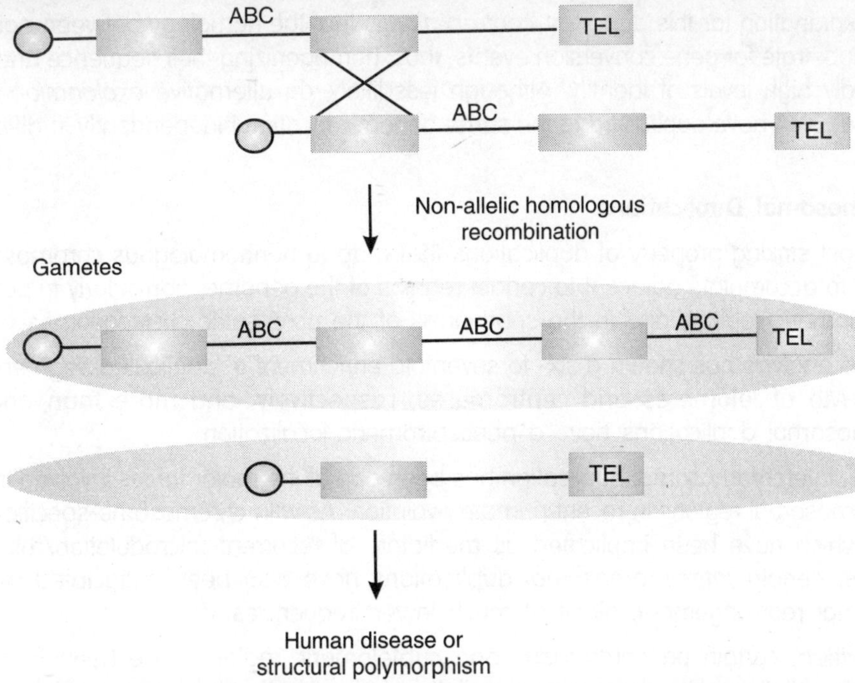

Figure 2.1: Homologous doublings mediate structural rearrangement. Misalignment of paralogous blocks of sequence during meiosis leads to unequal recombination and the deletion or duplication of the intervening sequence. These events may create structural polymorphisms or, if the genes (A,B,C) flanked by the duplications are dosage sensitive, genomic disease. O, centromere. TEL, telomere.

segments within the genome, and may also represent a significant impediment in the ability to distinguish true allelic *polymorphism* from paralogous genomic copies. Indeed, both gene and single nucleotide polymorphism annotation show significant improvements in accuracy when duplicated sequences are correctly defined within a genome assembly.

Thus, the correct definition of segmental duplications is an important aspect of achieving high-quality genomic sequence. Based on the neutral mutation rate in primates of approx 1.5×10^{-9} substitutions per site per year, the high levels of homology observed between many chromosome-specific duplications suggests they have emerged only recently during *evolutionary* history. Indeed, comparative analysis of different primate lineages has demonstrated that some are species-specific, *confirming* their *dynamic* nature.

In other cases, analysis of the levels of sequence divergence between pairs of duplication in different primates, such as those flanking the common Williams-Beuren syndrome deletion region in 7q11.23 and the large palindromic repeats on long arm of the Y chromosome, has provided evidence that they may also act as substrates for gene conversion. Although fluorescence *in situ* hybridization analysis shows the presence of these duplications in multiple primate species, thus, suggesting they originated before the separation of these lineages, estimates of their evolutionary age based on the rate of nucleotide divergence because of random mutation suggest a much more recent origin.

One explanation for this apparent *contradiction* is that the homology between pairs of repeats acts as a substrate for gene conversion events, thus, homogenizing their sequence and maintaining unexpectedly high levels of identity. Although less likely, an alternative *explanation* is that some segments of DNA have duplicated to the same genomic location independently in different primate *lineages*.

Interchromosomal Duplications

The most striking property of duplications that map to nonhomologous chromosomes is their propensity to accumulate adjacent to certain regions of the genome, particularly in pericentromeric and subtelomeric regions and in the short arms of the acrocentric chromosomes.

In silico analysis has shown a six- to sevenfold enrichment of duplicated segments within 100 kb and 1 Mb of telomeres and centromeres, respectively, and more than onethird of all interchromosomal duplications have a pericentromeric localization.

Indeed, interchromosomal duplication has been one of the major forces involved in remodeling these chromosomal regions in recent primate evolution. As with chromosome-specific duplications, many of which have been implicated as mediators of recurrent microdeletion/microduplication syndromes, certain *interchromosomal* duplications have also been associated with recurrent chromosomal rearrangement, albeit at much lower frequencies.

In addition, certain pericentromeric and subtelomeric regions have been found to exhibit polymorphic variation within the human population, indicating that the process of interchromosomal duplication is ongoing.

SPREDING OF HOMOLOGOUS DOUBLING

One of the most remarkable features of segmental duplications is their tendency to often cluster in close proximity to one another, particularly within peri- and subtelomeric regions. However, detailed mapping and sequencing studies show that the tendency for segmental duplications to accumulate within these regions is not a universal property of all human chromosomes.

Although some human chromosomes (e.g., 7, 9, 15, 16, 17, 19, 22, and Y) are significantly enriched for duplications, certain pericentromeric regions, such as those on chromosomes 3, 4, and 19, are almost completely devoid of duplications. Study of other sequenced organisms shows that this propensity for duplications to accumulate in clusters within specific chromosomal regions is not unique to humans.

Analysis of the working draft sequences of both *Rattus norvegicus* and *Mus musculus* shows that in both organisms, duplicated sequences are distributed *nonrandomly*, often occurring in large blocks with a distinct pericentromeric and subtelomeric bias, suggesting that the genomic distribution of segmental duplications may be a general property of mammalian *chromosomal architecture*.

Duplication of Pericentromeric Regions

Although the phenomenon of *pericentromeric* duplication is common to both men and mice, the properties of these duplicated regions differ. Typically, those found in mice are relatively small (<75 kb), whereas in humans they may be several 100 kb in length, and the degree of sequence

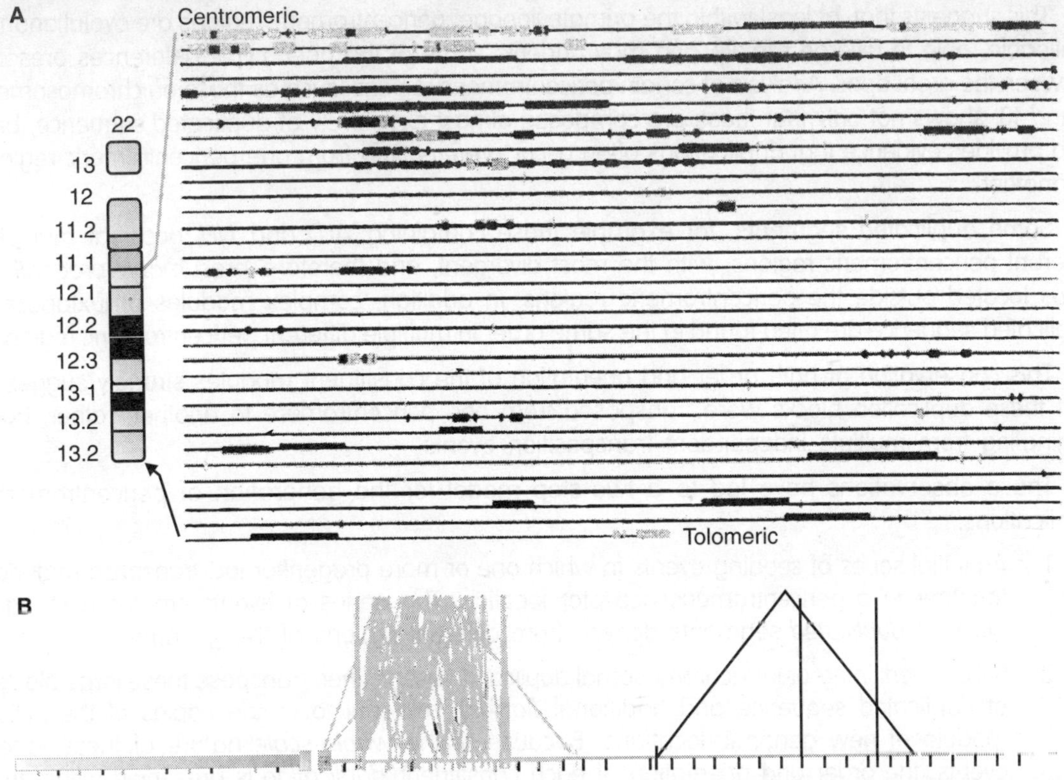

Figure 2.2: Homologous doubling on chromosome 22. (A) An overview of duplicated sequence on the long arm of chromosome 22 (27). A total of 715 duplications > 1 kb in length and >90% sequence identity were identified. Each horizontal line represents I Mb. Top left-hand corner is the most centromeric sequence contig and at the bottom right is the most telomeric sequence. Gray bars, interchromosomal duplications between nonhomologous chromosomes, black bars, intrachromosomal duplications. Overall, 9.1 '70 of the q arm is involved in large (>1 kb) duplications, with interchromosomal and intrachromosomal dupllcalions constituting 3.9 and 6.4% of the total sequence, respectively. Note the preponderance of interchromosomal duplications near the centromere and telomere. (B) A reduced view of chromosome 22 (each tick mark represents 1 Mb) showing the relationship of intrachromosomal duplications (black joining lines). This region is enriched for avariety of genomic disorders.

identity between pairs of duplications are often much lower (<97% in mice compared with >97% in humans). Although this latter property may be, in part, because the higher neutral *mutation* rate in mice, it suggests that many of the pericentromeric duplications in humans are *evolutionary* younger, and, therefore, more recently active, than those in mice. This is supported by the fact that certain *pericentromeric* duplications are unique to the human lineage, and are either completely absent or occur at different chromosomal locations in other *mammalian* species.

Indeed, the process of pericentromeric duplication appears to have occurred as punctuated events during primate evolution. Certain duplicated segments, such as those containing the genes *ALD* and *CTR,* are common to humans, *chimpanzees*, and *gorillas*, but are absent in orangutans, indicating their emergence predates this speciation event approx 12 million years ago. In contrast, analysis of duplications on chromosome 10 shows that these emerged at various times during primate evolution, ranging from 13 to 28 million years ago.

This suggests that, at least within the primate lineage, pericentromeric regions are evolutionarily malleable, able to diverge rapidly, and may underlie some of the phenotypic differences present between the great apes. Analysis of certain pericentromeric regions, such as those on chromosomes 2 and 10, shows not only that these are composed almost exclusively of duplicated sequence, but also provides evidence that duplications often occur *preferentially* from one pericentromeric region to another.

Some duplicated segments, for example those containing *ALD* and *NFI,* occur at multiple different pericentromeric regions, with the most divergent, and therefore *presumably progenitor, locus* located outside the pericentromeric regions. In addition, complex modules of juxtaposed duplicated sequence are often found in the same order in multiple different pericentromeric regions.

This conservation of both order and orientation of the constituent modules strongly suggests that these duplication blocks were transposed from one pericentromere to another, rather than originating from multiple independent transposition events.

These observations have led to a two-step model for the generation of pericentromeric duplications.

1. An initial series of seeding events in which one or more progenitor loci transpose material together to a pericentromeric acceptor location. This series of events creates a mosaic block of *duplicated* segments derived from different regions of the genome.
2. Subsequent inter- or intrachromosomal duplication events then transpose these large blocks of duplicated sequence and additional *flanking* material to create copies of the initial module at new genomic locations. Because of the *whole-scale* nature of these latter events, the order and orientation of each constituent *duplication* is preserved within the transposed block.

This model has since gained support from more recent studies, providing an explanation for the complex paralogous structure of many pericentromeric regions of the primate genome.

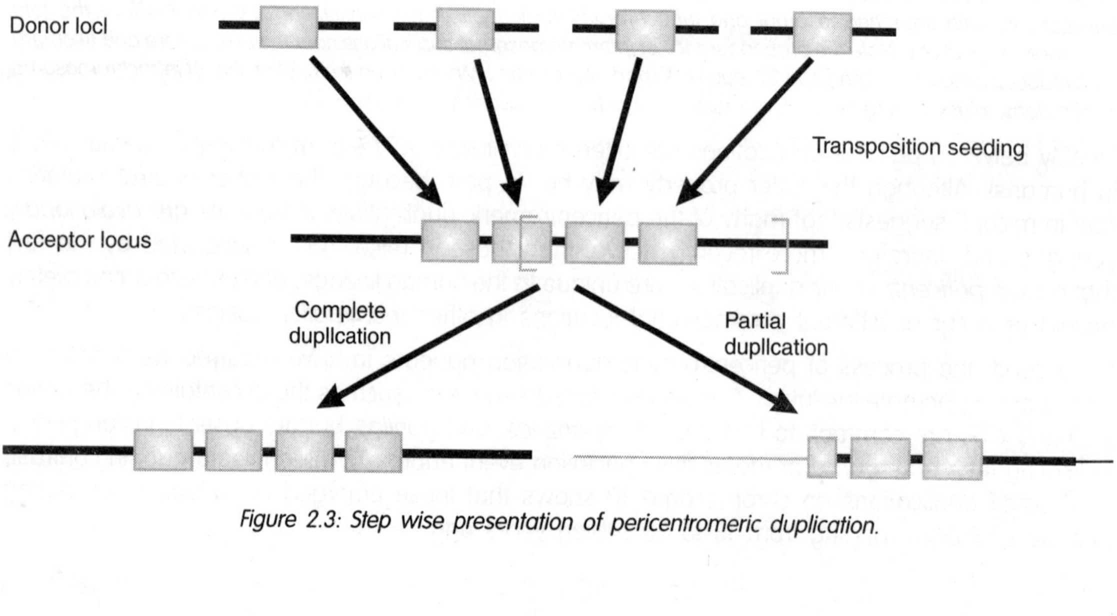

Figure 2.3: Step wise presentation of pericentromeric duplication.

Duplication of Subtelomeric Regions

Like pericentromeres, subtelomeric regions appear to preferentially accumulate duplications, and as with pericentromeres, many of these regions also show differences between species and within the human population. Olfactory receptor (OR) genes present a particularly striking example of this, and appear to have spread by recent subtelomeric duplication.

As a species, nearly half of all human subtelomeres contain one or more OR gene copies, although their distribution and copy number shows wide variations both between and within different ethnic groups, attesting to their recent *evolutionary* origins.

Similarly, members of the zinc-finger and *immunoglobulin* heavy-chain gene families are also located within multiple human subtelomeres and, like the OR genes, show marked variations in their copy number and distribution. Indeed, greater than one-third of all subtelomeric regions show high levels of polymorphism (*allelic frequency* >10%) in their duplication content.

Such an abundance of gene families in subtelomeric regions is a common feature of most *eukaryotes*, and may reflect a generally increased *recombination* and tolerance of *subtelomeric* DNA for rapid evolutionary change.

Uneven Spreading of Duplicated Sequences

A number of different explanations have been put forward in an attempt to explain the propensity for duplications to accumulate within *pericentromeric* and *subtelomeric* regions of the genome. Some authors suggest that pericentromeric regions may be the only regions of the genome that are able to accept duplicated material without deleterious *phenotypic* consequence.

Perhaps owing to a reduced gene density, these regions may have a greater tolerance for *duplicative* transposition events. However, contrary to the predictions of this model, there are many large euchromatic "*gene deserts*" outside of the pericentromeres that do not show a similar accumulation of *segmental* duplications.

This suggests that if an absence of selective constraint is the basis for this pericentromeric bias, lowered gene density is not the cause. Alternatively, the observation that large tracts of duplicated sequence appear to transpose near centromeres and telomeres may simply be an artifact resulting from the fact that *recombination* is suppressed in these regions.

Because of this reduced recombination rate, duplicated sequences may simply persist here for longer than in the rest of the genome, rather than being preferentially targeted to these regions. However, study of other sequenced organisms such as *Arabidopsis* and *Drosophila* shows no evidence for a similar accumulation of duplicated segments within pericentromeric regions, suggesting that this tendency may be a specific property of *mammalian* chromosomes. Could specific sequences present in pericentromeric and subtelomeric regions of mammalian chromosomes actively target duplication events to these regions?

Several authors have hypothesized the presence of hyper-recombinogenic sequences localized within pericentromeric regions that preferentially attract segmental duplications, and analysis of duplication *breakpoints* suggests that this is indeed the case. Unusual polyp urine/pyrimidine minisatellite-like sequences, termed pericentromeric *interspersed* repeats, have been identified at the junctions of many duplication breakpoints.

Analysis of duplications of *NF1, ABCD1,* and several paralogous sequences on chromosome 10 reveals the presence of flanking arrays of CAGGG, GGGCAAAAAGCCG, GGAA, CATTT, and HSREP522 motifs located at their boundaries. Although some of these sequences show homology to telomeric repeats and subtelomeric sequences, others have similarity to known recombinogenic sequences, such as *Escherichia coli* Chi recombination signals, G4 DNA, which has been implicated in meiotic pairing of homologous chromosomes, and those found in the *immunoglobulin* heavy-chain recombination regions that promote *nonhomologous* sequence exchange events.

These similarities suggest a potential role in facilitating the integration and dispersal of duplicated sequences within pericentromeric regions during recent mammalian evolution. Furthermore, phylogenetic analysis of one of these sequences showed that the repeat was both restricted to the primate lineages in which the duplications arose, and was present within the pericentromeric region before the integration of the duplicated sequences, consistent with a potential role in promoting duplicative integration.

A more recent genome-wide analysis of sequence features located at the junctions of duplications suggests that, in addition to certain types of satellite repeats, *Alu-Alu-mediated* recombination events also played a significant role in the recent proliferation of segmental duplications. Approximately one-third of all human segmental duplications terminate within *Alu* repeats that were active during the last 45 million years of primate evolution.

The primate-specific burst of *Alu* retrotransposition activity that occurred 35-40 million years ago suggests a model in which numerous sites of *Alu-Alu* sequence identity were created within the primate genome, providing a substrate for homologous recombination and an increased predisposition to duplicative transposition events.

This model provides a ready explanation for the several features of segmental duplications, including the disparity of segmental duplication between primates and other sequenced organisms, the timing of their *proliferation* within *primates,* and their bias towards gene-rich regions of the genome.

PHYLOGENY OF HOMOLOGOUS DOUBLING

The paradigm of segmental duplications as mediators of chromosomal rearrangement in a growing list of recurrent *microdeletion/microduplication* syndromes indicates their potential as catalysts of structural chromosome rearrangement.

However, with the recent availability of DNA sequence from several *mammalian* species, it is now clear that segmental duplications are also a major driving force in *karyotype* evolution and *speciation.* Comparative analysis of orthologous regions has provided strong evidence for a *duplicationdriven* mechanism underlying chromosomal evolution. In several cases, analysis of evolutionary breakpoints of translocation or *inversions* between humans and other primates has revealed the presence of duplicated sequences at the sites of *rearrangement.*

Indeed, *crossspecies* analysis using array *comparative genomic hybridization* (CGH) has shown a significant association of *segmental* duplications and sites of large scale rearrangement between the great apes, suggesting that the genomic *architecture* plays an important role in mediating *evolutionary* rearrangements.

Furthermore, recent comparative analyses of breaks in synteny between the genomes of humans and mice also show a highly significant enrichment of duplicated sequences at the sites of breakage.

Although definitive evidence showing that these duplicated sequences are the cause, and not a consequence, of the rearrangements, they strongly support a nonrandom model in which chromosomal evolution is driven by the duplication architecture of the genome. However, segmental duplications may also play an important role in genome evolution in other ways besides mediating large-scale rearrangement.

Many duplications contain either partial or complete gene sequences, and their proliferation can, therefore, create additional copies of transcripts that have the potential to evolve independently from one another. It has been suggested that such duplication removes the pressure for conservation of function that would normally constrain translational change, and that gene duplication may, therefore, be one of the major forces involved in generating proteorne diversity.

Although it is true that the majority of such duplication events will create nonfunctional pseudogenes owing to the loss of crucial exons, regulatory elements, or the accumulation of deleterious mutations, several important examples exist of genes that have acquired novel functions following duplication events.

These include the evolution of trichromatic color vision in the apes and monkeys and novel immunological functions of the eosinophil proteins *ECP* and *EDN,* demonstrating the ability of whole gene duplication to drive rapid phenotypic change.

Similarly, gene duplication provides an opportunity for the daughter genes to achieve functional specialization in response to environmental pressures, acting as novel substrates for adaptive evolution. Segmental duplications may also lead to the modification or creation of new genes by a number of different mechanisms. Novel exons may be inserted into an existing gene, a mechanism termed exon shuffling, as has been reported for the fibrinolytic factors.

Alternatively, the juxtaposition of segmental duplications containing different exonic structures has the potential to join two independent *transcriptional* units, thereby creating novel fusion transcripts lacking in functional constraints and able to evolve rapidly. Examples of these include *POM-ZP3, PMCHL1-PMCHL2, CECR7, KIAA0187.* and *USP6.* Segmental duplications can also frequently lead to the creation of novel splice variants in *existing* genes.

Indeed, analysis of alternatively spliced genes has shown that as many as 10% of all splice variants may have arisen from an initial intragenic duplication event. Finally, a recent survey of segmental duplications in the human genome found that approx 6% of transcribed exons are located in recently duplicated sequence.

Genes associated with immunity and defense, membrane surface interactions, *drug detoxification*, and growth and development were preferentially enriched, suggesting that gene duplication plays a major role in the evolution of these functions.

Thus, in contrast to the popular conception in which single nucleotide polymorphisms and small mutations are the major method of evolutionary change, larger genomic rearrangements may instead represent the driving force behind much of recent primate evolution, facilitated by the presence of abundant highly homologous duplicated sequences throughout the genome.

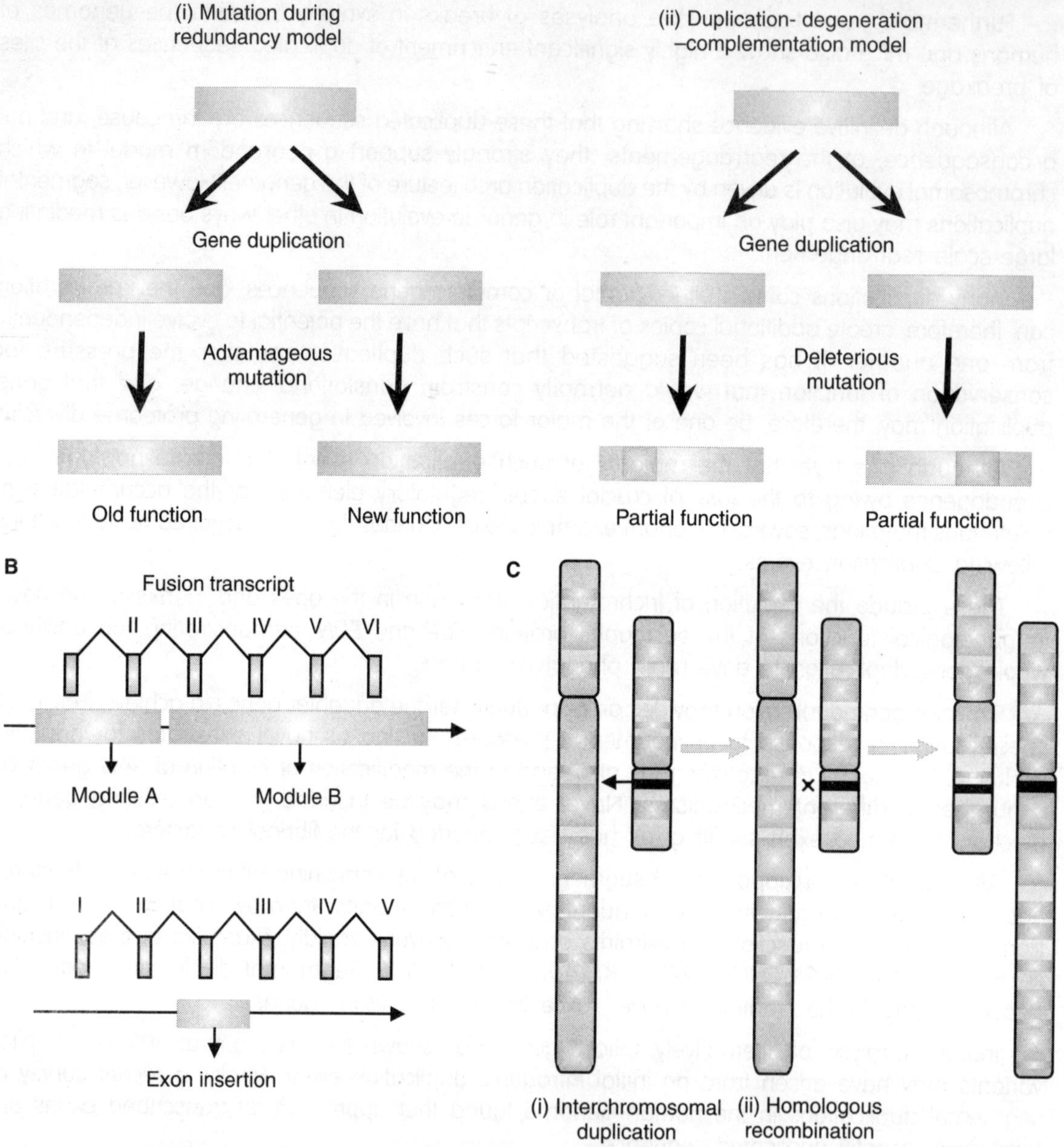

Figure 2.4: Mechanisms of phylogenetic changes : (A) gene evolution by complete duplication provides a period of relaxed selection, allowing the resultant copies to diverge and potentially acquire novel function. Two alternative models have been proposed: (1) the classical model of "mutation during redundancy" proposes that advantageous mutation in one duplicated copy leads to gain of function and (2) an alternative model, termed "duplication–degeneration–complementation," that suggests deleterious mutations occur in both copies, leaving each with partial functions that complement one another. (B) The creation of novel fusion transcripts by the juxtaposition of exons of independent origin, or alternatively insertion of novel exon(s) into an existing gene. (C) A hypothetical duplication-mediated chromosomal rearrangement that occurs as a result of an interchromosomal duplication followed by nonallelic homologous recombination.

VARIATION IN GENOMES

The role of duplications as mediators of recurrent genomic disorders is well established, with more than 25 separate syndromes now recognized. However, there is a growing body of evidence that suggests segmental duplications may also play a significant role in normal variation.

The existence of large genomic polymorphisms, originally termed heteromorphisms or *euchromatic variants*, has been recognized since the advent of high-resolution cytogenetic banding techniques (summarized at http://www.som.soton.ac.uk/research/geneticsdiv/ anomaly %20 register/).

Using more targeted molecular analyses, and more recently with the advent of high-throughput methodologies such as array CGH, large numbers of *submicroscopic* deletion/duplication and inversion polymorphisms have now been documented. As with the recurrent genomic *disorders*, many of these similarly seem to he mediated by the presence of *flanking* duplicated sequences.

In some instances these *polymorphisms* are associated with extreme divergences from one individual to another. For instance, copy number of *CYP2D6* has been found to vary from 0 to 13 per diploid genome, while the olfactory receptor genes show enormous variation in their *localization* within subtelomeric regions of human *chromosomes*.

The majority of polymorphic duplication and deletion events identified to date involve genes with functions in metabolism and immunity. Alterations in copy number of these genes often have profound effects on an individual's metabolic rate or resistance to environmental pathogens, and as such may be significant susceptibility factors for many common human diseases.

Although inversions are not usually associated with alterations in gene copy number and, thus, do not have a primary effect on phenotype, several of the polymorphic inversions identified to date confer a predisposition to further chromosomal rearrangement in subsequent generations.

Clusters of olfactory receptor genes at 4p16 and 8p23 are both sites of large common inversions which occur in the heterozygous state at significantly higher frequencies in the transmitting parents of individuals with the recurrent t(4; 8)(p16;p23) translocation, and inverted duplications, marker chromosomes and deletions of 8p23, respectively.

Similarly, a heterozygous inversion of the 1.6-Mb Williams-Beuren syndrome critical region at 7q 11.23 is observed in one-third of the parents who transmit a deletion of this same region to their offspring, a frequency much higher than that observed in the normal population.

Finally, an approx 4-Mb inversion in Yp11.2, present on one-third of normal Y chromosomes, is observed in nearly all cases of sex reversal with translocation between the X/Y homologous genes *PRKX* and *PRKY*.

Presumably these inversions predispose to secondary re-arrangement by switching the orientation of large homologous duplications on one chromosome *homologue*, allowing their subsequent alignment during synapsis, and, hence, facilitating unequal recombination.

Given the difficulty of *ascertaining* these submicroscopic inversions, it seems likely thct many other common inversion polymorphisms exist within the human genome.

Table 2.1: Summary of Polymorphic Deletions, Duplications, and Inversions Within the Human Genome"

Gene name(s)	Locus	Population frequency	Diploid copies	Size of variant segment	Associated phenotype
(a) Deletions/duplications					
GSTM1	1p13.3	>3%	1-3	18 kb	Altered enzyme activity
RHD	1p36.11	15-20%	0-2	approx 60 kb	Rhesus blood group sensitivity
CYP21A2	6p21.32	1.6%	2-3	35 kb	Congenital adrenal hyperplasia
LPA	6q25.3	94%	2-38	5.5 kb	Altered coronary heart disease risk
b*-Defensin gene cluster	8p23.1	Approx 90%	2-12	>240 kb	Immune system function?
IGHG1 region	14q32.33	12-74%	1-6	5-170 kb	Immune system function?
NF1/IGVH/GABRA5 pseudogenes	15q11.2-q13	80%	2-8	Approx 1 Mb	Unknown
CHRNA7/CHRFAM7A	15q13.3	10-80%	2-4	>270 kb	Unknown
IGVH/SLC6A8/CDM	16p11.2	Unknown	2-12	Undefined	Unknown pseudogenes
CCL3-L1/CCL4-L1	17q12	51/27%	0-6	>2 kb	Immune system function?
DNMT1	19p13.2	93%	2-5	>60 kb	Unknown
CYP2A6	19q13.2	1.7%	2-3	7 kb	Altered nicotine metabolism
IGL	22q11.22	28-85%	2-7	5.4 kb	Altered IgK:IgX in B lymphocytes
GSTT1	22q11.23	20%	0-2	>50 kb	Altered susceptibility to toxins and cancer
CYP2D6	22q13.1	1-29%	0-13	Undefined	Altered drug metabolism
Spermatogenesis genes	Yq11.2	3.2%	0-1	1.6 Mb	Low penetrance spermatogenic failure
OR genes subtelomeres	Multiple variable	Highly	7-11	>90 kb	Unknown

Table Contd.

Gene name(s)	Locus	Population frequency	Diploid copies	Size of variant segment	Associated phenotype
(b) Inversions					
NPHP1	2q13	21%		290 kb	Non-pathogenic
OR genes	4p16	12.5%		approx 6 Mb	Predisposition to t(4;8)(p16;p23) translocation
Williams-Beuren syndrome critical region	7q11.23	Unknown		1.6 Mb	Predisposition to deletion of WBS critical region (and atypical WBS phenotype?)
OR genes	8p23	26%		4.7 Mb	Predisposition to inv dup(8p), +der(8)(pter p23.1.:p23.2-pter) and del(8)(p23.1 ;p23.2)
EMD	Xq28	33%		48 kb	Non-pathogenic
Proximal Yp	Yp 11.2	33%		approx 4 Mb	Predisposition to *PRKX/PRKY* translocation

CONCLUDING REMAKS

There is a growing recognition that the duplication architecture of the mamalian genome plays a fundamental role in shaping the processes of evolution, genomic diversity, and phenotype. Because of the limitations of available technologies, until recent times athe major foucus of gentic research has been on alterations at either the cytogenetic band or DNA sequence level.

However, the availability of the complete genome sequence in combination with the adevent of techniques such as pulsed-field gel electrophoresis, fluroscence *in situ* hybridization, and array CGH now allows the study of an additional level of human variations, one below that visualized with the light microscope but above that detected by most sequence-based *methodologies*.

The prevalence of such variation is just beginning to be uncovered. Given the high gene content of duplicated sequences, it seems likely that many genomic polymorphisms may play a significant role in human *phenotypic* variation.

Chapter 3

Genomic Mutations

Genomes of living organisms consist of genes and noncoding DNA arranged in particular orders on one or more chromosomes. *Eukaryotes* usually have several pairs of linear chromosomes (humans have 23 pairs in each *somatic cell*), but prokaryotes typically have only one or a few circular (sometimes linear) chromosomes. The order of genes (or sets of genes) along a particular chromosome of a particular species is ordinarily conserved, and the list of genes or physical markers along a genome, together with the distances between these genes or markers, is called either a *genetic* or *physical map*, depending upon the units in which distances are measured. For each species, there is a particular genetic map that distinguishes it from other species.

For example, even though the mouse *(Mus museulus)* contains most of the same types of genes found in humans, the genetic map of the mouse differs from that of humans. This is illustrated in Figure elsewhere in this chapter for human *(Homo sapiens)* chromosome Hsa6, which contains blocks of DNA sequences corresponding to portions of six different mouse chromosomes.

Clearly, segments of DNA in the two species were rearranged with respect to each other during their independent descent from a common ancestor. If blocks of sequences are excised from one location and inserted into another, those sequences are said to have been *translocated.* On the other hand, it is possible that the chunks of sequence in a syntenic block remain together but are *permuted* in one organism relative to the other. An example of permutation is blocks 1-2-3 of Hsa6 in Figure elsewhere in this chapter, which appear in the order 3-1-2 in the mouse chromosome Mmu13. Sometimes sequences in the genome of an organism are duplicated elsewhere in the same genome (either on the same or on a different chromosome). For example, it is estimated that around 5% of the human genome (exclusive of transposable elements; consists of duplicated sequences.

COMPARATIVE STUDY

When comparing genomes of two organisms such as mice and humans, we sometimes

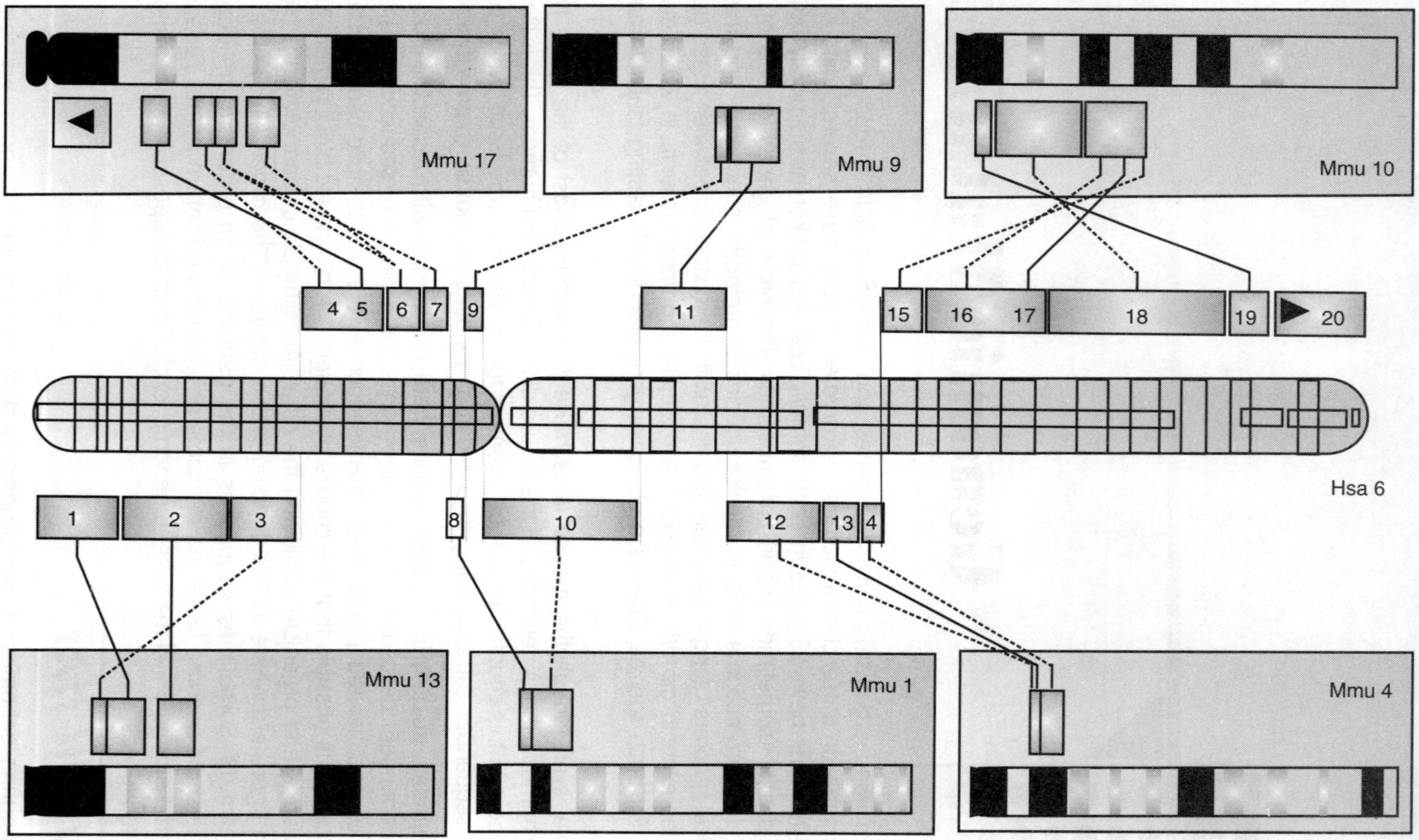

Figure 3.1: Syntenic blocks conserved between human chromosome Hsa6 and mouse chromosomes. Broken lines indicate regions that pit appear in inverted orders in the two organisms.

observe local similarities in the gene content and gene order. If two or more genes that occur together on a single chromosome in organism B also occur together on a particular chromosome in organism C, this occurrence is an example of *conserved synteny*.

As shown in Figure elsewhere in this chapter. The gene orders may or may not be conserved within syntenic blocks that are distributed among different chromosomes in a related organism. We will use a simple example to show how such patterns can arise by genome rearrangement. Figure elsewhere in this chapter shows four different ancestral "chromosomes," each distinguished by a different color.

Each ancestral chromosome is of identical size, and each has ten "genes" 0, 1, ..., 9. Each number/color combination represents a unique gene. This means that 2-green is not the same gene as 2-red. The telomeres are indicated by a single slash, /. We will model an evolutionary process indicated in the following diagram:

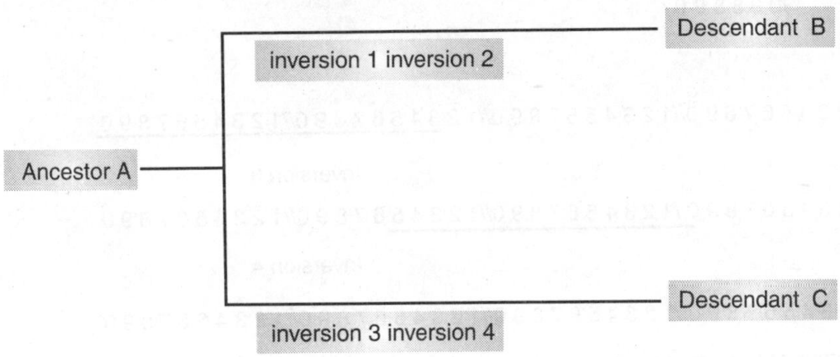

To model the genome rearrangements, we line up the A chromosomes in tandem and then invert different portions of this combined string in two independent steps to obtain chromosomes of Descendant B. The same process is repeated by inverting different segments to produce Descendant C.

By inversion, we mean that we take a portion of the sequence of genes and reverse their order relative to the surrounding genes. For example, if the middle three genes represented by *1 2 3 4 5* were inverted, the result would be *1 4 3 2 5*. Inversions of chromosomal regions are well-documented in most organisms for which there are genetic data. Conserved synteny occurs when two or more genes occur together on any chromosome of Descendant B *and* on any chromosome of Descendant C.

For example, 0-black and 9-black occur together on chromosome C4, and they also occur together on chromosome B2. 0-red and 5-green appear together on chromosome C3, but they do not appear together on any of the B chromosomes. This is *not a* conserved synteny. In the modelled example, C1 shares regions of conserved synteny (and syntenic blocks) with B1, B2, B3, and B4. Inversions of the type shown in this example can produce results like those illustrated in Figure elsewhere in this chapter.

Also Hsa6 has segments that are syntenic with six mouse chromosomes. The observed arrangements of genes on vertebrate chromosomes resemble what we would see if a process

Ancestor A:

/1 2 3 4 5 6 7 8 9 0//1 2 3 4 5 6 7 8 9 0//1 2 3 4 5 6 7 8 9 0//1 2 3 4 5 6 7 8 9 0

Inversion 1

/1 2 3 4 5 6 7 8 9 0//1 2 3 4 5 6 7 8 9 0//1 2 3 4 5 6 7 8 9 0//1 2 3 4 5 6 7 8 9 0

Inversion 2

/1 2 3 4 5 6 7 8 9 0//1 2 3 4 5 6 7 8 9 0//1 2 3 4 5 6 7 8 9 0//1 2 3 4 5 6 7 8 9 0

Chromosomes of Descendant B:

B1 /1 2 3 4 5 6 7 7 6 5 4 3 2 1 /
B2 /0 9 8 7 6 5 4 5 6 7 8 9 0 /
B3 /1 2 3 4 3 2 1 /
B4 /0 9 8 8 9 0 /

Ancestor A:

/1 2 3 4 5 6 7 8 9 0//1 2 3 4 5 6 7 8 9 0//1 2 3 4 5 6 7 8 9 0//1 2 3 4 5 6 7 8 9 0

Inversion 3

/1 2 3 4 5 6 7 8 9 0//1 2 3 4 5 6 7 8 9 0//1 2 3 4 5 6 7 8 9 0//1 2 3 4 5 6 7 8 9 0

Inversion 4

/1 2 3 4 5 6 7 8 9 0//1 2 3 4 5 6 7 8 9 0//1 2 3 4 5 6 7 8 9 0//1 2 3 4 5 6 7 8 9 0

Chromosomes of Descendant C:

C1 /1 2 3 4 5 6 7 8 3 4 9 8 7 6 5 4 3 2 1/
C2 /0 9 8 7 6 5 4 3 2 1/
C3 /0 5 6 7 8 9 0/
C4 /0 9 2 1/

Relationships between chromosome C1 and chromosomes B1–B4:

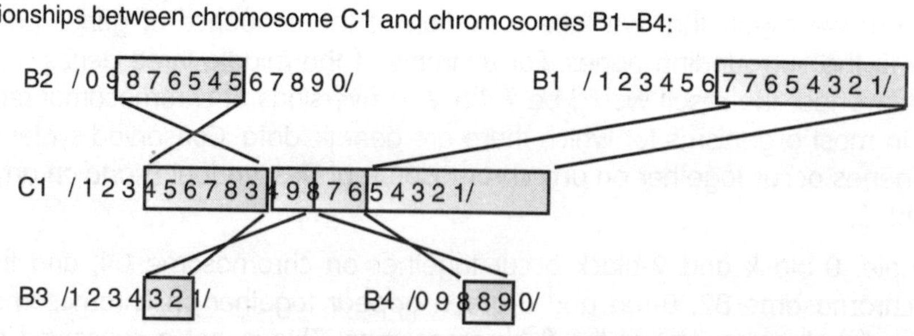

Figure 3.2: Model producing conserved synteny in descendant chromosomes B and C after independent inversion steps starting with ancestral chromosome A. Telomeres (/) divide the "genes" (numbered) into four chromosomes. Chromosomes are placed end-to-end before inversions (reversals) are performed to simulate translocation. Two inversion steps on the path leading to B and two inversion steps on the path leading to C are sufficient to generate syntenic relationships.

similar to the one modelled had occurred. In the example modeled, placing chromosomes end-to-end and then inverting segments corresponds to a translocation if the two breakpoints associated with the inversion fall within two chromosomes and to an inversion if they both fall within only one chromosome. The model just considered should not be taken as the actual mechanism for chromosome evolution. It merely illustrates how one type of rearrangement process can produce results similar to what is observed in contemporary eukaryotic chromosomes.

However, rearrangements like these have been observed in closely related contemporary species. For example, human chromosomes 1 and 18 have large inversions compared with the corresponding chimpanzee chromosomes, and chimpanzee chromosomes 12 and 13 together contain material that is located in human chromosome 2. Chromosome rearrangements thus may be used to help track evolutionary processes.

Abrration of Circular Genomes

As we mentioned above, bacteria often have circular chromosomes. Circular chromosomes are also found in mitochondria and chloroplasts. Remember that mitochondria and chloroplasts are organelles that are thought to have been reduced from eubacterial endosymbionts. These organelles contain many copies of their respective genomes. For example, human mitochondria contain approximately 10^3-10^4 DNA circles, each about 16,000 by long and containing 37 genes.

Circular chromosomes or organelle DNAs of related organisms may not have identical gene arrangements. Because of their circular structures, it is possible to invert segments of circular chromosomes by a single crossover event, which produces an *inversion,* a permutation of the original sequence with two novel junctions, or breakpoints.

When this occurs repeatedly, considerable "shuffling" of the gene orders may occur. One of the most interesting applications of computational biology is the reconstruction of phylogenies of organisms. This is often accomplished by using the actual DNA sequences of genes or gene segments. Human mtDNA sequences have been used to reconstruct the ancestor-descendant relationships leading to contemporary human populations.

Organellar DNAs are particularly useful sources of ancient DNA samples because the copy numbers of mtDNA sequences are thousands of times more abundant than particular sequences from nuclear DNA. But one problem with organellar DNA and other small circular DNAs is that they may have high mutation rates, which makes it hard to establish with confidence the time separating distantly-related gene homologs.

One way of dealing with this problem is to delineate blocks of conserved genes, or *conserved segments.* Evolutionary relationships are then analyzed in terms of these conserved segments rather than the actual DNA sequence. In this chapter, we will show how genome sequences can evolve by inversion (which we will be calling these *reversals*), and we will be estimating the *distance* between two genomes by using the number of reversals required to convert the order of conserved segments in one genome into the order found in the other.

A set of distances relating a collection of genomes can be used to construct phylogenetic trees, which are graphs of ancestor-descendant relationships between these genomes. These are discussed in detail in other chapter this book.

UNICHROMASOMAL GENOMES

Principles

To begin, we analyze rearrangements of linear, unichromasoinal genomes (viral, organellar, bacterial), with each segment representing a group of contiguous genes (a conserved segment). As noted, many of these genomes are circular but have an analog linear genome, as otherwise the analysis is *needlessly* complicated. Because each. conserved segment contains several genes, we can define an *orientation* for each segment. Before we deal with orientation, we first discuss the analysis of ordered elements, g_1, g_2,... g_n. In Section 5.3, each of these elements gi will be identified with conserved segments, and the permutation G will be identified with a particular genome.

The identity permutation of these elements is defined as the permutation in which the indices increase in numerical sequence:

$$I : g_1\ g_2\ g_3\ \dots\ gn$$

The elements may also appear as a permutation of the elements in the identity permutation. For example, if $n = 7$, permutation G might have elements in the following order:

$$G : g_3\ g_7\ g_2\ g_6\ g_4\ g_1\ g_5.$$

For simplicity, we represent this particular permutation by just listing the index numbers,

$$G : 3\ 7\ 2\ 6\ 4\ 1\ 5,$$

and we represent the identity permutation similarly:

$$I : 1\ 2\ 3\ 4\ 5\ 6\ 7.$$

How can we transform G into I ? One way is to pick up the numbers one or more at a time and drop them into the correct position. The equivalent biological process is translocation. Transformation of G to I by translocation is diagrammed as:

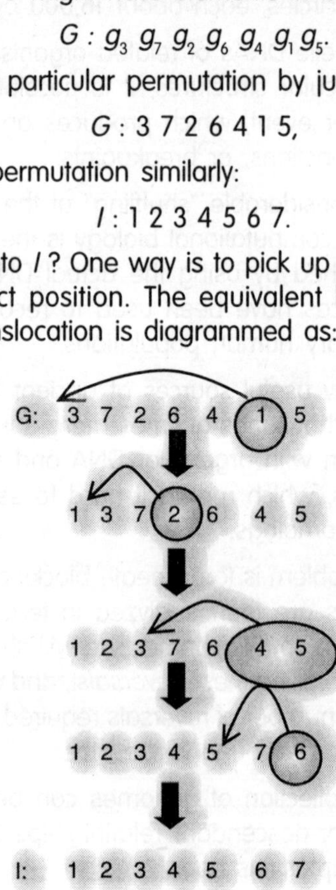

As indicated above, a common way for DNA molecules to rearrange is by inversion. The *equivalent* term in computer science is *reversal.* An example of a reversal performed on I is illustrated below, where the interval 3 4 5 is reversed:

1 2 | 5 4 3 | 6 7

The order of the elements in the box is reversed from the order in which they were found in the identity *permutation.* In general, if we are given a *permutation*:

$$g_1\ g_2\ g_3 \cdots g_{i-1}\ g_i\ g_{i+1} \cdots g_{j-1}\ g_j\ g_{j+1} \cdots g_n,$$

a reversal of the interval [*i, j*] yields the permutation

$$g_1\ g_2\ g_3 \cdots g_{i-1}\ [g_i\ g_{j-1} \cdot\cdot\ g_{i+1}\ g_j]\ g_{j+i} \cdot g_n$$

The order of the elements in brackets has been reversed. Adjoining elements may represent either an adjacency or a breakpoint.

$g_k g_j$ =* *Adjacency* if $j = k \pm 1$,

$g_k g_j$ =, *Breakpoint* if $j =^* k \pm 1$.

Since in the second string immediately above g_{i-1} is followed by g_j, there is one breakpoint after g_{i-1}. Similarly, since *gi* is followed by g_{j+1}, there is another breakpoint just before g_{j+1}. The *inversion* has produced two *breakpoints.* Let's now return to our previous example *G* = 3 7 2 6 4 1 5. We want this to represent a circular array of permuted elements.

This means that the first element in our linear representation is connected to the last element. If the first element is in the correct position, then we do not consider it to be preceded by a breakpoint, even if the last element is not yet in the correct position. (We don't want to count the same breakpoint twice.)

By our definition of breakpoint shown above, you can see that there are seven breakpoints. Instead of transforming *G* to *I* by transposition of elements, we can effect this transformation by successive inversions (reversals) of two or more elements (brackets enclosing the elements to be inverted in each step):

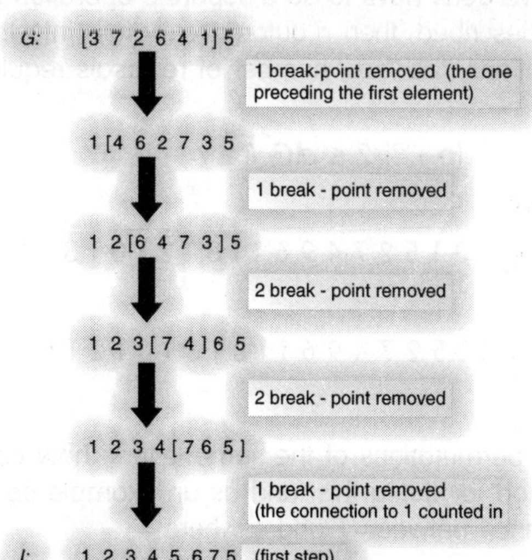

As we see from this example, we have removed the seven breakpoints that were in the original sequence. It should be clear that the maximum number of breakpoints that can be removed by a reversal is 2. Note that the reversals do not always remove two breakpoints.

We define $d(G, I)$ as the *minimum number of reversals* required to convert G into I. Notice that reversals destroy breakpoints during the transformation process. If the *number of breakpoints* in G is $b(G)$, then it should be clear that

$$d(G, I) \geq b(G)/2.$$

Also, each reversal can remove at least one breakpoint, so

$$d(G, I) \leq b(G)$$

and it follows that

$$b(G)/2 \leq d(G, I) \leq b(G).$$

Let's check this against the example that we just used. We should recognize that there is a breakpoint just before 3 in G (that element should have been 1 if there had been no breakpoint). Using the same reasoning, there is also a breakpoint after the last element. In fact, $b(G) = 8$, and $d(G, I) \geq 8/2$.

We know that it will take at least four reversals to transform G to I. In the illustration, we employed five reversals. What is the greatest number of reversals required to transform the most unfavorable *permutation* G to I (worst-case scenario)? What is the least number of reversals it would take to transform G to I?

We can easily estimate what the greatest number of reversals (worst case) would be. We can imagine that you start from this worst-case permutation by reversing the set of elements from the initial one in the permutation up to and including g_1. This puts g_1 in front. Then reverse all elements beginning with the second in the permutation up to and including g_2.

Continue the process until the identity permutation is achieved. In the worst case, we will have to do this for $n - 1$ elements. We don't have to do a separate operation for g_n, because if the first $n - 1$ elements are sorted as described, then n automatically falls into the last position. It can be shown that, for some permutations, the *least* number of reversals required is $(n + 1)/2$. Thus we have

$$(n + 1)/2 \leq d(G, I) \leq n - 1.$$

Examples of the worst case are

$$G_{13} = 3\ 1\ 5\ 2\ 7\ 4\ 9\ 6\ 11\ 8\ 12\ 10\ 13$$

and

$$G_{14} = 3\ 1\ 5\ 2\ 7\ 4\ 9\ 6\ 11\ 8\ 13\ 10\ 12\ 14.$$

Cycle Decomposition

Given X and Y, which are permutations of the same letters, how can we determine $d(X, Y)$? We describe a graphical method for doing this, and as an example (to keep the graphs simple) we estimate the reversal distance between F and I, where

F: 1 2 4 5 3 I: 1 2 3 4 5.

First, extend each permutation by adding 0 before the first element and $n + 1$ after the last element, where n is the number of elements. These changes do not alter the reversal distances between the permutations because the elements are in their correct positions.

$$F': 0\ 1\ 2\ 4\ 5\ 3\ 6 \qquad\qquad I': 0\ 1\ 2\ 3\ 4\ 5\ 6.$$

Each element index corresponds to a *vertex* of the graph, and lines connecting the vertices are called *edges*. First, we connect all adjacent vertices of F¢ with black edges:

Now, in the same way, connect all of the adjacent I' vertices with grey edges.

Γ (F′, I′):

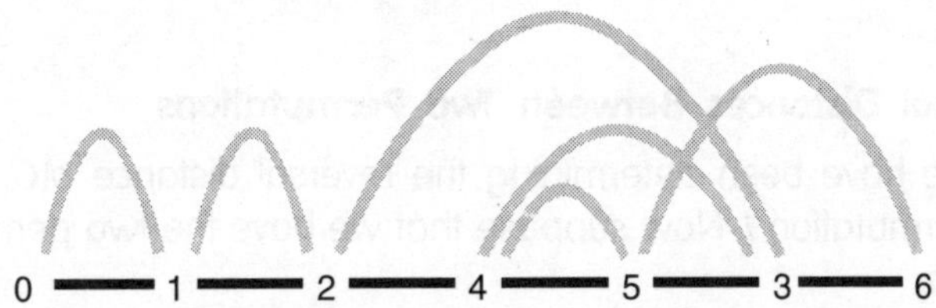

The diagram above is called a graph, Γ (F′, I′). This type of graph is said to be *balanced* because the number of black edges at each vertex is equal to the number of grey edges. II can be decomposed into *cycles* that have edges with alternating colors and that don't share edges with other cycles (i.e., they are *edge-disjoint*). We do not give an algorithm for this decomposition, which in general is a challenging problem. We define $c\,(F', I')$ as the maximum number of alternating, edge-disjoint cycles into which Γ(F′, I′) can be decomposed. There are four in this case, and they are shown in the following diagram:

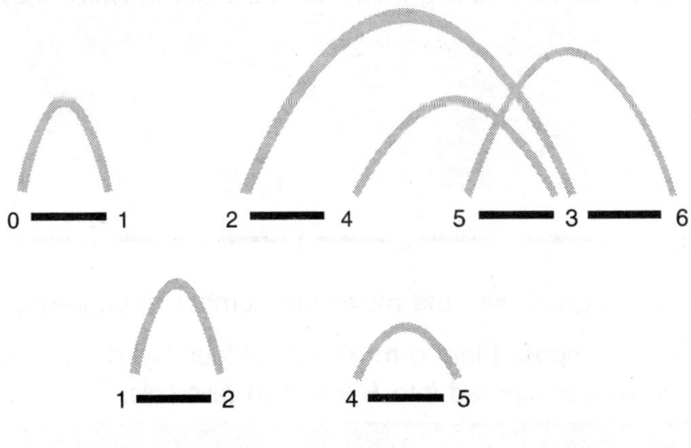

The reason that we did this is that for most of the usual and interesting biological systems, the equality holds for the next expression for reversal distance between two permutations X and Y:

$$d(X, Y) \geq n + 1 - c(X, Y) = \bar{d}\,(X, Y).$$

The general problem, just like the TSP, is NP-complete. In the case of permutation F, the number of elements n was equal to 5, and we found that $c(F, I) = 4$. This means that $\bar{d}\,(F, I) = 2$. In other words, we should be able to transform *F* into the identity permutation with just two reversals. They are shown below (again, we reverse the order of elements between the brackets):

F: 1 2 [4 5] 3

↓

1 2 [5 4 3]

↓

I: 1 2 3 4 5

Estimating Reversal Distances Between Two Permutations

Up to now, we have been determining the reversal distance *d*(*G*, *I*) between permutation *G* and the identity permutation *I*. Now suppose that we have the two permutations *F* and *H*, with the following elements:

F: 1 2 4 5 3 *H*: 1 4 3 5 2.

The distance between *F* and *H*, *d*(*F*, *H*) and *d*(*F*, *H*), is calculated just as was done above, except that this time the second set of edges will not be drawn connecting vertices in increasing numerical order but instead *in the order in which these vertices appear in H*.

For simplicity, we again represent each element by its numerical index and, as before, we extend each permutation by prefixing it with 0 and by adding $n + 1$ at the end (6 in this case). We connect the vertices of *F* with black edges, taking vertices in succession as they appear in *F*. Then we add grey edges between vertices, *taking them, 'an the order in which they appear in H*. The result is:

Γ (*F*, *I*): 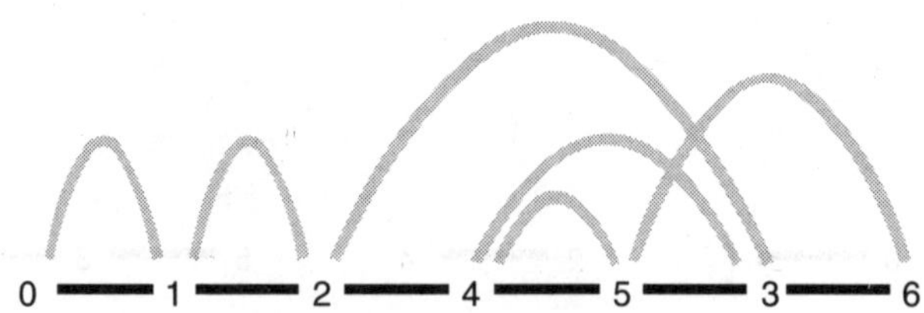

The decomposition of this graph into the maximum number of edge-disjoint cycles is:

Since Γ(*F*', *H*') can be decomposed into a maximum of four cycles, $\bar{d}(F, H) = 5 + 1 - 4 = 2$, and we expect that *F* can be transformed into *H* with two reversals.

ANALYZING GENOMES

In the previous section, we showed how to analyze permutations. When we start to analyze genomes, the elements are now genes (or sets of genes in an interval that contains no breakpoints), and these have *orientation*. Gene orientation might be indicated by transcription direction, which might be denoted as "+" if it is transcribed left to right in a particular map as drawn and "–" if it is transcribed right to left. Genes α, β, and γ might appear in one genome in an order +α + β + γ but in another genome in the reverse order –γ – β – α. These might represent conserved segment g_r which would be represented as $+g_i = +\alpha + \beta + \gamma$ in the former case and as $-g_i = -\gamma - \beta - \alpha$ in the latter case.

For example, consider the two genomes *J* and *K* below:

J : $+g_1 +g_5 -g_2 +g_3 +g_4$ K : $+g_1 - g_3 +g_2 +g_4 - g_5$

These could be represented graphically as (again, employing only the subscript labels):

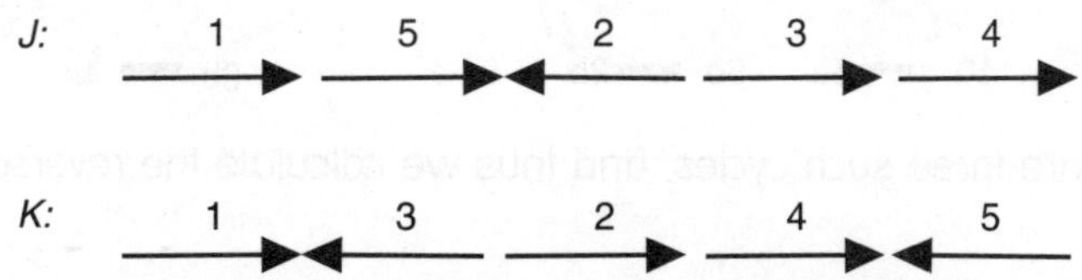

As before, we are seeking the reversal distance *d(J, K)*, which is the minimum number of reversals needed to convert *J to K*. But now when we do reversals to sort the segment orders, we must also keep track of the orientations of the conserved segments to make sure that both gene order *and* orientation are correct.

Just as we did before, we calculate *d(J, K)* by constructing an edgecolored graph and then counting the maximum number of alternating cycles into which the graph can be decomposed. But we must modify the procedure to keep track of the orientations of the conserved segments. (As noted above, the genome rearrangement problem with unsigned segments is NP-complete. Although signed segments would appear to make the problem more difficult, actually there is an $O(n^2)$ *algorithm*! Further discussion of this is beyond the scope of this chapter.) To track the orientation of conserved segment *i*, label its left end (tail) ia and its right end (head) *ib*. If genome *J* has *n* conserved segments, there are now 2*n* elements representing their ends, thus *specifying* their location and orientation.

We prefix the list of elements in *J* with 0 and add element $n + 1$ to the right end as before to yield a total of 2*n* + 2 elements. Genomes J and K are each represented by a particular permutation of the same 2*n* + 2 elements, and the approaches described in the previous section can now be applied. Our example above now becomes

J: 0 *la lb 5a 5b 2b 2a 3a 3b 4a 4b* 6.

K : 0 *la lb 3b 3a 2a 2b 4a 4b 5b 5a* 6.

Now we create an edge-colored graph, but we only draw edges where adjacent conserved segments are connected (i.e., we won't draw edges between the ends of the same conserved segment, *ia* and *ib)*. We designate edges between conserved segments in *J* with black lines and edges between conserved segments in K with grey lines.

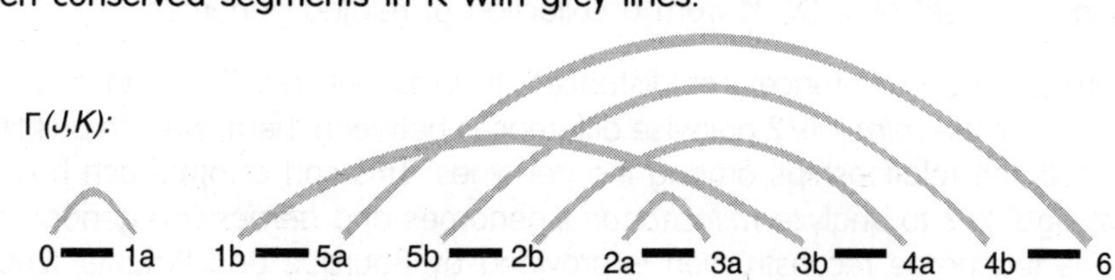

The decomposition of this graph into edge-disjoint alternating cycles is:

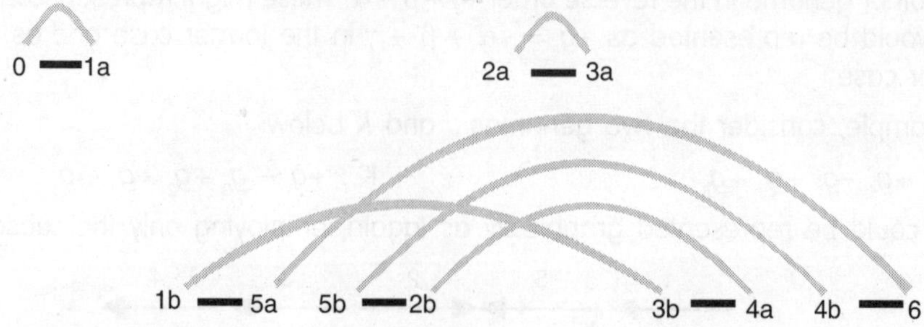

We find that there are three such cycles, and thus we calculate the reversal distance between *J* and *K*:

$$(J, K) = n + 1 - c\,(J, K) = 5 + 1 - 3 = 3.$$

We then expect to be able to transform *J* to *K* by using three reversals. These are shown in the following diagram (with [ ] indicating set of conserved segments being reversed):

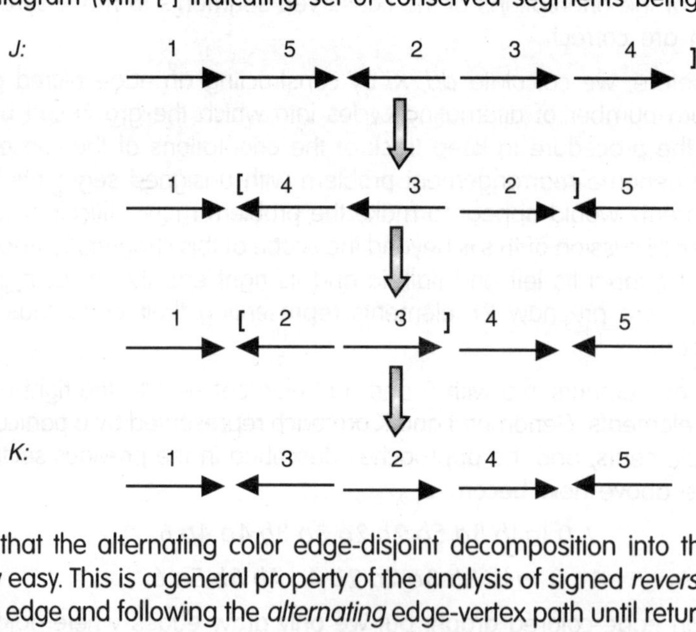

We notice that the alternating color edge-disjoint decomposition into the maximum number of cycles is now easy. This is a general property of the analysis of signed *reversals*. It is accomplished by choosing an edge and following the *alternating* edge-vertex path until returning to the beginning.

There are no choices to make. A challenging problem became easy with the addition of biological detail. Let's remember again at this point why we are going through all of this. What we are extracting is the set of $\bar{d}\,(X, Y)$ from a collection of related genomes.

These are pairwise differences, or distances. It turns out that if we have a collection of *m* genomes and all of the *m(m* – 1)/2 pairwise differences between them, we can build a phylogenetic tree that reveals the relationships among the genoines. This sort of approach has been used by different investiga. tors to analyze *mitochondrial* genomes and *herpesvirus* genomes. A discussion of approaches togenome reconstruction is provided by Bourque and Pevzner (2002).

Finding reversal distance *d(X, Y)* between genome *X* and genome *Y*

Given two related genomes containing the same set of conserved segments, but in different, orders and orientations:

Step 1: Label the ends of each conserved segment *i* as *ia* and *ib*, for all $I <^{*}{}_{i} <^{*} n$, where *n is* the number of conserved segments. These labeled ends will be the vertices of the graph.

Step 2: Write down in order the ends of the oriented conserved segments in the sequence that they occur in genome *X*.

Step 3: Add 0 at the beginning of the permutation representing genome *X*, and add *n* + 1 at the end of the permutation.

Step 4: Repeat steps 2 and 3 for genome *Y*.

Step 5: Connect vertices from *different* fragments with a black line if they are joined in genome *X*. Connect vertices from *different* fragments with a grey line if they are joined in genome *Y*. (No connections will be made for *is* and *ib.)*

Step 6: Decompose the graph generated in step 5 into the maximum number of edge-disjoint alternating cycles. This number is *c(X, Y)*.

Step 7: Compute the reversal distance, the minimum number of reversals of one or more conserved segments needed to transform one genome into the other: $d(X, Y) = n + 1 - c(X, Y)$.

SIGNIFICANCE OF GENOMIC STUDIES

The illustrations above applied to unichromosomal linear genomes, which can be generalized to circular genomes of *eukaryotic organelles* and many bacteria. Relating multichromosomal organisms by genome rearrangements is much more complicated, although this has been done for humans and mice. We will illustrate these more complex problems without going into detail. Genome rearrangements are also discussed in other chapter of this book. We continue to use the mouse and human genomes as examples.

Synteny & Conserved Segments

In the first and second sections of this chapter, we discussed conserved synteny and conserved segments. We need to be a bit more precise to *accommodate* the data. According to one definition, a conserved segment is a region found in two genomes in which homologous genes retain the same order and relative map position in both genomes.

Does it follow that conserved segments contain *no* gene rearrangements? It is estimated that there are about 25,000 mammalian genes in genomes of about 3×10^9 bp. This means that, on average, conserved segments that include three genes would exceed 300,000 bp. Is it possible that there have been rearrangements in intervals of size 200,000 to 1,000,000bp that would not be detected in conserved segments? The availability of sequenced genomes makes possible more *precise* measures of conserved *synteny* and conserved segments, based upon a higher density of *physical markers*.

Regions of local similarity between two sequenced genomes can be readily identified (sequence similarity and alignment are discussed in the next chapter). Suppose that subsequence f_i in genome

F is similar to subsequence gi in genome G, and that (a) alignment of f_i to *G* yields g_i as the highest-scoring alignment and that (b) alignment of g_i to *F* yields *fi* as the highest-scoring alignment (i.e., f_i and g_i are reciprocal "best hits"). The subsequences *fi* and *gi* can be used as *sequence anchors* or *landmarks* to aid in comparing the two genomes.

In principle, there can be many more sequence anchors than there are genes and, unlike genes, their identification is uninfluenced by errors in gene annotation (e.g., incorrectly identifying a gene in *F* as an ortholog of a gene in G-see in other chapter of this book for how orthologs are identified).

In a recent study relating the mouse and human genome organizations, nearly 560,000 genome anchors were employed. Whereas conserved segments were defined above in terms of genes, synteny blocks can also be defined in terms of anchors.

A *syntenic block* is a set of sequence anchors that appear together (but not necessarily in the same order) in the genomes from two different organisms. Because there are so many more anchors than there are genes, the synteny blocks are based upon much finer *partitioning* of the chromosome than are conserved *segments* defined by genes.

Conserved segments may contain short inversions or other types of *microrearrangements* that are not detected by mapping with sequence anchors. Synteny blocks can be used to define *macrorearrangements* relating two chromosomes by *approaches* similar to those we have discussed above.

Two Dimensional Presentation

The methods we have discussed for single chromosomes have relied on making what were, in effect, one-dimensional diagrams. When comparing two chromosomes or genomes, it is possible to represent the relationships by using *two-dimensional* diagrams, with each axis representing the synteny blocks for each organism. The projection of the two-dimensional graph onto either axis produces a *unidimensional* diagram of the type that we have employed up to now.

Figure elsewhere in this chapter shows an example of a two-dimensional diagram comparing the human and mouse *X* chromosomes. The problem to be solved is exactly the same as the one that we have discussed in the unichromosomal case: determining the distance between the chromosomes by enumerating the number of reversals required to convert the order of synteny blocks found in mice into the order found in humans.

This is of course just a part of the larger problem, identifying the total rearrangement distance between the complete genomes. Exercise 9 illustrates cycle decomposition in two dimensions. To demonstrate how the observations recorded in Figure elsewhere in this chapter could have arisen, we perform a simple simulation.

Simulating Genome Rearrangements

Using R, provide an ordered set of 100 anchor loci for genome x. This might represent the ancestral genome. We will not perform any rearrangements on x.

```
x<-c(1:100)
```

Next we provide an ordered set of 100 anchor loci for genome y:

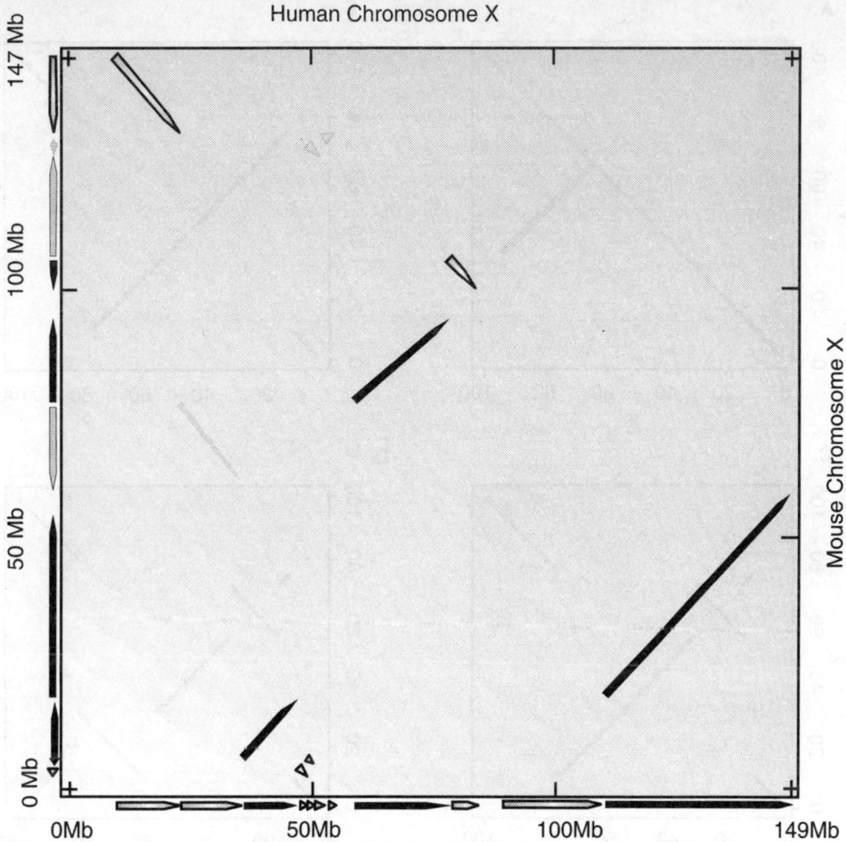

Figure 3.3: Synteny blocks shared by human and mouse X chromosomes. The arrowhead for each block indicates the direction of increasing coordinate values for the human X chromosome.

```
y<-c (1:100)
```

We now transform genome y into genome y4 through a series of four reversal steps of our choice, saving each intermediate genome yl, y2, and y3. y2 is produced by applying a second reversal to yl, which resulted from the first reversal, and so on.

```
yl<–y
yl [11: 751 <–yl [75:11]
#Reversal of interval 11 to 75 inclusive in y
y2<-yl
y2[15:301<-y2[30:151
#Reversal of interval 15 to 30 inclusive in yl
y3<-y2
y3[55:801<-y3[80:,551
#Reversal of interval 55 to 80 inclusive in y2
```

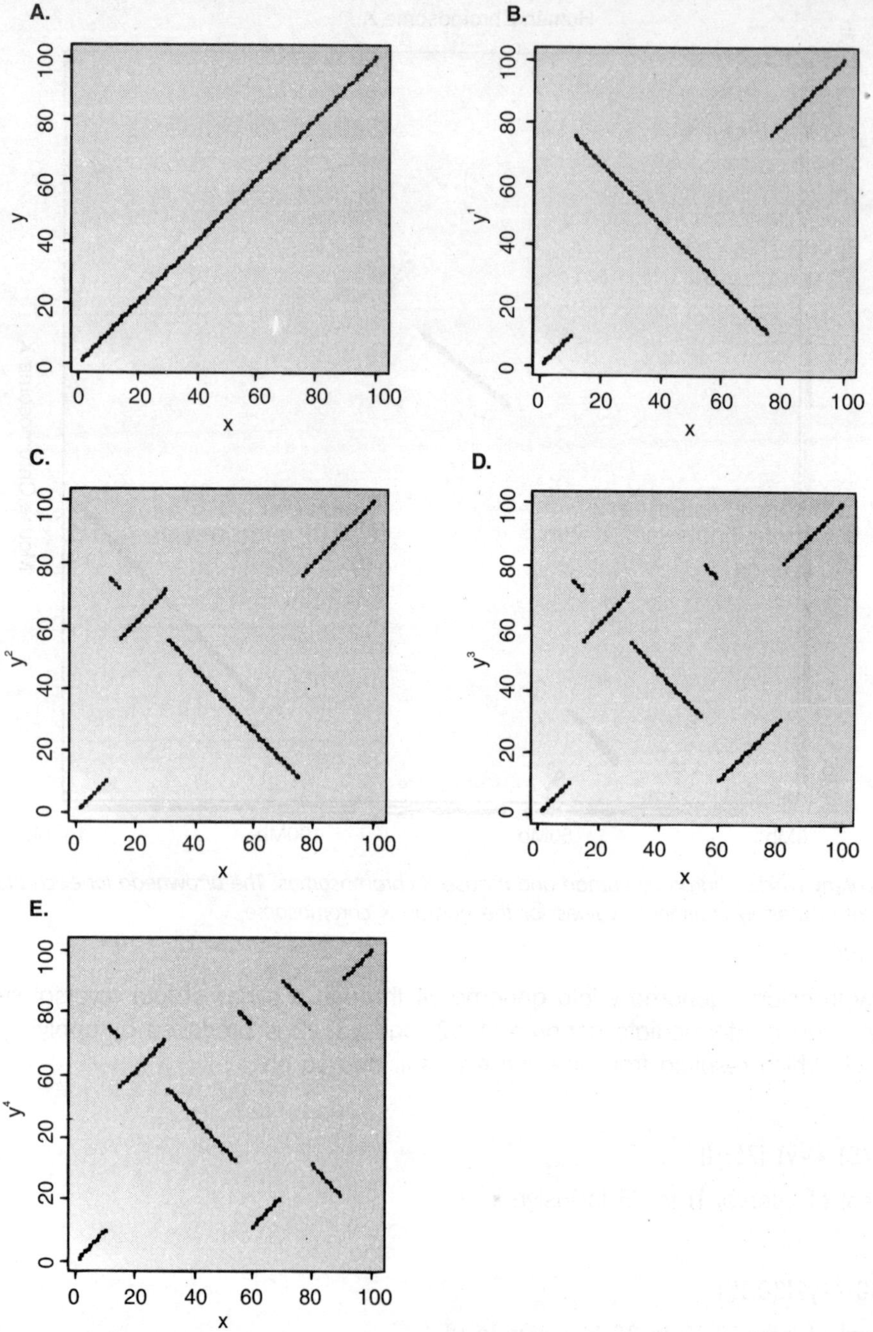

Figure 3.4: Simulation of rearrangement of a single chromosome by reversals. Anchor loci are plotted along axes. Panel A represents chromosomes x and y in the unrearranged states. Chromosomes y1, y2, y3, and y4 in panels B through E are related to chromosome y by one reversal relative to the previous genome (e.g., y1 has one reversal compared to y, y2 has one reversal compared to y1, etc.).

```
y4<-y3
y4[70:90]<-y4[90:70]
#Reversal of interval 70 to 90 inclusive in y3
```

Now we plot the results on a single plot with three rows of two columns (mfrow=c (3,2)) and using "." as the print character for each point (pch=" .").

```
par(pin=c(2.5,2.5), mfrow=c(3,2),pch=".")
plot(x,y, main="A.")
plot(x,yl, main="B.")
plot(x,y2, main="C.")
plot(x,y3, main="D.")
plot(x,y4, main="E.")
```

The results of this simulation are plotted in Figure elsewher in this chapter. We can see step by step how a graph such as the one shown in Figure elsewhere in this chapter can be generated by reversals.

We note in passing that these graphs are analogous to the "dot plots" that will be discussed in other chapter of this book. In that chapter, the plots are of k-word matches in two different sequence strings. Here, the plots are of synteny blocks in two different genomes. *Interpretations* in terms of diagonals are very similar in both cases.

Genomic similarities & differences

Mouse and human genomes have been compared by Pevzner and Tesler (2003), who also present an extremely complicated multichromosomal breakpoint graph. They identified 281 synteny blocks (within stated gap and size criteria), and they concluded that the order of the mouse genome synteny blocks can be converted into the order seen in the human genome by 245 genomc rearrangements: 149 inversions, 93 translocations, and 3 fissions.

Mice and humans share a common *ancestor* estimated to have lived about 83 million years ago. Since that time, there have been about 1.5 rearrangements occurring per million years of evolutionary time separating the two species (245 rearrangements ÷ (2 × 83 million years); the factor of 2 reflects the fact that evolution is occurring in both lineages). As more mammalian genomes are sequenced, comparisons of their sequence organizations will provide additional measures of genetic distance, which can be used together with DNA sequence differences to delineate evolutionary relationships.

Chapter 4

Chromosomal Aberrations

[dup(17)(p11.2p11.2) and 17p Rearrangement]

An approx 4-Mb genomic segment on chromosome l 7p11.2 commonly deleted in 70-80%, of patients with the *Smith-Magenis* syndrome (SMS) is flanked by large, complex, highly identical (approx 98.7%), and directly oriented, proximal (approx 256 kb) and distal (approx 176 kb) low-copy repeats (LCRs), termed SMS-REPs. These LCR copies mediate *nonallelic homologous* recombination (NAHR), resulting in both SMS deletion and the reciprocal *duplication* dup(p11.2p11.2). A third copy, the middle SMS-REP (approx 241 kb) is inverted and located between them.

Several additional large LCR17ps have been identified fomented by *breakpoint* mapping in patients with deletions ascertained because of an SMS phenotype. LCRs in proximal 17p constitute more than 23% of the analyzed genome sequence, approx *fourfold* higher than *predictions* based on virtual analysis of the entire human genome.

LCRs appear to play a significant role not only in common recurrent deletions and *duplications*, but also in other rearrangements including unusual sized (i.e., uncommon, recurrent and nonrecurrent) *chromosomal* deletions, *reciprocal translocations*, and marker chromosomes. DNA sequence analysis from both common and unusual sized recurrent SMS deletions and common dup(l7)(p 11.2p11.2) reveals "recombination hotspots" or a remarkable positional preference for strand exchange in NAHR events.

Large *palindromic* LCRs, mapping between *proximal* and middle SMS-REPs, are responsible for the origin of a recurrent somatic *isochromosome* i(17q), one of the most common recurrent structural abnormalities observed in human *neoplasms*, suggesting genome architecture may playa role in mitotic as well as meiotic *rearrangements*.

LCRs in proximal 17p are also prominent features in the genome evolution of this region whereby several serial segmental duplications have played an important role in *chromosome* evolution accompanying primate speciation. The gene- and LCR-rich, human genomic region 17p11.2-p12 is rearranged in a variety of different *constitutional*, *evolutionary*, and cancer-associated structural chromosome *aberrations*, and thus is an excellent model to *investigate* the role of genome architecture in DNA rearrangements.

HEREDITARY NEUROPATHIC DISEASES

Similar to Charcot-Marie-Tooth disease type IA (CMTIA) and *hereditary neuronopathy* with pressure palsies (HNPP), the LCR-mediated NAHR mechanism is responsible for two other constitutional genomic disorders in *proximal* chromosome l7p: SMS and the dup(17)(p11.2p11.2) syndrome.

Smith-Magenis Syndrome

SMS (MIM 182290) is a multiple *congenital anomalies* and mental *retardation* disorder associated with an interstitial deletion within chromosome 17p11.2. Clinical *characteristics* include minor craniofacial and skeletal *anomalies* such as *brachycephaly, frontal bossing, synophrys, midfacial hypoplasia,* short stature, and brachydactyly, neurobehavioural *abnormalities* such as aggressive and self-injurious behaviour and sleep disturbances, ophthalmic, otolaryngological, cardiac, and renal *anomalies.* As defined by fluorescence *in situ* hybridization (FISH) and by a unique *de novo* junction fragment identified in pulsed-field gel electrophoresis, 70-80% of SMS patients harbor i common approx 4-Mb deletion within 17p11.2.

Approximately 20-25% SMS patients have either smaller or larger sized deletions. Recently, *premature termination* codon mutations in the retinoic acid inducible-1 gene, *RAII,* which maps within the SMS critical region, have been found in five SMS-like patients without deletion, suggesting *RAII haploinsufficiency* causes SMS.

Bioinformatics analyses of the dosage sensitive *RAII* genc, and comparative *genomics* between human and mouse orthologs, revealed a zinc finger like-PHD domain at the carboxyl terminus that is conserved in the *trithorax* group of chromatin-based transcriptional regulators, suggesting that RAI1 might be involved in chromatin *remodeling.*

These findings suggest *RAI*1 is involved in *transcriptional* control through a *multi-protein* complex, and its function may be altered in individuals with SMS. Interestingly, despite a common deletion size, the only constant objectively defined features among patients with SMS are sleep disturbances, low adaptive functioning, and mental retardation. There is no *pathognomonic* clinical feature, no characteristic *cardiovascular* defect, renal anomaly, *otolaryngological,* nor *ophthalmic* abnormality in SMS.

Pharmaceutical Effect of the dup(17)(p11.2p11.2) Syndrome:

The common SMS deletion region is duplicated in patients with a milder, *predominantly neurobehavioural* phenotype and the *reciprocal* chromosome duplication—dup(17)(p11.2p 11.2). Similar to SMS patients with common deletion, subjects with common duplication dup(17)(p11.2p11.2) also *manifest* marked variability in the physical features and behavioural profile.

Clinical findings include dysmorphic craniofacial features, hypotonia and failure to thrive, oropharyngeal dysphasia, *neurocognitive* impairment, and behavioural problems including *autistic,* aggressive, and *self-injurious* behaviour. Structural cardiac anomalies including aortic root enlargement, have been identified.

However, the frequency of organ system developmental abnormalities appears to be less than that observed for patients deleted for this *genomic* interval (i.e., SMS). Sleep disturbances are seen in all *patients* yet the findings are distinct from those of deletion 17p11.2. It is predicted that the *incidence* of dup(17)(p11.2p11.2) may be equal to that of SMS given the *reciprocal* nature of the common rearrangements responsible for the conditions.

However, as this duplication is difficult to detect by routine *cytogenetic* analysis, many of these patients are currently probably not ascertained. Systematic clinical evaluation of a cohort of patients with dup(17)(p11.2p11.2) will be necessary to determine the features most characteristic of this *microduplication syndrome*. The relatively high frequency of constitutional *genomic* disorders in proximal chromosome l7p is further substantiated by the identification of an individual with two distinct *megabase-sized* DNA rearrangements of this *genomic* interval.

These included both *a de novo* dup(17)(p11.2p11.2) and an inherited HNPP deletion on the other homolog. These rearrangements were associated with mild delay and a family history of *autosomal* dominant carpal tunnel *syndrome*.

SMS-REPS, LCR CHROMOSOMAL DELETION

Physical mapping studies have demonstrated that the SMS common deletion interval is flanked by large (approx 200 kb), highly identical (>98%), LCR gene clusters termed *proximal* and distal SMS-REPs that during either maternal or paternal gametogenesis act as substrates for NAHR. To delineate the genomic structure (*size, orientation, sequence identity, gene content*) and evolution of the SMS-REPs, we constructed and sequenced a complete approx 5-Mb bacterial and P1-derived artificial chromosome (BAC/PAC) contig in 17p11.2-p12.

Our analysis revealed that both the proximal SMS-REP (approx 256 kb) and the distal copy (approx 176 kb) are located in the same orientation and derived from a *progenitor* copy, whereas the middle SMS-REP (approx 241 kb) is inverted and appears to have been derived from the *proximal copy*.

This architecture likely explains why the common SMS deletions occur between *proximal* and distal SMS-REPs. There are four regions of significant stretches of identity between the proximal and the distal SMS-REPs. The sum of these high sequence *similarity* regions is approx 170 kb (169,905 bp), and the identity is greater than 98% with the *exception* of the D region (>95%). The largest conserved segment is 126 kb in size.

Two large sequence blocks (between A and B, and between C and D) in the proximal SMS-REP are absent in the distal SMS-REP. Two smaller blocks, flanking areas of the B region in the distal SMS-REP are absent in the proximal SMS-REP. FISH analysis using SMS-REP-specific BAC clones as probes revealed strong *hybridization* signals on *metaphase* chromosomes 17p11.2 and three strong signals on the *interphase* chromosomes. However, SMS-REP-specific BACs also showed weaker hybridization signals in interphase analysis and metaphase spreads; these map to chromosome 17p13.1, 17p12, 17q11.2, 17q12, 17q21.2, and 17q23.2.

In *concordance* with FISH results, BLAST analysis revealed that other approx 11-30 kb "SMS-REP-like" paralogous sequences, *fragments* of SMS-REPs, are localized on 17p13.1 (approx 28 kb), 17p12 (approx 11 kb), 17q11.2 (approx 30 kb), 17q12 (approx 11 kb), 17q2L2 (approx 25 kb), and 17q23.2 (approx 28 kb). Sequence analyses show that all three SMS-REPs within the SMS common deletion are not present in the mouse *syntenic* region.

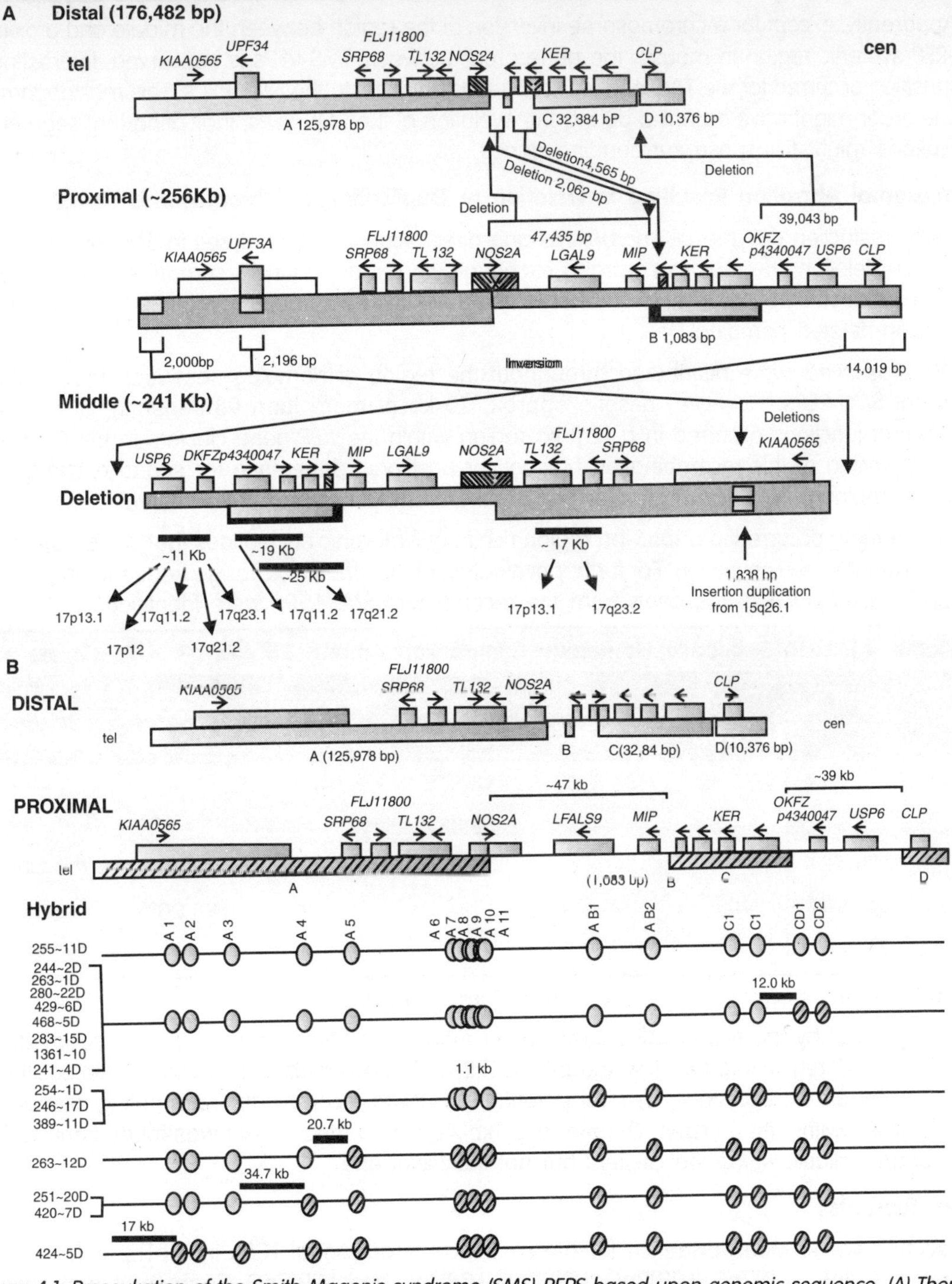

Figure 4.1: Preseukation of the Smith-Magenis syndrome (SMS)-REPS based upon genomic sequence. (A) There are four regions of sequence identity more than 95% between the proximal and the distal SMS-REPs (A, B. C, and D). The A, B. and C sequence blocks have more than 98% identity between distal and proximal REPs, whereas the D regions show more than 95% identity.

Apparently, except for a chromosome inversion of the region between the middle and proximal SMS-REP syntenic region in mouse, the gene order between SMS-REPs is conserved. Interestingly, transposition occurred for the *TACI* and *KCNJ12* genes adjacent to the SMS-REPs. This *rearrangement* of gene order might have occurred during the evolution of the SMS-REPs, indicating that segmental duplications might transpose surrounding genes.

Chromosomal Abrration Resulting in Deletion or Duplication of Chromosome 17p11.2

Using restriction enzyme *cis-morphisms* and direct sequencing analyses in SMS patients with a common deletion, the regions of strand exchange were mapped in 16 *somatic-cell* hybrids that retain the deletion chromosome which effectively isolates the recombinant SMS-REP from the copies on the non-deleted homolog.

The *crossovers* were distributed throughout the region of homology between the proximal and distal SMS-REPs. However, despite approx 170 kb of more than 98% identity, 50% of the recombinant junctions occurred in a 12.0-kb region within the *KER* gene clusters in the C region. DNA sequencing of this recombination hotspot, or positional preference for strand exchange, in seven recombinant SMS-REPs narrowed the crossovers to an approx 8-kb interval.

Four of them occurred in a 1655-bp region rich in polymorphic nucleotides that could potentially reflect frequent gene conversion. For further evaluation of the strand exchange frequency in patients with SMS, *novel junction fragments* from the recombinant SMS-REPs were identified.

Table 4.1: DNA Sequence Homology Comparison Among Different LCR17p Copies

	LCR17pB	LCR17pC	LCR17pD	Proximal CMT1A-REP	Middle SMS-REP	Distal SMS-REP
LCR17pA	98.70%	88.45%	98.57%	—	—	—
LCRI7pC	—	—	88.03%	—	—	—
Distal CMTIA-REP	—	—	—	98.15%	—	—
Proximal SMS-REP	—	—	—	—	98.20%	98.13%
Middle SMS-REP	—	—	—	—	—	98.29%

SMS, Smith-Magenis syndrome; CMT1A, Charcot-Marie-Tooth disease type 1A.

As predicted by the reciprocal-recombination model, junction fragments were also identified from this hotspot region in patients with dup(17)(p11.2p11.2), *documenting reciprocity* of the positional preference for strand exchange. Several potential *cis-acting* recombination-promoting sequences were identified within the hotspot. Of note, a 2.1-kb AT-rich inverted repeat was found flanking the *proximal* and middle *KER* gene clusters but not the distal one.

LCR17p Repeats

Recently, we provided evidence for the existence of additional LCRs in 17p11.2-p 12 termed LCRI7pA, LCR17pB, LCR17pC, LCR17pD, LCR17pE, LCRI7pF, and LCR17pG. Segmental *duplications* constitute more than 23% of approx 7.5 Mb of genome sequence in proximal 17p, approx fourfold higher than predictions based on virtual analysis of the entire human genome. Based on the genomic sequence information *spanning* these LCRs, we determined the size, structure, *orientation*,

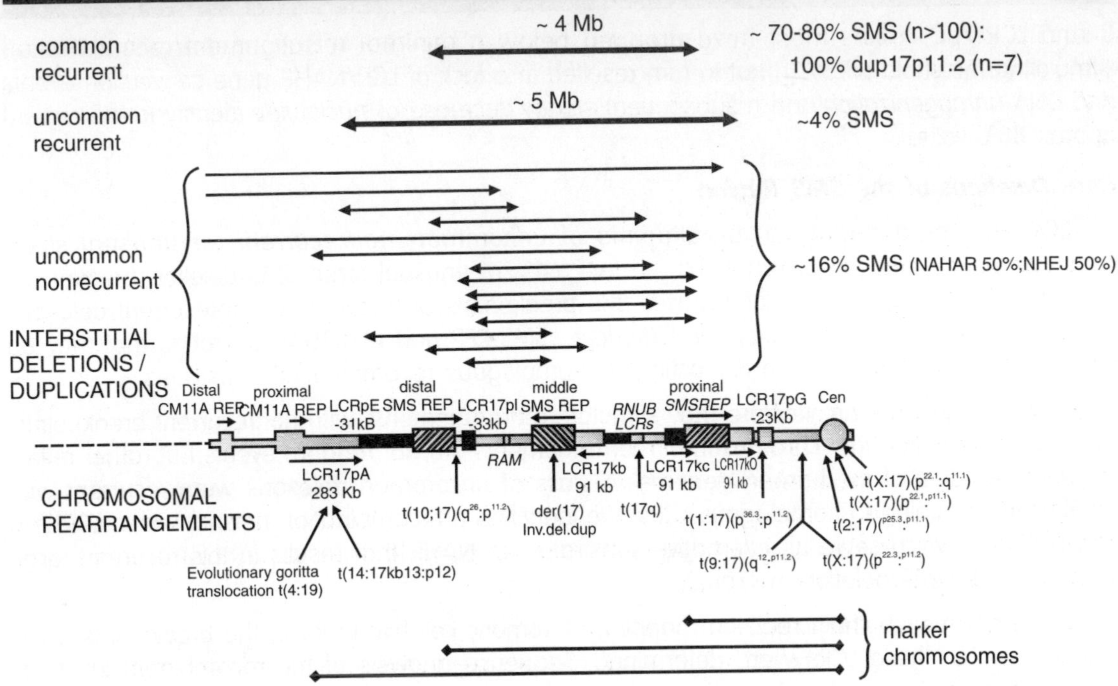

Figure 4.2: Schematic diagram of breakpoints for DNA rearrangements in proximal chromosome 17p. (Top) Interstitial deletions and duplications are shown as horizontal arrows. Recurrent, common (approx 4Mb) and unusual sized (approx 5 Mb) deletions are responsible for 70-80% and 4% of SMS cases, respec-tively. They both utilize low-copy repeats (LCRs) as substrates for nonallelic homologous recombination (NAHR). In approx 16% of SMS cases, uncommon nonrecurrent deletions have been found. Approximately half of them arise owing to NAHR mechanism between repetitive sequences and half through nonhomologous end-joining. (Bottom) LCR-associated chromosome translocations. isochro-mosome 17q, and marker chromosomes are depicted. The LCRI7p structures are depicted in colors to better represent their positional orientation with respect to each other; the shaded rectangles and horizontal arrowheads represent the orientation of the LCRs.

and extent of *homology* for and between each copy. BLAST comparisons of LCR17p repeats revealed that LCR17pA is composed of three subunits, which we *termed* LCRI 7pA/B (approx 232 kb, *homologous* and in an inverted *orientation* with respect to LCR 17pB) and adjacent to it on the centromeric side two overlapping sequences LCR 17pA/C (approx 79 kb), and LCR l 7pA/D (approx 115 kb) (both homologous and directly oriented with respect to LCR17pC [approx 91 kb] and LCR17pD [approx 118 kb], respectively).

Interestingly, the LCR17pC and LCR17pD copies directly flank the proximal SMS-REP. It is remarkable that despite different sizes, locations, orientations, and, most importantly, times of origin, proximal and distal CMT 1 A-REPs, *proximal, middle,* and distal SMS-REPs, LCR17pA/B and LCR17pB, and LCR17pA/D and LCR17pD, have retained very similar (approx 98.1-98.7%) *nucleotide* identity. Our studies show that in contrast to LCR 17pA/D-and LCR17pD-mediated NAHR, resulting in recurrent SMS *chromosome* deletions, no deletions with *breakpoint* within similarly sized LCR 17pA/C and LCR 17pC have been found.

We propose that during primate evolution, the DNA sequence homology between LCR 17pA/

C and LCR17pC copies must have dropped below a minimal misalignment/recombination stimulating threshold (<95%?) that in turn resulted in a lack of LCR/NAHR gene conversion events and DNA *homogenization* and a subsequent *steady* decrease of *nucleotide* identity to the current approx 88% value.

Rare Deletions of the SMS Region

LCR 17ps predominate at breakpoints of uncommon *nonrecurrent*, or unusual sized chromosome *deletions* in proximal 17p. In fact, 64% of unusual sized SMS deletion breakpoints occur in LCRs (19, 29, 30). In some cases, the breakpoints of uncommon nonrecurrent deletions map to two different nonhomologous LCRs (e.g., SMS-REP and LCR17pA), indicating that they do not mediate the rearrangement by acting as homologous recombination substrates.

However, genome architecture may stimulate rearrangements with nonrecurrent breakpoints, supporting the notion that chromosomal rearrangements are not random events but rather reflect structural features of the human genome. A class of uncommon deletions were recurrent with breakpoints mapping to homologous LCRs. We identified a recombination hotspot within LCR17pA and LCR17pD, which serve as alternative substrates for NAHR that results in this recurrent large (approx 5 Mb) SMS deletion in 17pl1.2.

Using polymerise chain reaction mapping of somatic cell hybrid lines, the breakpoints of six deletions within these LCRs were determined. Sequence analysis of the recombinant junctions revealed that all six strand exchanges occurred within a 524-bp interval, and four of them occurred within an *AluSq/x* element.

This interval represents only 0.5% of the 124-kb stretch of 98.6% sequence identity between LCRl7pA and LCRl7pD. These findings indicate that alternative LCRs can mediate rearrangements, resulting in haploinsufficiency of the SMS critical region, and reimplicate homologous recombination as a major mechanism for genomic disorders.

Rare Nonrecurrent 17p11.2 Chromosomal Rearrangements

To examine recombination mechanisms involved in uncommon, nonrecurrent rearrangements, we sequenced the products of strand exchange in hybrids retaining such rare deletion chromosomes from patients with SMS. Two of the four deletions are a product of *Alu-Alu* recombination, while the remaining two deletions result from a nonhomologous end-joining *mechanism*.

Of the breakpoints studied, 3/8 are located in LCRs, and 5/8 are within repetitive elements, including *Alu* and MER5B sequences. These findings suggest that higher-order genomic architecture, such as LCRs, and smaller repetitive sequences such as *Alu* elements, can mediate chromosomal deletions via homologous and nonhomologous mechanisms. These data further implicate homologous recombination as the predominant mechanism of deletion formation in this genomic interval.

Human Neoplasia

i(17q) is one of the most common recurrent structural abnormalities observed in human neoplasms including chronic myeloid leukemia and in solid tumors such as childhood primitive *neuroectodermal* tumors. The i(17q) breakpoints were mapped in 11 cases with different *hematological malignancies* and the genomic structure of the involved region was determined. Our results revealed a complex genomic architecture in the i(l7q) breakpoint *cluster* region characterized by large (approx 38-49 kb), *palindromic,* LCRs, strongly suggesting that somatic rearrangements are not random events but rather reflect *susceptibilities* owing to the *genomic* structure.

Genetic Models of SMS and dup(17)(p11.2p11.2)

By examining the deleted regions in SMS patients with unusual-sized deletions, we refined the minimal SMS critical region (SMCR) to an approx 1.1-Mb *genomic* interval that contains 20 genes. The number, order, and *orientation* of genes within the SMCR are highly conserved in mouse chromosome 11, 32-34 cM. *To* identify the causative gene(s) in SMS, using chromosome engineering, we generated chromosomes carrying either the deletion/deficiency (*Df(11)* or duplication (*Dp(11)* of an approx 2-Mb genomic interval containing the mouse region syntenic to the SMCR.

These models partially reproduce the craniofacial and behavioural phenotype in humans. The *Df(11)17/+* mice exhibit *craniofacial abnormalities, seizures, marked obesity, abnormal circadian rhythm,* and are hypoactive. The *Dp(11)17/+* animals are underweight, *hyperactive,* and have impaired *contextual* fear conditioning.

Most of thc phenotypes in the *Df(11)17/+* mice including craniofacial abnormalities, seizures, obesity, and some behavioural abnormities are rescued in *Df(11)17/Dp(11)17* mice, suggesting the existence of a dosage-sensitive gene(s) in this approx 2-Mb genomic interval. To refine regions responsible for different SMS phenotypic features, three lines of mice (*Df(11)17-1, Df(11)17-2,* and *Df(11)17-3*) with approx 590-kb deletions were generated using *retrovirus* mediated chromosome engineering for constructing nested deletions.

Heterozygous mice with these smaller deletions manifest craniofacial anomalies and obesity. The identification of *RAI1* point mutations in SMS patients suggest that haploinsufficiency of RAI1 causes the *neurobehavioural, craniofacial,* and *otolaryngological phenotypes* in SMS. *Rai1,* the mouse homology of *RAI1,* is located in the smaller deletion.

The heterozygous mice with disruption of *Rai1* gene also exhibit a similar craniofacial phenotype and obesity, suggesting that *Rai1 haploinsufficiency* results in the *craniofacial* abnormalities and obesity in the smaller deletion mice. Importantly, the severity and penetrance of the *craniofacial* phenotype were significantly reduced in the smaller deletion mice and *Rai1* heterozygous mice in comparison to *Df(11)17/ +* mice, indicating that genes or regulatory regions in the larger deletion that are not within the smaller deletion influence both the *penetrance* and *expressivity* of the phenotype.

We observed phenotypic variation in these mouse models in both the same and different genetic backgrounds, perhaps reflecting the variations in SMS patients. Our studies on mouse models suggest that in SMS even a single clinical endophenotype such as craniofacial features is affected by multiple genes, in which *RAI1* is the major causative gene, whereas other genes or regulatory regions within the SMCR, or located elsewhere, modify the phenotype.

The molecular basis of dup(l7)(p11.2p11.2) syndrome is still unknown. Examining whether the *Rai1* mutant allele can rescue the phenotype in *Dp(11)17/+* mice will help understand the role of *RAI1* in this disorder.

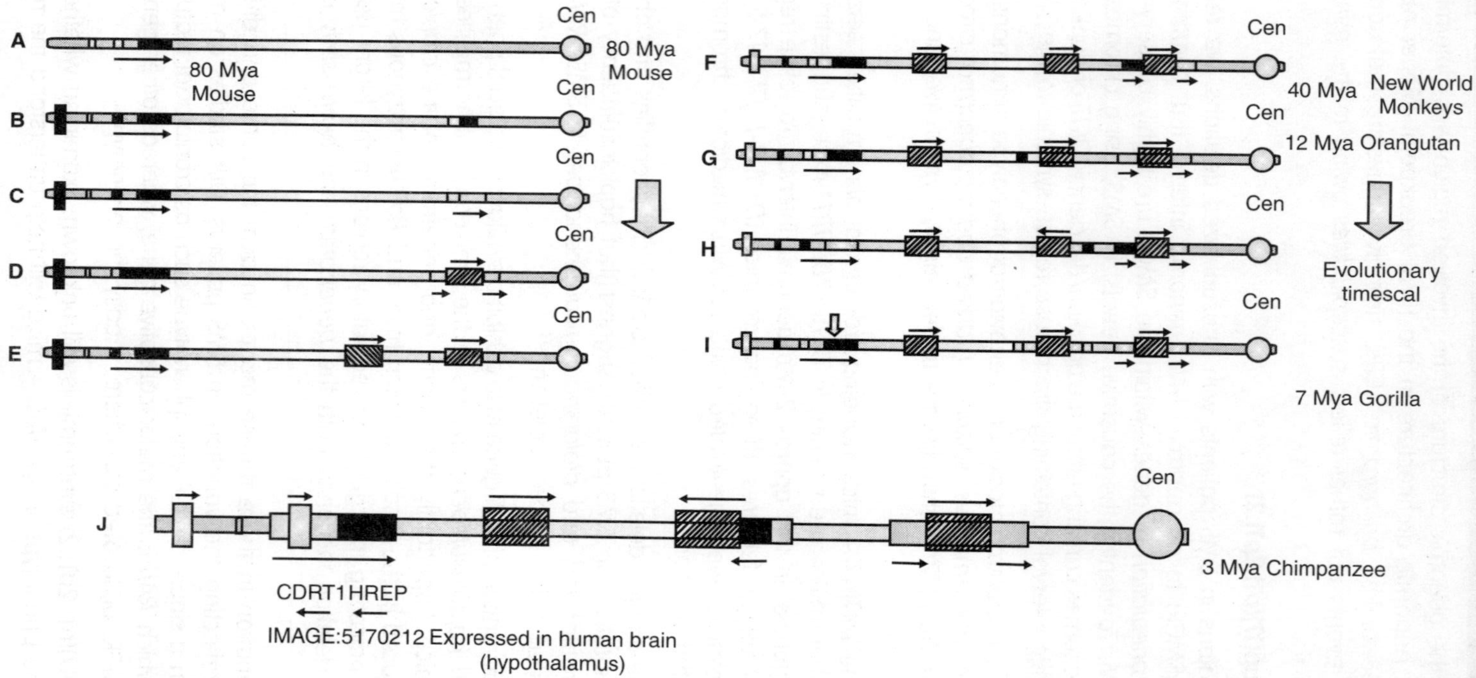

Figure 4.3: Rearrangements of proximal 17p low-copy repeat (LCR) during primate evolution. Proximal chromosome 17p is depicted by two thin horizontal lines with the centromere (white circle) *shown to the right. LCRs are shown as horizontal rectangles with the same color or black-and-white graphic, representing highly homologous sequence. (A) In the mouse genome only LCR17pA is present. (B,C) The proximal portion of LCR 17pA was duplicated more than 25 million years ago (Mya). Note that LCR17pC and LCRI 7pD represent two overlapping portions of LCR17pA. (D) The proximal SMS-REP split the LCR l7pC and LCR 17pDcopies resulting in three directly adjacent large LCRs. Two tandem duplication resulted in directly oriented middle (E) and distal (F) Smith-Magenis syndrome (SMS)-REPs. (G) After the divergence of orangutan and gorilla 7-12 Mya, the distal portion of LCR17p was tandemly duplicated creating a directly oriented LCRl7pB copy. (H) Both middle SMS-REP and LCRl7pB, adjacent to it, were inverted. (I) Following that, at the junction between L.CR 17pA/B and LCR 17pD copies, the evolutionary translocation t(4; 19) occurred in a pre-gorilla individual, 7-12 Mya (vertical open arrow). (J) Finally, between gorilla and chimpanzee, 3-7 Mya the proximal CMTI A-REP, present only in human and chimpanzee. resulted from the insertional duplication of the distal copy. To the right is shown a time line of mammalian, mainly primate evolution with Mya, as indicated.*

Serial Homologous Doubling During Primate Evolution

Using multifaceted approaches, consisting of FISH studies in fibroblast and lymphoblast cell lines from several primate species and computational analyses of human genome sequence (molecular clock analysis), we *investigated* the complex structure and evolution of the genome *architecture* in an approx 7.5-Mb region of proximal 17p.

Cumulative data from studies of different segmental duplications in proximal 17p suggest a potential model that most parsimoniously explains how the complex genomic architecture evolved. The complex genome architecture in proximal 17p results from a series of consecutive segmental duplications during primate evolution.

Both repetitive sequences and transposable elements were identified at the breakpoints of genome rearrangements, suggesting a potential role in their generation. Segmental duplications appear to influence genome evolution by a number of different mechanisms including the creation of novel fusion/fission genes at the LCR insertion site, as well as potentially enabling an increased mutation rate owing to LCR-mediated genomic inversion associated with reduced recombination.

In essence, such genome architecture enables shuffling of the genome and generating new gene functions, both at segmental duplication breakpoints and by divergence of duplicated copies, providing ample diversity for selection to act upon during evolution.

CONCLUDING REMARKS

The proximal region of human chromosome 17p contains multiple LCRs, many of which have been identified at the brekpoints of DNA rearrangements associated with genomic disorders. The predominant mechanism for deletion and reciprocal duplication is NAHR utilizing homlogous LCR as recombination substrates.

Genome rearrangements can occur either during meiosis or mitosis, accompany gemome evolution, and can cause genomic disorders. Chromosome rearrangements are not random events but insted reflect *underlying* genome architecture.

Chapter 5

Chromosomal Structural Aberrations

Paternally transmitted chromosomal *abnormalities* may lead to birth defects and genetic diseases in *offspring*. However, the etiologies of numerical and chromosomal defects in sperm and their association with subsequent abnormal reproductive outcomes remain generally unknown. Extensive testing in *rodent* models has shown that various environmental toxicants, when given to males or females before or after *mating*, can have profoundly deleterious effects on reproduction, such as infertility, lethality during development, malformations, as well as cancer among offspring. This has raised concern that certain environmental, occupational, or medical *exposures* can have detrimental effects on the genetic *integrity* of human male germ cells.

Of special concern are long-term *chronic* low-dose exposures to environmental *mutagens* such as smoking and air pollution that affect large numbers of individuals, or short exposures of more limited populations to very high doses of mutagens such as cancer chemotherapies. *Paternally* mediated abnormal reproductive outcomes may be a consequence of abnormal reproductive physiology, predisposing genetic factors, past and present male environmental exposures, or random errors that occur during *spermatogenesis.*

Elucidating the relative contribution of these factors from *epidemiological* surveys of affected offspring is extremely difficult because the sample sizes of offspring with specific defects are generally too small and prenatal selection against defective embryos vary among different types of genetic and chromosomal defects.

This has provided an incentive for devel-oping effective *biomarkers* to detect genomic damage directly in sperm where greater statistical power can be attained by detecting changes in the frequencies of defective sperm of only a few well characterized individuals. Fluorescence *in situ* hybridization (FISH) is an efficient approach for labeling DNA of chromosomes in interphase cells including sperm. Its effec-tiveness has improved with the availability of chromosome-specific DNA probes for every human chromosome and with the increased emphasis on the importance of scoring criteria.

Since its introduction, FISH assays evolved from a one- to a two-, three-, and four-chromosome assay using multiple probes specific for each chromosome. Emphasis has also shifted from using any chromosome for which an effective DNA probe was available to selected use of chromosomes with clinical relevance in human *aneuploidy syndromes.*

Human sperm FISH has identified several potential risk factors for chromosomally abnormal sperm including certain lifestyles factors and *environmental/occupational* exposures. Sperm FISH also has the intrinsic advantage of being *broadly* applicable to any laboratory and domestic species for which chromosome-region-specific probes are available and several *multicolour* FISH assays have also been developed to detect numerical abnor-malities in *laboratory* animals.

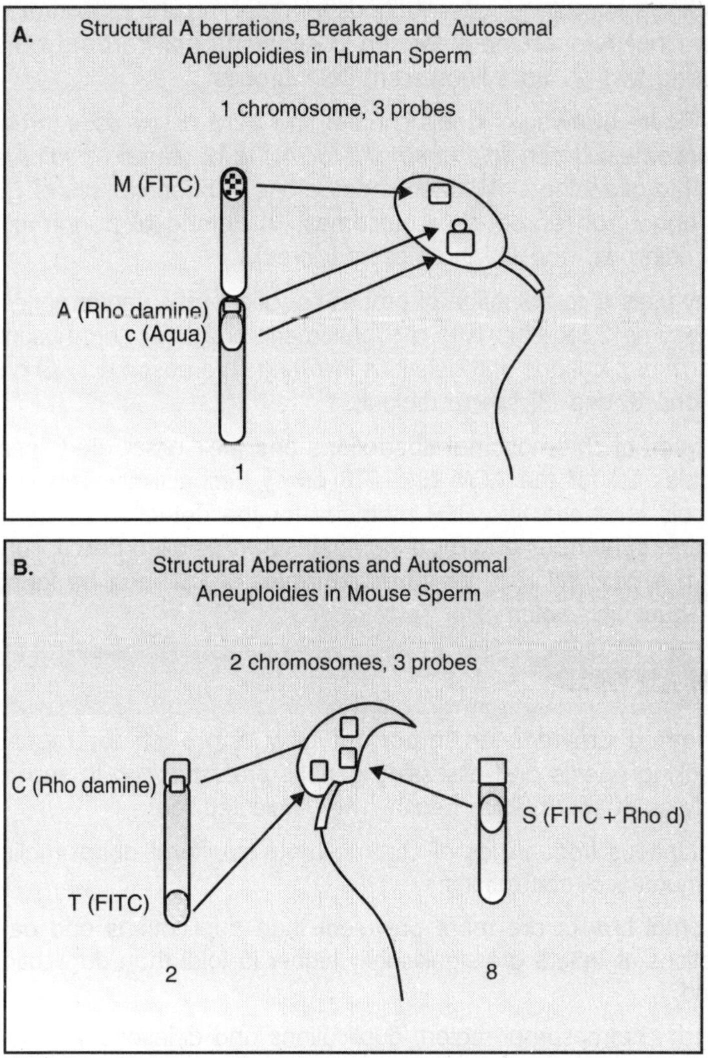

Figure 5.1: Sperm FISH labeling strategies for detecting chromosomal structural aberrations in sperm of humans and mice. (A) ACM assay for the detection of deletions and duplications of chromosome 1p, breaks in the classical satellite region of 1q, as well as numerical abnormalities in human sperm. (B) CT8 assay for detection of deletions and duplications of chromosome 2 and numerical abnormalities of chromosomes 2 and 8.

Sperm FISH assays for laboratory animals may provide a platform for systematic tests of the genetic damage to germ cells of the myriads of chemicals present in the environment, and for *prioritizing* human epidemiological studies of paternally mediated abnormal reproductive outcomes. An important recent extension of sperm FISH methods has been the development of assays for detecting human (*ACM assay*) and rodent (CT8 assay) sperm carrying chromosomal structural aberrations.

Unlike trans-mitted germinal aneuploidy, which originates from meiotic nondisjunction predominantly during oogenesis, transmitted *de novo* structural aberrations tend to originate predominantly during spermatogenesis. Methods for detecting chromosomal aberrations in sperm have in common that two or more loci on a single target chromosome are interrogated simultaneously using two or more fluorescent DNA probes.

As shown in Figure elsewhere in this chapter, the ACM assay uses three probes for specific regions of chromosome 1: 1 cen (*alpha satellite* or A), 1q.12 (*classical satellite* or C), and 1p36.3 (midisatellite or M) to allow the simultaneous detections of three classes of chromosomal defects associated with abnormal reproductive outcomes: (1) numerical abnormalities, (2) segmental duplications and deletions, and (3) chromosomal breaks.

The CT8 assay uses a combination of probes specific for the centromeric (C) and telomeric (T) regions of chromosome 2 together with a subtelomeric probe for chromosome 8 to detect three types of damage: (1) duplications and deletions involving chromosome 2, (2) aneuploidies involving chromosomes 2 and 8, and (3) sperm diploidy.

The specific types of chromosomal aberrations and their associated sperm FISH phenotypes are shown in Tables 5.1 for the ACM and CT8 assay, respectively. This chapter describes the laboratory protocols for these two FISH methods for the detection of chromosomal structural aberrations in sperm of humans and mice together with a detailed description of the criteria used for collecting data. A description of the basic principles of FISH can be found in the chapter by Pacchierotti and Sgura (this volume).

THE ACM METHOD

The ACM method provides an important new approach for measuring exposure to chromosome-breaking agents and assessing genetic *predisposition* to such damage. As shown in Table 5.1, analyses of sperm from healthy men indicated that

(1) the spontaneous frequencies of chromosomal structural abnor-malities are higher than those of numerical aberrations;

(2) chromosomal breaks are more prevalent than duplications and deletions; duplications and deletions of 1q36.3 are significantly higher (5-fold) than duplications and deletions of 1 cen; and

(3) within each chromosomal region, duplications and deletions tend to occur at the same rate.

The ACM assay has been used recently to show that *oligozoospermic* men have higher frequencies of sperm with chromosomal structural aberrations than *normozoospermic* sperm and that there is a gradual increase in the *frequencies* of sperm with chromosomal structural aberrations

Table 5.1: Chromosomally Abnormal Sperm Detected by the Human ACM Assay.

Sperm chromosomal defects	FISH phenotype	Baseline frequencies
Segmental aneuploidies		
1 pter duplication	ACMM	1.0 ± 0.3
1 pter deletion	ACO	0.5 ± 0.2
1 cen-1g12 duplication	ACACM	2.2 ± 0.4
1 cen-1 q 12 deletion	OOM	1.6 ± 0.3
Chromosomal breaks		
Breaks between 1 cen and 1g12	A.CM	5.4 ± 0.7
Breaks within 1 q 12	AC.CM	5.5 ± 0.9
Total structural aberrations		16.2 ± 1.3
Numerical abnormalities		
Disomy I or diploidy	ACACMM	12.9 ± 2.1
Nullisomy 1	000	0.3 ± 0.2

as men age. The following subheadings describe the materials, *reagents*, and protocols for performing the ACM assay.

Basic Requirements

Preparation of Semen Smears

1. 3-9 μL of fresh or frozen semen.
2. Glass slides with frosted end.
3. 100% Ethanol-200 proof.

Decondensation of Sperm

1. 40 mL of dithiothreitol (DTT, Sigma-Aldrich, St. Louis, MO) solution in a Coplin jar placed in ice.
2. 40 mL of lithium 3,5-diiodosalicylate (LIS, Sigma-Aldrich) in a Coplin jar at room temperature (RT).
3. Tris (*hydroxymethyl*) *aminomethane hydrochloride* (Tris-HCl, Sigma); adjusted to pH 7.8.
4. *Autoclaved* distilled water (dH_2O).
5. Circulating water bath set to 77-78°C.
6. 70% formamide/2× SSC: 315 mL of formamide (Shelton Scientific, Peosta, IA), 45 mL 20× SSC (87.65 g of NaCl, 44.1 g of Na citrate, 400 ML of dH_2O; adjust pH to 7.0 with HCl or NaOH; adjust volume to 500 mL with dH_2O), 60 mL autoclaved dH_2O, then adjusted to pH

7.0 with 2 *N* HCl and to volume 450 mL with extra autoclaved dH_2O.

7. 70%, 85%, 100% Ethanol in Coplin jars.

Random Priming

1. Template DNA: human 1 cen unlabeled pSDZI-1; human 1 classical satellite pUC 1.77; I midi D 1 Z2.
2. 2.5× Random primers (Invitrogen).
3. 10× Buffer: 10 μL of Tris-HCl, pH 7.5, 5 L of Na2EDTA (stop buffer), 485 μL of sterilized H_20.
4. Bioprime labeling kit with Klenow, sterilized H_2O, dNTP mix (Invitrogen).

Mixture Preparation

1. Master mix: 5.5. mL of formamide, 0.5 mL of 20× SSC, 1 dextran sulfate, heat at 72°C until completely dissolved, adjust pH to 7.0, and adjust the volume to 7 mL with dH_2O. Aliquot in 1.5-mL *Eppendorf vials* and store at -20°C.
2. Herring sperm (Invitrogen).
3. Labeled probes generated from random priming reactions.

Hybridization

1. Slide warmer set to 37°–42°C.
2. Glass cover slips 22 × 22 mm, No. 2 (Corning).
3. Rubber cement (Starkey, La Grange, IL).
4. Prewarmed humidity chamber.
5. Incubator with temperature set to 37°C.

Washing

1. Circulating water bath set to 77-78°C.
2. 60% Formamide/2× SSC: 30 mL of formamide, 5 mL of 20× SSC, 10 mL of autoclaved dH_2O, then adjust pH to 7.0 with 2 N HCl and adjust volume to 50 mL with dH2O.
3. 2× SSC: Dilute 100 mL of 20× SSC with 800 mL of *autoclaved* dH_2O. Adjust the pH to 7.0 with 2 N HCl. Add more water until the volume reaches 1000 mL. Filter sterilize. Store at RT.

Antibody Staining

1. 40 μL of PNM: Add 5 g of nonfat dry milk to approx 100 mL of PN buffer and add 20 μL of 0.02% sodium azide (*Sigma-Aldrich*). Incubate at 37°C for 60 min. Leave on a lab bench overnight. Remove supernatant in two 50-mL, centrifuge tubes and aliquot into 1.5-mL Eppendorf tubes. Store at 4°C.
2. 1 μL of Pacific Blue-streptavidin (stock concentration 2.5 μg/mL; Molecular Probes, Eugene, OR).
3. 0.5 μL of anti-digoxigenin (DIG) -fluorescein isothiocyanate (FITC) (stock concentration 0.2 mg/mL; Boehringer Mannheim, Indianapolis, IN).

Double staining

1. Vectashield (Vector, Burlingame, CA).
2. Cover slips 22 × 22 mm, No. 1 (Goldseal).

Techniques for the Human ACM Assay

Preparing smear

1. Soak slides in 100% ethanol for at least 2 days. Layer the slides in a jar or a container with a lid in a *criss-cross* fashion to expose as much of the slide as possible to the alcohol. Fill the jar with ethanol until all slides are covered. When removing the slides, wear *appropriate* personal protective equipment (PPE), wipe all slides *vigorously* with a *Kimwipe*, and place them in a slide box.
2. Using a diamond stylus, cut a piece from another clean slide in which the width is slightly less than the slide's width (using another slide for measuring); after marking the surface, break the slide by *bending* inwards allowing the piece to break off carefully. Be careful not to touch the edge that will be in contact with the sample. As an alternative, cover slips may be used instead; however, they can break more easily during smearing. Wear appropriate PPE.
3. Apply 7 μL of semen sample onto a clean glass slide using a pipet. This volume may vary (i.e., 3-9 μL) depending on the sample's sperm *concentration*.
4. Smear the sample using a cut piece of a previously prepared slide. Holding the frosted end in one hand so that the slide is slightly inclined and the other end *resting* on the *workbench*, place the cut slide piece (or cover slip) directly on the sample drop; the clean untouched factory edge only should be in contact with the slide at this point. Slowly move the edge of the cut piece/cover slip so it comes in contact with the sample; the sample will *automatically* distribute along the length of the edge. After the sample is distributed along the edge, slide the edge toward the frosted end, allowing the sample to smear about 2.5-4 cm.
5. Lay the smeared slide flat on a clean paper towel. When the sample dries, store the smeared slides into a slide box with the cover slightly off to allow *air-drying* for at least 24 h. It is highly recommended to leave slides out for several days.

ACM Probe Preparation

This process uses random priming to generate labeled probes for *hybridization*. The random octamers are annealed to denatured DNA template and extended by Klenow *fragments*. During the process, fluorescently labeled-11-dUTP is incorporated during a 10- to 40-fold *amplification* to produce probes for use in FISH.

Alpha-Rhodamine

1. Prepare *rhodamine* dNTP mix (for five reactions, prepare the following: 2 μL of 25 mM dCTP, 2 μL of 25 mM dGTP, 2 μL of 25 mM dATP, 1.4 μL of 25 mM dTTP, 15 μL of 1 mM rhodamine-6-dUTP).
2. Mix 20 μL of random primer, 0.2 μL of template DNA (pSDZI-11 cen rhodamine), and 23.8 μL of *sterilized* water in a *polymerase* chain reaction (PCR) tube.

3. Program *thermal cycler* to denature for 10 min at 99.9°C. remove the vial. and place immediately in ice for 5 min.
4. Add 5 μL of 10× buffer, 5 μL of dNTP mix, and 1 μL of Klenow from Bioprime Kit to denatured template. Total volume = 55 μL.
5. Gently mix and centrifuge for 30 s.
6. Program thermal cycler to incubate at 37°C for 3 h, 70°C for 10 min, and maintain at 4°C.
7. Transfer probe to a 1.5-mL *Eppendorf tube*, label, and store in a -20°C freezer.

Classical Biotin

1. Mix 20 μL of random primer, 0.3 μL of template DNA (pUC 1.77-*classical biotin*), and 23.7 μL of sterilized water in a PCR tube.
2. Program thermal cycler to *denature* for 10 min at 99.9°C, remove the vial, and place immediately in ice for 5 min.
3. Add 5 μL of 10× buffer, 5 μL of dNTP mix from the Bioprime Labeling Kit, and 1 μL of Klenow from *Bioprime* Kit to denatured template. Total volume = 55 μL.
4. Gently mix and *centrifuge* for 30 s.
5. Program thermal cycler to incubate at 37°C for 3 h, 70°C for 10 min, and maintain it at 4°C.
6. Transfer the probe into a 1.5-mL Eppendorf tube, label, and store in a -20°C freezer.

Midi-Digoxigenin

1. Prepare DIG dNTP mix (for five reactions, prepare the following: 2 μL of 25 mM dCTP, 2 μL of 25 mM dGTP, 2 ML of 25 mM dATP, 1.4 μL of 25 mM dTTP, 15 μL of 1 mM DIG- 11-dUTP).
2. Mix 20 μL of random primer, 0.1 μL of template DNA (D 1 Z2 midi-digoxigenin), and 23.9 μL of sterilized water in a PCR tube.
3. Program thermal cycler to *denature* for 10 min at 99.9°C, remove the vial, and place immediately in ice for 5 min.
4. Add 5 μL of 10× buffer, 5 ML of dNTP mix, and 1 μL of Klenow from the *Bioprime* Kit to *denatured* template. Total volume = 55μL.
5. Gently mix and centrifuge for 30 s.
6. Program the *thermal cycler* to incubate at 37°C for 3 h, 70°C for 10 min, and maintain it at 4°C.
7. Transfer the probe into a 1.5-mL Eppendorf tube, label, and store at -20°C.

Probe Mixture

1. Prepare probes using the following amounts per slide: 10.5 μL of master mix; 1.0 μL of herring sperm DNA; 1.5 μL of D 1 Z5 A. rhodamine; 1.0 μL of pUC 1.77-C. biotin; 1.0 μL of D1Z2 M. digoxigenin.
2. Mix by gentle pipetting and centrifuge briefly (2 s) using a tabletop centrifuge.
3. Keep probe mix in –20°C until ready for denaturation.

DECONDENSATION

1. Place slide smears in DTT (4 mL of 1 M Tris-HCl, pH 7.8, + 36 mL of dH_2O + 0.0617 g of DTT) on ice for 30 min. DTT should be prepared fresh every time. Do not denature more than four slides per Coplin jar. DTT disrupts *disulfide bridges* between protamine molecules.
2. Transfer slides to LIS (4 mL of I M Tris-HCl pH 7.8 + 36 mL of dH_2O + 0.0633 g of LIS) at RT for 75-90 min. LIS penetrates sperm membranes and swells nuclei. If slides are kept in LIS for less than 75 min, dimmer signals are produced. If slides are kept in LIS for more than 90 min, diffuse signals are produced.
3. Dry slides upright up at RT on absorbent paper for 2-3 h. Over-drying (3+ h) may allow spontaneous *reformation* of *disulphide* bonds and *recondensation* of sperm.
4. To prepare for hybridization, warm denaturation solution (70% formamide/2× SSC) in circulating water bath set to 77-78°C (20-30 min before use).
5. Denature slides for 2 min in 70% formamide/2× SSC at 77-78°C in a circulating water bath.
6. Keep slides in ice-cold ethanol series of 70%, 85%, and 100% for 2 min each.
7. Air dry slides upright on absorbent paper at RT for 30 min.
8. After drying, check sperm *density* under *microscope* using phase contrast and etch an area with a consistent sperm density to *hybridize*. Choose an area big enough for a cover slip. Continue hybridization only if you have enough cells.
9. Immerse the cover slip in ethanol for a few seconds and clean dry with a Kimwipe.
10. Set the slide warmer to 37°C.
11. Retrieve probes from freezer, tap and centrifuge a few seconds using a benchtop centrifuge.
12. Denature probe mix for 6 min (it can be done for up to 10 min) at 77-78°C in circulating water bath and *plunge* immediately into ice for no more than 5 min.

Hybridization

1. Place the slide on a slide warmer and immediately drop denatured probe mix onto the *etched* area and add a cover slip. Avoid air bubbles.
2. Seal the cover slip with rubber cement.
3. Put slides in a *prewarmed humidity* chamber. A plastic tip box with half an inch of water at the bottom works perfectly for this task.
4. Allow the slides to incubate at 37°C over two nights.

Washing

1. Peel the rubber cement off carefully with forceps. Do not allow the cover slip to move around.
2. Place the slide in a Coplin jar containing 2× SSC at RT to allow cover slips to fall off.
3. Wash slides in two changes of fresh 2× SSC at RT for 3 min each.

4. Drain slides but do not allow them to completely dry.
5. Add 1 μL of Pacific Blue-streptavidin to 40 μL of PNM. Mix well and transfer to the slide.
6. Apply a plastic cover slip and incubate for 30 min in the dark at RT.
7. Drain excess liquid and wash two times in Coplin jars containing 2× SSC at RT for 3 min each.
8. Warm circulating water bath and washing solution (60% *formamide*/2× SSC) to 45°C. Thirty minutes is enough to time to allow solution to warm up to the desired temperature.
9. Wash slides (no more than two at a time) in 60% formamide-2× SSC at 45°C for 4 min.
10. Wash slides two more times in *Coplin jars* containing 2× SSC solution at RT for 5 min each.
11. Add 0.25 μL of Pacific Blue-streptavidin and 0.5 μL of anti-DIG FITC to 100 μL of PNM. Mix well and aliquot 40 μL of the mixture to each slide.
12. Apply a plastic cover slip and incubate using the humidity chamber in the dark 30 min at RT.
13. Drain excess liquid and wash again two times in the same Coplin jars containing 2× SSC solution for 3 min each.

Double Staining

1. Drain the slides well but do not allow them to dry completely. Apply 10 L of *Vectashield antifade* on the etched area and place a No. 1 thickness 22 × 22-mm cover slip on the slide.
2. Keep at 4°C until ready for analysis.

THE MOUSE SPERM CT8 ASSAY

The mouse CT8 assay is the first robust rodent screen for potential male germ cell *aneugens* and *clastogens* and it has been used to show that:

(1) the baseline frequencies of sperm carrying structural *aberrations* involving chromosome 2 are more common than sperm aneuploidy for chromosomes 2 and 8 combined;

(2) diploid sperm were the most common *anomaly* found in mouse sperm; and

(3) sperm with duplications or deletions of the *centromeric* and *telomeric* region of chromosome 2 occurred at similar frequencies.

Comparison of the spontaneous frequencies of sperm with chromosomal structural aberrations between humans and mice showed that healthy human males appear to produce approx 6 times higher frequencies of sperm with chromosomal aberrations than mice. Recently, the CT8 assay was used to show that exposure of male mice to *chemotherapeutic* relevant doses of etoposide resulted in major increases in the frequencies of sperm carrying chromosomal aberrations in both meiotic pachytene (27- to 578-fold) and *spermatogonial* stem-cells (8- to 16-fold), while aneuploid sperm were induced only after treatment of *meiotic cells* (27-fold) with no *persistent* effects in stem cells. The following sections describe the materials, reagents, and protocols for performing the CT8 assay.

Table 5.2: Chromosomally Abnormal Sperm Detected by the Mouse CT8 Assay.

Sperm chromosomal defects	FISH phenotype	Baseline frequencies
Segmental aneuploidies		
2pter duplication	CTT8	0.1 ± 0.3
2pter deletion	C08	0.1 ± 0.3
2cen duplication	CCT8	0.1 ± 0.3
2cen deletion	OT8	0.3 ± 0.5
Numerical abnormalities		
Disomy 2	CCTT8	0.3 ± 0.7
Nullisomy 2	OO8	0.1 ± 0.3
Disomy 8	CT88	0.2 ± 0.4
Nullisomy 8	CTO	0.6 ± 0.8
Diploidy	CCTT88	2.4 ± 1.8

Basic Requerements

Smear Preparations

1. 2.2% Sodium citrate (isotonic solution).
2. 1.5-mL microcentrifuge tubes.
3. Incubator,
4. Ethanol-cleaned glass microscope slides.

Decondensation

1. 40 mL of DTT (Sigma-Aldrich, St. Louis, MO) solution in Coplin jar placed in ice.
2. 40 mL of Tris-HCl (Sigma-Aldrich); adjust pH to 7.8.
3. Autoclaved dH_2O.
4. Circulating water bath set to 77-78°C.
5. 70% Formamide/2× SSC: 315 mL of formamide (Shelton Scientific, Peosta, IA), 45 mL of 20× SSC, 60 mL of autoclaved dH_2O, then adjust pH to 7.0 with 2 *N* HCl and adjust volume to 450 mL with extra autoclaved dH_2O.
6. 70%, 85%, 100% Ethanol in Coplin jars.

Priming

1. Template DNA: mouse chromosome 2 centromere probe (Research Genetics, 460-H-4), mouse chromosome 2 telomere probe (Research Genetics, 121-E-1). clones 4a and 5e form mouse chromosome 8.
2. 2.5× Random primers (Gibco, Bethesda, MD).

3. 10× buffer.
4. Bioprime Labeling Kit with Klenow, Sterilized H_2O, dNTP mix (Invitrogen).

Mixture Preparation

The following quantities are for preparing probes necessary to hybridize 4 slides. These probe quantities may need to be adjusted depending on the quality of each batch of probe.

1. 30 μL of mouse Cot-1 DNA (Invitrogen).
2. 4 μL of mouse 2cent-DIG.
3. 2 μL of mouse 2tel-bio.
4. 1 μL of mouse 8-bio (clone 4a).
5. 1 μL of mouse 8-bio (clone 5e).
6. 0.5 μL of mouse 8-DIG (clone 4a).
7. 0.5 μL of mouse 8-DIG (clone 5e).
8. 2 μL of herring sperm DNA (Invitrogen).
9. 4.1 μL of 3 M sodium acetate.
10. 112.75 μL of ice-cold 200 proof ethanol.

Hybridization

Same as previous Hybridization.

Washing & Staining

1. Circulating water bath set to 45°C.
2. 50% formamide/2× SSC: 25 mL of formamide (Shelton Scientific), 5 mL of 20× SSC, 10 mL of autoclaved dH_2O, then adjust pH to 7.0 with 2N HCL and adjust volume to 50 mL with dH2O.
3. 2× SSC: Dilute 10 mL of 20× SSC with 800 mL of autoclaved dH_2O. Adjust the *pH* to 7.0 with *2 N HCl. Add* more water until volume reaches 1000 mL. Filter sterilize. Store at room temperature.
4. 100 mL of PN buffer.
5. Dual detection reagent (1 μL of fluorescein avidin DCS; 1 μL of anti-digoxigenin-rhodamine, 498 μL of PN, store at 4° C).

Double staining

Same as previous Doublestaining.

Techniques for the Mouse CT8 Assay

Smear Preparation

1. Set the incubator to 32°C.
2. Fill 1.5-mL microcentrifuge tube with 300 μL of 2.2% sodium citrate, and prewarm to 32°C.
3. *Euthanize* mice with C02 following the *guidelines* for the care and use of *laboratory animals*

for research. Isolate the cauda epididymides from both testes. Hold each cauda with *tweezers* and cut small slits into the cauda using iris scissors. Be careful to keep the cauda as one piece.

4. Place both cauda in the *microcentrifuge* tube with 300 μL of 2.2% sodium citrate.
5. Incubate at 32°C for 10 min to allow sperm to swim out of the cauda.
6. Remove the cauda. The sperm suspension can be used immediately for making *smears* or stored at -20°C. Pipet 5 μL of the sperm suspension onto a clean glass slide. Gently smear the sperm suspension over an approx 22 × 22 mm 2 area of the glass slide using the side of a pipet tip. Alternately, a glass slide may be used to make a smear as long as an adequate cell density is attained. Allow the smears to air dry at room temperature for at least 24 h and store the slides at –20°C in nitrogen gas.

Decondensation

1. Make I *M* Tris-HCl stock solution: 157.6 g of Tris-HCl and dissolve in approx. 800 mL of dH_2O; adjust pH to 7.8 by adding NaOH. Bring up to total volume of I L. Autoclave before use.
2. Make 10 mM DTT (made fresh for each use):
   a. Make 0.1 *M* Tris-HCl: Measure 36 mL of dH_2O, pour into a Coplin jar placed in ice. Add 4 mL of I M Tris-HCl stock (from step 1). Mix well.
   b. Make 10 mM DTT: 0.0617 g of DTT from Sigma and mix with 0.1 M Tris-HC1 solution in the Coplin jar by stirring with spatula until the powder completely dissolves. Place the Coplin jar on ice.
3. Place the glass slides into the Coplin jar containing ice cold 10 mM DTT (step 2) for 30 min.
4. Remove slides from *Coplin jar* and *drain*. Briefly dip slide in dH_2O at room temperature and drain. Dry slide at room temperature for at least 30 min. Slides are ready for *hybridization* when thoroughly dried.

Priming

Random Priming With Digoxigenin- Modified Nucleotides

1. Mix 25-500 ng of DNA template with dilute buffer or water to make a final volume of 19 μL in a *microcentrifuge* tube. Add 20 μL of the 2.5× *random primer* solution, mix well and *denature* for 5-10 min at 100°C. Immediately cool on ice.
2. Add 5 μL of reaction buffer, 5μL of DIG/dNTP mix, mix well and briefly centrifuge.
3. Add 1 μL of Klenow fragment. Mix gently but thoroughly and centrifuge for 30 s.
4. Incubate at 37°C for 3 h.
5. Heat to 70°C to deactivate the enzyme (it is not necessary to add stop buffer). Store at -20°C until use.

Random Priming With Biotin-Modified Nucleotides

1. Mix 25-500 nag of DNA template with dilute buffer or water to make a final volume of 19μL in a *microcentrifuge* tube. Add 20 μL of the 2.5× random primer solution, mix well

and denature for 5-10 min at 100°C. Immediately cool on ice.

2. Add 5 µL of the kit 10× dNTP mix, mix well, and briefly centrifuge.
3. Add 1µL of Klenow fragment. Mix gently but thoroughly and centrifuge for 30 s.
4. *Incubate* at 37°C for 3 h.
5. Heat to 70°C to *deactivate* the enzyme (it is not necessary to add stop buffer). Store at -20°C until use.

Probe Mixture

1. Prepare the probe mixture with amounts of reagents as indicated elsewhere in this chapter. These quantities (per slide to be *hybridized*) may need to be adjusted depending on the quality of each batch of probe.
2. Keep at -80°C for 1 h to overnight.
3. Centrifuge for 30 min at 180g, pour off the *supernatant*, and dry the *pellet*.
4. Reconstitute probe mix with 3µL of water and 7 µL of CEP *hyhridization* buffer (per slide).
5. Denature probe mix 10 min at 78°C, and *preanneal* at 37°C for 30 min.

Pretreatment and Denaturation

1. Flood slides with 3:1 methanol-acetic acid and air-dry.
2. Place slides in DTT on ice (4 mL of IM Tris-HCl, pH 7.8, + 36 mL of dH_2O + 0.0617 g of DTT) for 30 min.
3. Dip in dH_2O and dry completely at RT (minimum of 30 min).
4. *Denature* smears in 70% formamide/2x SSC at 78°C for 6 min.
5. *Dehydrate* slides using a series of ice-cold 70%/85%/100% ethanol (2 min each).
6. Dry completely at RT; check under a microscope for sperm density and mark area to be *hybridized.*

Hybridization

1. Clean glass cover slips with 100% ethanol.
2. Prewarm slides on a slide warmer set to 42°C for about I min.
3. Add 10 µL of the probe mixture and cover the slide with a 22 × 22 mm^2 glass cover slip.
4. Leave the slide on the slide warmer for a few minutes and cover the edges with rubber cement. Make sure that all the edges are covered by rubber cement. Add more rubber cement if needed.
5. Incubate over 2 nights at 37°C in a moist *prewarmed* box.

Washing and Detection

1. Carefully remove the rubber cement with forceps.
2. Wash for 5 min in 50% formamide/2× SSC at 45°C. Repeat two additional times.
3. Wash for 5 min in 2× SSC at 45°C.

4. Wash for 5 min in PN at 45°C.
5. Wash for 5 min in PN at room temperature.
6. Drain the slide and without allowing it to dry, add 30µL of dual detection reagent, cover with a plastic cover slip, and keep at RT in a moist box for 40 min.
7. Final wash for 3 min in PN at room temperature (twice).

Double Staining

1. Apply 10 µL of DAPI (0.01 µg/mL) in Vectashield on marked area and apply cover slip.
2. Keep at 4°C until scoring.

OPTICAL STUDY

The previous subheadings have described the laboratory protocols for producing slides with good hybridization quality for both the ACM and CT8 assays. Hybridization quality, that is, high bright and compact FISH signals and low background, is probably the most important factor affecting the success of a sperm FISH study.

However, equally important are the visualization of the FISH signals and *protocols* for data collection that were developed to reduce to a minimum the influence of technical factors on the experimental outcome (see Pacchierotti and Sgura, this volume, for information about *fluorescence microscopes* and filter settings).

In the next few subheadings we will focus on several technical factors that are critical to the reliability of the sperm FISH assay such as:

(1) the development of strict scoring criteria to determine whether a sperm has an abnormal number of spots;
(2) blinding of scorers and procedures for data collection; and
(3) *harmonization* of scoring criteria among different scorers.

Objectives

The subjective evaluation of FISH signal is an issue that can result in great variability in scoring results among different laboratories and within laboratories. Our group has invested significant effort in developing strict scoring criteria that reduce scorer-to-scorer variability.

Strict *adherence* to these scoring criteria is essential for generating reproducible data. The first step in the scoring criteria is to determine whether a sperm should be scored. In order to score a sperm, it must meet all of the following criteria (these criteria apply to both human and mouse sperm):

1. The entire cell must be visible. Do not count a cell unless the total cell border can be seen. Do not count cells that are partially covered by other cells. Example: do not count cells that exist in large clumps where *overlapping* cells might obscure the fluorescence signals.
2. The cell must appear intact. The contents of the cell should remain within the cell borders. Sometimes overly *decondensed* cells will spill their nuclear content or their fluorescence domains and are not *acceptable* for scoring. Example: do not count cells where the

fluorescence domains are outside the main body of the sperm head.

3. The cell should be in a *well-hybridized* area of the slide. Usually on hybridized slides, there are whole regions that hybridize efficiently, as well as areas that are less well hybridized. It may be necessary to pass over some areas if the *hybridization* is poor.
4. The cell size should be no less than 5 pm and no greater than 15 pm. An eye-piece with a reticle can be used for this task. *Nondecondensed* cells do not hybridize reliably and to keep from biasing the data, cells less than 5 μm on the *reticle* are not included in the *denominator*. A tally of such cells is kept but their fluorescence domains are not scored. Cells that overly decondense give very diffuse *hybridization* signals and cannot be reliably scored.
5. The cell should have no background. There may be some background signals on the slide. An area with high background makes it difficult to distinguish true signals from the background and should not be scored.

Only sperm that pass all these criteria can be scored for their FISH phenotype. Next, we describe the decision tree for the ACM assay, followed by the decision tree for the CT8 assay.

Analyzing Results

Before analyzing the FISH signals the scorer should determine whether the cell qualify for scoring by using the phase contrast to evaluate the cell's outline (intact or spilled), presence of one or multiple tails. Also, it should be established that it is a single cell rather than two overlapping cells. To be scored, the following criteria must be met.

1. The entire cell is visible with no part hidden from view.
2. The cell border appears intact.
3. The cell has a tail or tail attachment site.
4. All the hybridization signals are inside the cell
5. The cell size is no less than 5 or greater than 15 notches by the reticle.
6. The cell is in an area of the slide in which a majority of the cells have hybridized.

If any of these *criteria* are not met, this cell is not acceptable for scoring. If contents are spilled or the cell is too small or too large, this information should be tallied, but the *hybridization* signals are not scored. If the cell is acceptable for scoring, the next step in the decision tree is to determine the FISH *phenotype*.

Break Between the Alpha (A) and Classical (C)

To be defined as a break between the A and C probes the following criteria must be met:

1. The A and C signals must be at least one A signal domain width apart.
2. Both signals must be of normal size based on the hybridization signals seen for these probes over the same region of the slide.
3. The signals must be entirely separate and not attached by any threads.
4. The cell has a tail or a tail attachment site.
5. There are no *hybridization* signals either in surrounding cells or within this cell that would

raise the question that this is an artifact.

If all of these *criteria* are met, the cell is scored as having a break between the A and C region; otherwise the cell is not scored.

Break Within the Classical (C) Probe?

To be defined as a break within the C probe, the following criteria must be met:

1. Only one C signal is associated with the A signal.
2. If both C signals are of the same size they must be separated by a width as large as the individual signals; if they are of unequal size, they must be separated by at least the width of the larger C domain.
3. Both signals must have approximately the same intensity.
4. The signals must be entirely separate and not attached by any threads.
5. The cell has a tail or a tail attachment site.
6. There are no hybridization signals either in surrounding cells or within this cell that would raise the question that this is an artifact.

If all of these criteria are met, the cell is scored as having a break within the C region; otherwise the cell is not scored.

Duplications or Deletions

To be defined as a duplication of the C region, the following criteria must be met:

1. The two C signals within the cell must have approximately the same size.
2. Both signals must have approximately the same intensity.
3. Both signals should be approximately of normal size based on the hybridization signals seen for this probe over the same region of the slide.
4. The signals must be at least a full domain's width apart.
5. The signals must be entirely separate and not attached by any threads.
6. The cell has a tail or a tail attachment site.
7. There are no hybridization signals either in surrounding cells or within this cell that raise the question that this is an artifact.

If all of these criteria are met, the cell is scored as having a duplication of the C region; otherwise the cell is not scored.

To be defined as a deletion of the C region the following criteria must be met:

1. The absence of the C signal is confirmed by using the single filter specific for this signal.
2. There are no debris near or around cell.
3. The cell is not damaged.
4. The cell has a tail or a tail attachment site.

If all of these criteria are met, the cell *is* scored as having a deletion of the C region; otherwise the cell is not scored. A similar process is used to determine whether there are *duplications* and *deletions* of the A and M signals.

To be defined as a duplication of the AC probes, the following criteria must be met:

1. The two signals of each probe must have approximately the same size.
2. The two signals of each probe must have approximately the same *intensity*.
3. The two signals of each probe must be approximately of normal size based on the hybridization signals seen for these probes over the same region of the slide.
4. The signals must be at least a full domain's width apart.
5. The signals must be entirely separate and not attached by any threads.
6. The cell has a tail or a tail attachment site.
7. There are no hybridization signals either in surrounding, cells or within this cell that raise the question that this is an artifact.

If all of these criteria are met, the cell is scored as having a duplication of the A and C regions; otherwise the cell is not scored.

To be defined as a deletion of the AC regions, the following *criteria* must be met:

1. The absence of the A and C signals is confirmed by using triple, single, and DAPI filters.
2. There are no debris near or around cell.
3. The cell is not damaged.
4. The cell has a tail or a tail attachment site.

If all of these criteria are met, the cell is scored as having a deletion of the AC regions; otherwise the cell is not scored.

Diploid Sperm

To be defined as a diploid sperm the following criteria must be met:

1. There are two A signals plus two C signals plus two M signals.
2. The two signals of each probe must have approximately the same size.
3. The two signals of each probe must have approximately the same *intensity*.
4. The two signals of each probe must have approximately normal size based on the hybridization signals seen for this probe over the same region of the slide.
5. The two signals of each probe must be at least a full domain's width apart.
6. The signals must be entirely separate and not attached by any threads.
7. There are no hybridization signals either in surrounding cells or within this cell that raise the question that this is an artifact.
8. The cell is really one cell and not two *overlapped* cells (this should be checked carefully by phase contrast).

If all of these criteria are met, the cell is scored as being diploid; otherwise the cell is not scored.

Signals Within the Cell

Occasionally sperm with no *hybridization* signals will be seen. In these cases, the absence of the signals should be confirmed using all filters. In addition, under phase contrast it should be checked that the cell is within size requirements, that there is no debris overlapping sperm, and that there are no cells overlapping. After checking the above criteria, this cell is scored as 000.

Analyzing Results

Before analyzing the FISH signals the scorer should determine whether this cell qualifies for scoring by using phase contrast to evaluate the cell's outline (*intact* or *spilled*), hooked shape of the head, and *presence* of one or *multiplae tails*.

Also, it should be determined that it is a single cell rather than two *overlapping* cells. In order to be scored, the following criteria must be met:

1. The entire cell is visible with no part hidden from view.
2. The cell border appears intact.
3. The cell has a hook.
4. All the hybridization signals are inside the cell.
5. The cell is in an area of the slide in which a majority of the cells have hybridized.

If any of these criteria are not met, this cell is not acceptable for scoring. If contents are spilled or the cell is too small or too large, this information should be *tallied*, but the hybridization signals are not scored. If the cell is acceptable for scoring, the next step in the decision tree is to determine the FISH *phenotype*.

Duplications or Deletions of the Centromeric Region

To be defined as a duplication of the C region, the following criteria must be met:

1. The two C signals within the cell must have approximately the same size.
2. Both signals must have approximately the same intensity.
3. Both signals should be approximately of normal size based on the hybridization signals seen for this probe over the same region of the slide.
4. The signals must be at least a full domain's width apart.
5. The signals must be entirely separate and not attached by any threads.
6. The cell has a hooked shape.
7. There are no hybridization signals either in the *surrounding* cells or within this cell that raise the question that this is an artifact.

If all of these criteria are met, the cell is scored as having a duplication of the C region; otherwise the cell is not scored.

To be defined as a deletion of the C region the following criteria must be met:

1. The absence of the C signal is confirmed by using the single filter specific for this signal.
2. There are no debris near or around cell.

3. The cell is not damaged.
4. The cell has a hooked shape or a tail.

If all of these criteria are met, the cell is scored as having a deletion of the T region; otherwise the cell is not scored.

Duplications or Deletions of the Telomeric Region

To be defined as a duplication of the T region, the following criteria must be met:

1. The two T signals within the cell must have approximately the same size.
2. Both signals must have approximately the same intensity.
3. Both signals should be approximately of normal size based on the hybridization signals seen for this probe over the same region of the slide.
4. The signals must be at least a full domain's width apart.
5. The signals must be entirely *separate* and not attached by any threads.
6. The cell has a hooked shape.
7. There are no *hybridization* signals either in surrounding cells or within this cell that raise the question that this is an *artifact.*

If all of these criteria are met, the cell is scored as having a duplication of the T region; otherwise the cell is not scored.

To be defined as a deletion of the T region the following criteria must be met:

1. The absence of the T signal is confirmed by using the single filter specific for this signal.
2. There are no debris near or around cell.
3. The cell is not damaged.
4. The cell has a hooked shape or a tail.

If all of these *criteria* are met, the cell is scored as having a deletion of the T region; otherwise the cell is not scored.

Aneuploid Chromosome 2

To be defined as being *disomic* for chromosome 2 the following criteria must be met:

1. The two C and two T signals within the cell have approximately the same size.
2. Both signals of each probe have approximately the same intensity.
3. Both signals of each probe are approximately of normal size based on the *hybridization* signals seen for these probes over the same region of the slide.
4. The two signals of each probe are at least a full domain's width apart.
5. The two signals of each probe are entirely separate and not attached by any threads.
6. The cell has a hooked shape.
7. There are no other hybridization signals either in surrounding cells or within this cell that would raise the question that this is an artifact.

If all of these criteria are met, the cell is scored as being disomic for chromosome 2; otherwise the cell is not scored.

To be defined as being nullisomic for chromosome 2 the following criteria must be met:

1. The absence of the C and T signals is confirmed by using the single filter specific for each signal.
2. There are no debris near or around cell.
3. The cell is not damaged.
4. The cell has a hooked shape or a tail.

If all of these criteria are met, the cell is scored as being nullisomic for chromosome 2; otherwise the cell is not scored.

Aneuploid for Chromosome 8

To be defined as being *disomic* for chromosome 8 the following *criteria* must be met:

1. The two 8 signals within the cell have approximately the same size.
2. Both signals have approximately the same intensity.
3. Both signals are approximately of normal size based on the hybridization signals seen for these probes over the same region of the slide.
4. The two signals are at least a full domain's width apart.
5. The two signals are entirely separate and not attached by any threads.
6. The cell has a *hooked shape*.
7. There are no other hybridization signals either in surrounding cells or within this cell that would raise the question that this is an artifact.

If all of these criteria are met, the cell is scored as being disomic for chromosome 8; otherwise the cell is not scored.

To be defined as being nullisomic for chromosome 8 the following criteria must be met:

1. The absence of the 8 signal is confirmed by using the single filter specific for each signal.
2. There are no debris near or around cell.
3. The cell is not damaged.
4. The cell has a hooked shape or a tail.

If all of these criteria are met, the cell is scored as being *nullisomic* for chromosome 8; otherwise the cell is not scored.

Diploid Sperm

To be defined as a diploid sperm the following criteria must be met:

1. There are two C signals plus two T signals plus two 8 signals.
2. The two signals of each probe must have approximately the same size.
3. The two signals of each probe must have approximately the same intensity.

4. The two signals of each probe must have approximately normal size based on the hybridization signals seen for this probe over the same region of the slide.
5. The two signals of each probe must be at least a full domain's width apart.
6. The signals must be entirely separate and not attached by any threads.
7. There are no hybridization signals either in surrounding cells or within this cell that would raise the question that this is an artifact.
8. The cell is really one cell and not two overlapping cells (this should be checked carefully by phase contrast).

If all of these criteria are met, the cell is scored as being diploid; otherwise the cell is not scored.

Signals Within the Cell

Occasionally sperm with no hybridization signals will be seen. In these cases, the absence of the signals should be confirmed using all filters. In addition, under phase contrast it should be checked that the cell is within size requirements, there is no debris overlapping sperm, and there are no overlapping cells. After these criteria are checked, the cell is scored as 000.

Data Collection

Scorers and Coding Procedures

Despite the *implementation* of strict scoring criteria as described in the previous subheading, the scoring of sperm FISH signals still remains a subjective analysis and therefore subject to scoring biases. Extreme care should be taken to completely blind the scorers to the origin of the samples being scored. Several *procedures* can be used to achieve this goal.

- First, it is very important that the scorer is not able to identify the origin of the slide based solely on sperm *concentration*. Chemical exposure may result in a *significant* reduction in the number of sperm in the semen. In these situations, a third person should take particular care that a section of the slide with the highest *concentration* of sperm is chosen for hybridization so that slides from control and exposed samples do not differ greatly in sperm concentration.
- Second, the person coding the slides for scoring should ensure that all *experimental* groups and controls are represented in the set of five or six slides given to the scorer. Any marking on the slide should be removed and an *adhesive sticker* with the code should be applied on the frosted part of the slides.
- Third, the method used for tallying the number of sperm scored should provide information to the scorer only of the number of sperm scored and not of how many sperm with abnormal FISH *phenotypes* have been found.

We routinely score 10,000 sperm per sample (in two scoring sessions of 5000 sperm each). The advantage of analyzing large numbers of sperm in a relatively short time by sperm FISH confers a relatively high level of *sensitivity* and *statistical* power to these assays, so that small increases can be detected by analyzing sperm from a small number of donors per *experimental* group.

As *previously* mentioned, sets of five or six coded slides are given to the scorer. Once the scorer has finished analyzing 5000 sperm on each of the coded slides, these are returned to the coder, who removes the sticker, and applies a new *sticker* with a second code. Once the slides are *recoded*, they are given back to the scorer, who will score a second set of 5000 sperm.

The scoring of the first set of 5000 sperm is *routinely* started on the left side of the cover slip, and the second set of 5000 starting from the right side of the cover slip. To make sure that two independent sections of the slide are scored, a mark with a *permanent* marker is made on the side of the slide to indicate where the scoring was stopped. In addition, *coordinates* of the location where scoring was stopped can be recorded.

If during the scoring of the second set of 5000 sperm the scorer comes too close to the mark, the scoring should be stopped and another slide from the same donor/animal should he used to complete the second set of 5000 sperm.

Cytoscore

Collection of data is done using the Cytoscore© program that was developed at Lawrence Livermore National Laboratory to keep track of the number of sperm scored and their FISH phenotype. The CytoScore© program allows the user to input, in addition to the FISH phenotype, other information, such as size and shape of sperm head, whether a second opinion was provided, and any additional information that the *scorer* may feel is necessary.

For each set of data, the CytoScore© creates three files:

(1) total count, which shows the total data scored. At the beginning of the file, information about the date the data was collected, the scorer, microscope used, project, and additional microscope settings are reported. Following is the number of sperm analyzed and the number and types of abnormal cells;

(2) split summary, which report the same information contained in total count; however, here the data are reported for each set of 500 sperm. Reporting the data in this way may provide information of the quality of the hybridization. In case of homogenous hybridization, the abnormal cells should be divided evenly among the various splits. If this is not the case, it indicates that there may be regions of the slide that are better (or worse) than others;

(3) anomaly data, which report all cells that were called abnormal together with the location of the cell on the slide, plus a set of additional information including comments inserted by the scorer or the person giving a second opinion. The files generated by the CytoScore software can be exported into Excel spreadsheets for statistical analyses. The first step in this process is to decode the samples and compare the two sets of 5000 sperm from the same slide using a X^2 test. If the comparison is good the data become part of the study; if the comparison is not good, i.e., there are major variations between the two sets, the data are discarded and the slide must be rescored.

Scoring Harmonization

The goal of scoring harmonization is to establish standard scoring criteria among scorers, studies, and laboratories in order to maximize consistency of analyses and minimize technical variation. The approach of the harmonization process is to: (1) help the new scorer create

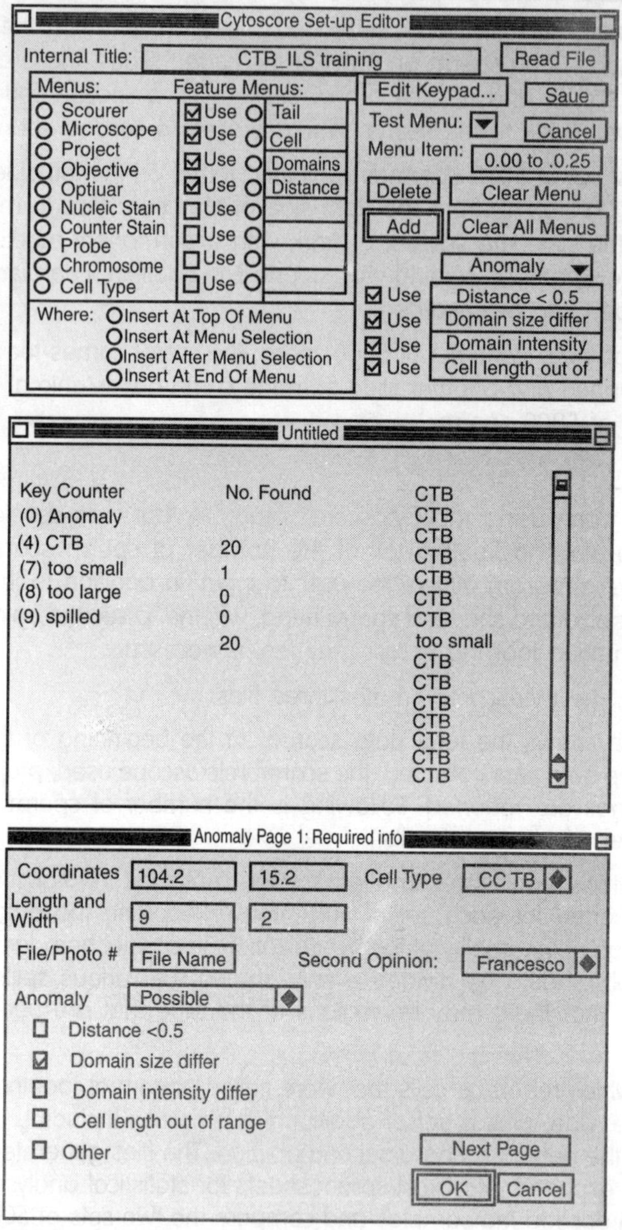

Figure 5.2: Examples of Cytoscore screenshots. (A) Cytoscore set-up editor showing the menu listing and features that are stored for each sperm that is scored. (B) Cytscore scoring data window showing the active keys that were selected during the set up of the CytoScore generator (left column), the total number of sperm scored (center column), and the last 16 sperm that were scored with the most recent at the top (right column). Pressing the 4 key adds a CT8 sperm to the total count. Note that the number of anomalies found is not listed. (C) Cytoscore anomaly page. Every time the zero key is pressed, this window pops-up and requires the scorer to provide information about the anomaly. In the example presented, we have a CCT8 sperm that was classified as possible because the 2 CC signals differ in size.

visualization images on the situations described in the scoring criteria and (2) conduct a scoring *harmonization* trial with samples with established *frequencies* of *abnormal sperm*.

Criteria Training

The objectives of this training are to familiarize the new scorer with all components of the scoring criteria before observing slides, emphasizing the requirement to consistently use multiple filters, including phase contrast.

In addition, the new scorer needs to work side by side with an experienced scorer who presents examples of sperm FISH phenotypes, especially cells that may introduce judgment difficulties, such as background noise, split signals, clusters of spots that remain connected by signal filaments, and cells with signals on the outside of the cell. Key aspects for the trainee to learn are to:

1. Recognize the difference between hybridization artifacts and abnormal cells.
2. Learn to recognize fields that are inappropriate for scoring because of:
   a. Unhybridized cells or uneven hybridization.
   b. Excessive clustering and overlapping of cells.
   c. Debris obscuring cells to be scored.
3. Keep track of areas that have already been scored in order to prevent duplication by documenting the starting and stopping coordinates.

Active discussion among scorers is also a key aspect of the scoring training. When an atypical cell is observed, both scorers examine the sperm identified as having an abnormal *phenotypc* and discuss any reason for *suspecting* an abnormality with respect to the established scoring criteria.

It is *beneficial* to show examples of cells that are definitely normal versus normal cells that have various artifacts interfering with the detection of the hybridization signals, including background noise, spilled cells, and split signals. Finally, comparing slides and observing cells described by thc othcr scorers helps to create visual images of the situations that are listed in the *scoring criteria*.

Harmonization Plan

The objective of the scoring *harmonization* exercise is to test the new scorer's ability to score sperm FISH samples using an established scoring criteria and replicate findings obtained by more experienced scorers. This is generally accomplished in three phases. During phase 1, the new scorer and at least one *veteran* scorer will score 10,000 cells on two reference samples. *Scoring* is conducted side by side, which each-scorer showing to the other every abnormal cell *encountered*.

Discussion of the reasons why each abnormal cell is classified as such is an important aspect of this training and should continue until a *consensus* is reached. After reaching 5000 sperm, the scorers will exchange slides and score 5000 cells in the same area that was already scored, with continued *discussion* of every abnormal cell encountered.

Once the second set of 5000 sperm is analyzed, comparisons between the two sets of 10,000 sperm from each scorer are made using X^2 analyses. If there are no statistically significant differences, the scorers proceed to phase 2; if statistically significant differences are found, the scorers review the scoring criteria and repeat phase 1. Phase 2 will be conducted in the same manner as phase 1, except that this time each scorer will show the other only those cells that he or she believes are *questionable*.

At the end of the scoring, *comparisons* will be made between the results of the new scorer and those of the *veteran scorer*. If there are no significant differences, the scorers will proceed to phase 3. If there are significant differences the scorers will revert back to the side-by-side scoring exercise outlined in phase I using a new slide. In the last phase of the scoring harmonization plan. both scorers will score a common set of four coded slides. After a set of 5000 sperm on each slide are scored, the slides will be recoded and an additional 5000 cells will be scored. Finally, *comparisons* will be made between the first and second set of 5000 sperm scored for each slide by each scorer.

Comparisons will also be made between the trainee's data and data previously collected. If there are significant differences the scorers will revert to phase 2. If there are no significant differences, the scoring harmonization exercise was successful and the trainee will have generated data that are in line with *historic data* generated from those samples by experienced scorers and is ready to generate data from new samples.

POINTS TO REMEMBER

1. Properly stored samples can be used for FISH for several years after collection. We have obtained good hybridization results from samples that had been stored at -80°C for more than 15 years. Fresh samples should be frozen as soon as possible to prevent sperm from settling at the bottom of the vial.
2. For low-density sperm count donors, a pipet tip can be used to make a spiral smear. The ideal volume is 8-10 μL depending on density observed microscopically after the smear is made. Pipet the appropriate volume, and then place the pipet at the center of the glass slide.

   From there, the pipet tip is moved in a circular motion while a small volume is pipetted. After the smear has been made, the smear will appear as a spiral similar to a snail's shell. It is recommended to allow the smears to air-dry for at least 48 h before *hybridizing* or storing them in the freezer. If low sperm density is still a problem, the semen sample aliquot can be left in the refrigerator for a day or two to allow sperm to settle to the bottom before making smears.
3. If the slides are not immediately hybridized, place the slide box in a plastic bag and then fill the bag with nitrogen air. As much as possible of the oxygenated air should be removed. Seal the bag and place it in a freezer at -20°C. In preparation for hybridization, allow the bag to stand at room temperature for about 30 min before opening it and removing the slides.
4. Whenever a new probe is labeled it should be checked for efficiency and overall quality. A poor probe may be difficult to detect during a study, and it may not be noticed until several slides have been scored.
   a. Use a control slide to test the probe. Hybridize the probe alone or with other good

probes. Some probes may behave differently when combined with others.

b. Score at least 500 cells on the slide, looking only at the new probe, and only in regions that are well hybridized. The probe should be present in greater than 99% of the cells.

c. Note the quality of the probe. It should be bright with minimal signal splitting and very little or no noise.

If the probe is efficient, clean, bright, and tight, then the probe may be used for FISH.

5. Do not take the Klenow out of the freezer until ready to use. Once you have used the Klenow, return it immediately to the freezer.
6. Denature no more than two slides at a time in the 70% formamide-2x SSC solution. Each slide will drop the temperature of the solution about 1-1.5°C.
7. Prepare fresh every time. Keep at –4°C for at least 30 min before using.
8. On a humid day, 1 h of drying time provides better results.
9. This step can be done while slides are drying.
10. This should be done under yellow light or partial darkness because these stains are sensitive to white light. Also, prewarming cover slips on the warming plate a few minutes before adding the denaturing probes helps maintain the temperature at 37°C during the first few minutes of hybridization.
11. Keep an empty pipet tip box with half an inch of water at the bottom in the incubator at all times so that it will be ready to serve as a humidity chamber any time it is needed. Periodically add water to replace that which has evaporated.
12. The probe mixture for the ACM assay is a combination of directly labeled (A) and indirectly labeled (C and M) probes; therefore detections steps are needed for the C and M probes. Amplification of the C signal is done in two steps with an intermediate washing step with formamide. This step is necessary to reduce nonspecific staining for this probe.
13. If cover slips do not fall off in less than 5 min, then the cover slips were not sealed adequately or there was not enough water in the humidity chamber. Leave cover slip in 2× SSC until cover slips fall off even if it takes longer than 5 ruin.
14. If signals are not bright enough, this step can be done for 45 min.
15. Small numbers of sperm may be removed from the slide during the denaturation and washing process and remain in the solution. For this reason, denaturing and washing solutions should be changed weekly if heavily used. They should also be changed whenever different treatment sets are hybridized.
16. Equally important is the use of a fluorescence microscope that is equipped with the proper filter sets to detect the light emitted by each fluorochrome. See Table elsewhere in this chapter in Pacchierotti and Sgura (this volume).
17. This is especially critical when scoring human samples because of the *occasional* presence of somatic cells in the semen. The hooked shape of the mouse sperm is the main

discriminating factor for *identifying* sperm from somatic cells.

18. During the scoring of a large number of samples, it is encouraged to include *periodically* one slide already scored in the set of coded slides given to the scorer. This will help to keep track of the consistency of the scoring criteria used.

Chapter 6

Whole Genome Sequencing

Whole-genome sequencing is fueling an information revolution that is changing the face of biology. The ability to determine the complete complement of DNA of a wide range of organisms is allowing us to ask in *unprecedented* detail, fundamental questions about the *molecular* basis of life. First, what is the genetic capacity of an organism? Second, what does an organism do with its *genetic* capacity?

Third and very importantly, how susceptible is this genetic capacity to evolutionary change, what types of changes occur, and which are of *biological* significance? With respect to the genomes of bacterial pathogens, we are interested in how answers to these questions about genetic capacity help to explain *pathogenicity*. In using the genome to understand pathogenicity we are faced with attempting to address the difficult but fundamentally important question of "what creates a pathogen?". We cannot hope to understand the intricate details of the molecular basis of the pathogen-host relationship without integrated *epidemiological, population*-level, *phylogenetic, ecological,* and, perhaps most importantly, experimental studies (preferably in an animal model). These approaches, and the impact that genome sequencing is having on these fields of research, are covered in other chapters.

Here we focus on what kinds of questions can be posed and answered using *in silico* studies of wholegenome sequences. To do this, we first give an overview of the current status of pathogen genome-sequencing efforts, and then we present three key examples of how insights derived from the analysis of a large number of bacterial genomes is helping to place the study of pathogens into context. For the remainder of the chapter, we discuss how genomes are shedding light on the numbers and types of virulence determinants in bacteria, and further, on the molecular basis of three aspects of *bacterial biology* that play an important role in *pathogenicity—physiology,* adaptation to a preferred niche, and the ability to evolve through the *acquisition* of foreign DNA.

HISTORICAL BACKGROUND

The first bacterial genome sequence to be completed (in 1995) was that of *Haemophilus*

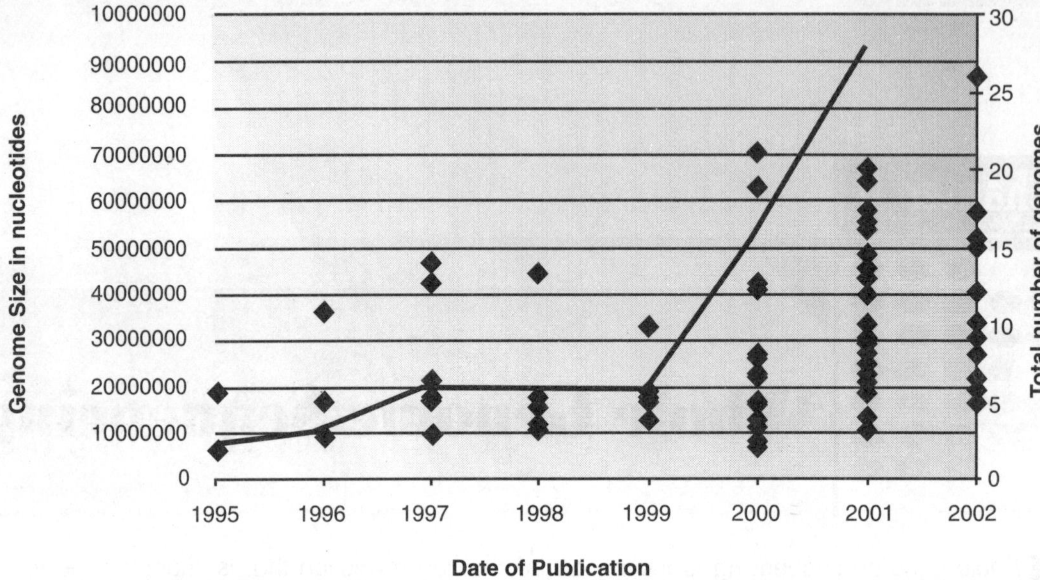

Figure.6.1: Graphical presentation of bacterial genome sequences.

influenzae, a human commensal with the potential to cause a variety of diseases including meningitis. The sequencing of this 1.8-millionbase-pair genome was a conceptual and technical breakthrough. This *milestone* was made possible by the shotgun sequencing approach, which removed the need for *painstakingly* constructed genetic maps.

The shotgun approach relies on sequencing enough clones with unique DNA fragments to increase the probability that all bases in the genome have been sampled past an acceptably high *threshold* (ideally 100%). The challenge lies in the assembly of hundreds of thousands of overlapping partial DNA fragments of 300-500 nt (followed by the manual closure of the inevitable remaining gaps in the sequence).

This advance opened the door for the sequencing off genomes even from species which are poorly characterized, e.g., those of intracellular pathogens that have fastidious growth requirements, like *Treponema pallidum.* The success of sequencing the *H. influenzae* genome also reflected the application of new computational methods, and demonstrated that a bacterial genome could be sequenced in 6 months for less than 50 cents a base.

In short order, proposals for tens of genomes were funded, and the pace of genome sequencing rapidly increased thereafter. As of this writing, we now have more than 80 complete genome sequences, and the largest sequencing laboratories can finish the raw sequencing of complete *bacterial genomes* in hours.

The Joint Genome Institute (JGI) recently held a "microbial marathon," sequencing 15 bacterial genomes in a single month (http://www.jgi.doe.gov/). In 2001, 28 genomes were published, an average of just over two per month. Not only is the number of completed bacterial genomes increasing, but so too is the size of genome sequences being determined. In 2002, the genome of *Streptomyces coelicolour,* the largest bacterial genome sequenced to date, was published.

This species has more genes than the eukaryotic yeast *Saccharomyces cerevisiae* and is a key producer of natural antimicrobials. Of the first 80 bacterial genomes, published between 1995 and summer 2002, 16 are for members of the domain Archaea, a vast taxonomic grouping of bacteria with no known pathogenic representatives.

A total of 64 strains of Eubacteria, representing 41 genera, were completed, and 23 of these genera contain species that have the capacity to be human pathogens. Of these, 29 species (37 strains) represent obligate, facultative, and opportunistic pathogens. These pathogens include the causative agents of some of the most prevalent human diseases (tuberculosis), and agents of historically important pandemics and recurrent epidemics (plague, cholera, typhoid, and typhus).

Other species are the etiological agents of *nosocomial-* and comnmunity-acquired infections *(Staphylococcus aureus)* and opportunistic infections *(Pseudomonas aeruginosa)*. We are also beginning to sequence *pathogens* of other species, thereby adding to the value of experimental models of *human* disease.

Table 6.1: The Taxonomic Diversity Found in the First 80 Published Genome Sequences.

	Genomes (strains)	Species	Genera
Archaea	16	16	11
Eubacteria	64	52	41
Eubacterial human pathogens	37	29	23
Total	80	68	52

This collection contains genomes that belong to species of both bacterial domains and represent a total of 68 species from 52 genera. Among Eubacteria, we have genomes from 29 species (37 strains) with the capacity to causc disease in humans.

Standard Work Flow

The standard workflow of a genome-sequencing project is depicted in Figure elsewhere in this chapter. The first crucial step is the selection of the target species. Our ability to study pathogenicity is largely dictated by past and future decisions about which organisms to sequence. Earlier sequencing efforts were distinctly biased towards human pathogens, but even now the desire to learn more about pathogen biology continues to be the single most significant impetus behind the sequencing of bacterial genomes.

Pathogenic species have been chosen for a variety of reasons. For fastidious intracellular obligate pathogens like *T. pallidum* and species of *Chlamydia* in which classical genetics has proved very difficult, genome sequences provide crucial new tools for studying their biology. The investment in the five chlamydial genomes sequenced thus far shows the perceived importance of this approach.

Chlamydial species were also selected because they are deeply separated from other eubacteria. *Escherichia coli K12* was originally sequenced to complement long-term molecular

Table 6.2: Genomes Sequenced From Strains That Have the Capacity to Cause Disease in Humans

Pub	Date	Species	Causative agent of
1	1995	Haemophilus influenzae	Meningitis, otitis media
2	1995	Mycoplasma genitalium	Urethritis
3	1996	Mycoplasma pneumoniae	Pneumonia
4	1997	Helicobacter pylori	Gastric ulcers, cancer
5	1997	Borrelia burgdorferi	Lyme disease
6	1998	Mycobacterium tuberculosis	Tuberculosis
7	1998	Treponema pallidum	Syphilis
8	1999	Chlarnydia trachomatis	Genital tract infections, trachoma
9	1999	Rickettsia prowazekii	Typhus
10	1999	Helicobacter pylori	(As above)
11	1999	Chlamydia pneumoniae	Pneumonia and bronchitis
12	2000	Ureaplasma urealyticum	Opportunistic infections during pregnancy
13	2000	Campylohacter jejuni	Gastroenteritis
14	2000	Chlamydia pneumoniae	(As above)
15	2000	Neisseria meningitidis	Meningitis and septicemia
16	2000	Neisseria meningitidis	(As above)
17	2000	Chlamydia pneumoniae	(As above)
18	2000	Vibrio cholera	Cholera
19	2000	Pseudomonas aeruginosa	Sepsis in immunocompromised hosts
20	2001	Escherichia coli	Gastroenteritis and urinary tract infections
21	2001	Mycobacterium leprae	Leprosy
22	2001	Escherichia coli	(As above)
23	2001	Pasteurella multocida	Cellulitis
24	2001	Streptococcus pyogenes	TSS, rheumatic fever, scarlet fever, pharyngitis
25	2001	Staphylococcus aureus	Boils, abscesses, sepsis
26	2001	Staphylococcus aureus	(As above)
27	2001	Mycobacterium tuberculosis	(As above)
28	2001	Streptococcus pneumoniae	Pneumonia, septicemia, and meningitis
29	2001	Rickettsia conorii	Mediterranean spotted fever
30	2001	Yersinia pestis	Plague
31	2001	Salmonella typhi	Typhoid fever
32	2001	Salmonella typhimurium	Gastroenteritis
33	2001	Listeria monocytogenes	Listeriosis
34	2002	Brucella melitensis	Malta fever
35	2002	Clostridium perfringens	Gas gangrene
36	2002	Fusobacterium nucleatum	Dental caries
37	2002	Streptococcus pyogenes	(As above)

genetic studies of a model laboratory organism. It now provides an invaluable basis of comparison for the study of two more recently sequenced pathogenic strains of *E. coli* that are responsible for outbreaks of serious human disease.

The serotype B strain of *Neisseria meningitidis* was sequenced in the hopes that access to a complete genome sequence would accelerate the development of a vaccine. Although pathogenic, mycoplasmas were primarily selected because they have extremely small genomes.

As such, they afford a model with which to study the minimal set of genes required for a free-living, replicating cell. The selection of an index strain is the next critical step. This decision has important implications because there is tremendous genetic and phenotypic diversity within bacterial species. Most species that colonize humans and are capable of causing disease also contain strains that are nonpathogenic, or even beneficial.

Among pathogenic strains in a species there are often a significant variety of distinct serotypes, biovars, or otherwise genetically distinct strains that cause different types of disease, and show different tissue and even host tropisms. Finally, the DNA of the genomes of strains within a single species can differ by up to 20%, and often it is debatable whether we really understand what a bacterial species is. Once the DNA of a genome has been sequenced and assembled, the annotation process begins.

The quality of *annotation* is critical, as it impacts on downstream analyses based on the genome sequence and the quality of our public *electronic* databases. The basic features of this process were established in the groundbreaking analysis of the first complete bacterial genome sequence, but sequence annotation remains an active area of research.

The analysis of a final annotated gene list is a first true look at the genetic capacity of an organism, and both the discovery of unexpected genes and the lack of expected genes are valuable outcomes of this process. Also valuable is an *appreciation* of the number of genes that can be *ascribed* a putative function vs those whose function remains undefined.

The ability to assign functions to a significant number of genes in a genome, especially a large proportion of those highly conserved genes with housekeeping roles, further makes it possible to construct a metabolic and biochemical schematic of a cell. Similar analysis (presence/absence, shared/novel) of the pathways represented in this *schematic* likewise provides clues about the "*lifestyle*" of an organism. Equally important, a schematic of the major landmarks and features of a bacterial genome sequence, which can be constructed from a complete annotation, can have a profound impact on our understanding of *biological* potential.

For example, genes found in operons and clusters often have related functions, while the annotation of the various classes of intergenic regions and repetitive sequences in a genome provides information about how organisms regulate their repertoire of genes, how genetic diversity is generated, and how long-term evolutionary processes, like the reduction of *genome* size, proceed. The publication of a genome sequence, no matter how *informative*, is only a first step towards new and productive lines of *investigation*.

This holds true whether the genome is a first window into the biology of a poorly characterized species or is yet another tool, made more powerful when used in parallel with data collected over decades of study of a species like *E. coli*. When properly analyzed, genome sequences should

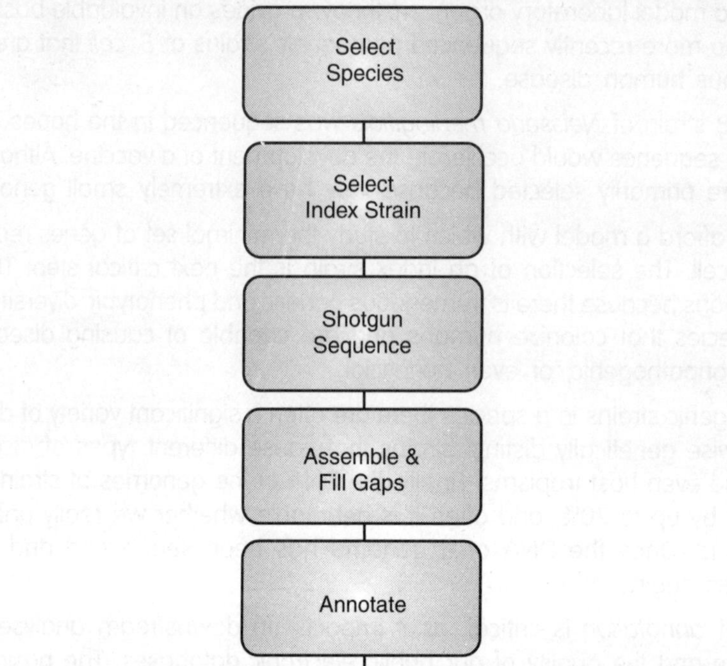

Figure 6.2: The standard work flow of a bacterial genome-sequencing project.

raise more questions than they answer. *Scrutiny* of a genome sequence will elucidate the genetic basis of particular traits, but the *longer-lasting* contribution of each new genome is a list of candidate genes that is made available for the future study of a much larger variety of *phenotypic* traits.

Powerful comparative genomic and experimental postgenomic technologies, direct outgrowths of the successes of whole-genome sequencing, now provide the promise of allowing researchers to elucidate the information contained in genome sequences to determine genotype-phenotype relationships explicitly and explore what organisms really do with their genetic potential.

Prokaryotic Genomes

The collection of pathogenic and nonpathogenic organisms sequenced over the past 8 yr represents a wealth of genetic and ecological diversity, and provides excellent starting material for comparative genomic studies.

The fact that we now have genome sequences for a large number of bacteria, half of which have the capacity to cause disease in humans, has highlighted a core set of *fundamental* biological properties of bacterial genomes that place the in-depth study of bacterial *pathogens* into proper context.

Orphan Genes

Perhaps the most striking general insight to emerge from whole genome sequencing (of both bacteria and eukaryotes) is the number of genes without known relatives-also referred to as "orphans" or "ORFans".

Wholegenome sequences are of particular value because they contribute a copy of all the genes found in a genome,. and not just the most intensively studied genes. We already have around 225,000 ORFs from bacterial genome sequences alone, and yet, despite this wealth of data, there continue to be large numbers of genes that are known only from a single species.

Some authors have suggested that the compilation of this long list of unshared, orphan genes is the biggest surprise to come from genome sequencing thus far. Orphans are not a phenomenon restricted to species for which only distantly related genome sequences are available, but may also be found in strains within a species.

For example, when two pathogenic strains of *E.coli* were compared with strain K12, a total of 1632 ORFs were found, 1387 of them *strain-specific.* With respect to pathogens, the detection of

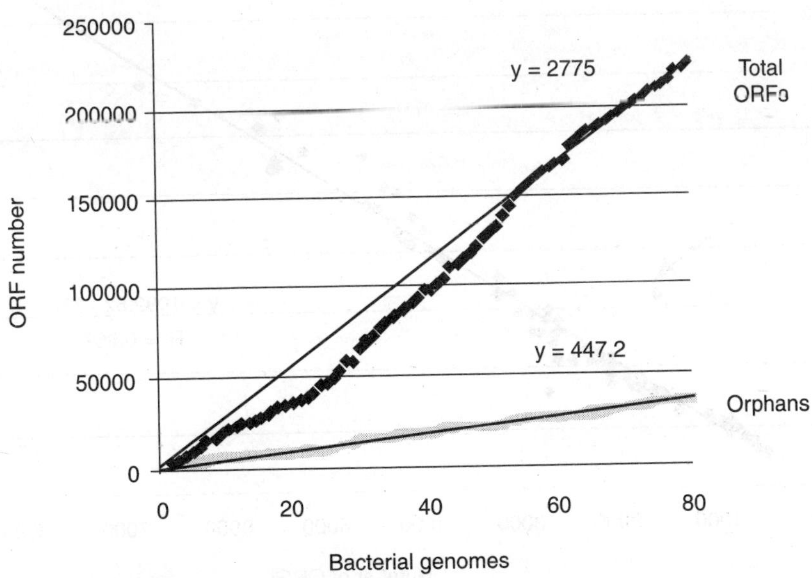

Figure 6.3: Showing the ORF number in bacterial genomes.

orphans and taxonomically restricted genes is of special interest because some may be *responsible* for their virulence behaviour.

Relatianship Between Gene Number and Gene Size

Another striking observation that has emerged from the sequencing of tens of genomes is the relationship between genome size and coding capacity. We have genome sequences that represent a significant portion of the total range of genome sizes observed in bacteria. From the trendline in it can be seen that on average there are 1000 genes per million base pairs of a bacterial genome sequence.

This relationship holds true for the smallest genomes of pathogenic mycoplasmas and the symbiont *Buchnera aphidacola,* and also for the largest sequenced genomes of generalist soil bacteria like *P. aeruginosa.* It is due to the fact that overall gene length (approx 1000 bp) and the proportion of noncoding DNA (85-94% of the total genome is usually coding DNA) is generally conserved across bacterial taxa. The intracellular pathogen *Mycobacterium leprae,* though, is a striking exception. This species deviates significantly because it is in the process of shedding large numbers of genes.

Caught in the process of genetic decay, *M. leprae* represents a rare glimpse into the biological phenomenon of "reductive genome evolution"-a key discovery to come from the genome sequencing of highly host-adapted intracellular pathogens. Only 49.5% of the *M. leprae* genome encodes for functional proteins, while 27% contains recognizable genes that have decayed into

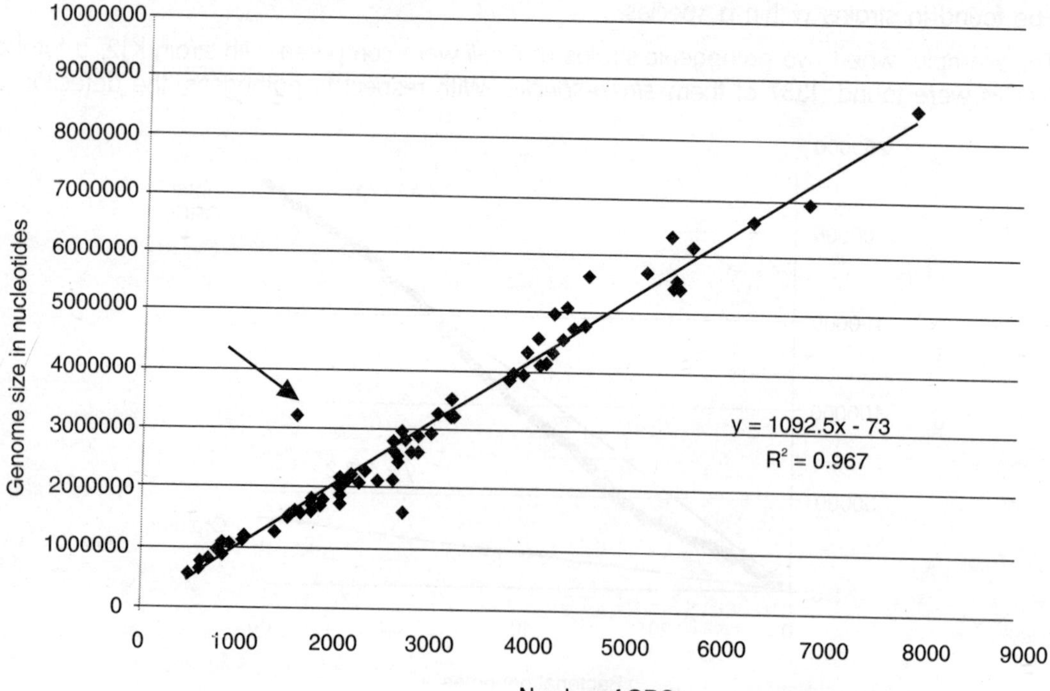

Figure 6.4: Cap relationship between ORFs and genome size.

nonfunctional pseudogenes through the accumulation of mutations. The remaining 23.5% is purely noncoding. When the 1,116 pseudogenes found in the genome are included, *M. leprae* nonetheless fits the expected ratio of ORFs to genome size. The larger genome of the closely related *Mycobacterium tuberculosis* contains functional copies of many of these pseudogenes and has a standard ratio of ORFs to genome size.

Phylogeny

Genome sequencing has further underscored the high frequency and biological importance of the common ability of bacteria to acquire foreign DNA. Horizontally acquired sequences can be detected in a single genome in association with mobile elements or through significant departures from genome-wide average values for features such as G+C content, dinucleotide frequencies, and codon composition.

Comparative genomic analysis has further provided ample evidence of significant genetic transfer between strains within a species to the level of taxa that belong to different bacterial domains. Transfer events can involve the acquisition of large numbers of small sequences over time or the acquisition of large sequences that, when they carry a set of genes involved in virulence, are termed *pathogenicity islands* (PAIs).

This feature of bacterial genomes was originally described in *E. coli* in the latter half of the 1980s. Subsequently, in part owing to the analysis of whole genomes, pathogenicity islands have been discovered to be one of the major "themes" of bacterial pathogenesis, due both to their prevalence in bacterial populations and the importance of the traits they confer on their host genomes.

The confirmation of high rates of horizontal transfer between bacterial genomes, by contrast, is also encouraging a growing appreciation of the concept of the *genetic backbone,* comprising a core set of indigenous genes that remain stable over time. Genetic backbones are now actively being defined by homology comparisons for various evolutionary groups including serogroups, species, and even bacterial domains.

REQUIREMENT OF GENES TO BECOME A PATHOGEN

The challenge in analyzing the genome and genetic capacity of any bacterial pathogen lies in distinguishing the genes responsible for disease from all the other genes in its genome. Access to the wealth of data provided by complete genome sequences underscores two important facts-first, the inherent difficulty of defining what a virulence determinant is, and second, that pathogenicity is often due to a complex combination of genes working together.

Accordingly, it has been suggested that there are three classes of proteins responsible for virulence. These are true virulence factors, which are found only in pathogens, genes that regulate "true virulence genes," and "life-style" genes, such as those that are required to colonize the host. Questions like "How many genes are required for virulence?" can be answered only by experimentation, but the analysis of a genome can play an important role by providing a list of candidate virulence determinants.

Given a draft list of candidate virulence determinants we can ask further questions, like "What proportion of the total coding capacity of the genome do these genes represent?" and "What classes of genes are involved?" Answers to these questions differ among pathogens with different lifestyles. Experimental studies have demonstrated that in enteric bacteria like *E. coli,* becoming a pathogen can be as simple as acquiring a single region of DNA (by horizontal transfer), which is integrated into a "permissible" genetic background full of genes that allow (benign) colonization of the human host.

The intracellular enteric pathogen *Salmonella enterica* requires at least 60 genes to be a pathogen, and many of these genes are shared with nonpathogenic strains or have housekeeping roles. To put the types and frequencies of virulence determinants in a genome into proper context, we need to understand an organism's physiology, ecological niche, and susceptibility to the acquisition of foreign DNA.

Physiology and preferred niche are intimately linked-biochemical capacities evolve in response to the environment (in the case of pathogens, the host), and physiology determines where and how an organism can live. The probability of acquiring foreign DNA and the niche are also intimately related.

The ecological niche of a bacterium should determine the frequency at which a pathogen has a chance to acquire foreign DNA and the probability that this DNA will be from another species, genus, or even a more distantly related taxon. Pathogens that live inside host cells are expected to have limited opportunities to acquire foreign DNA, while enteric and mucosal pathogens, which live in complex microbial communities, should experience frequent opportunities for genetic exchange.

Pathogenicity

Identifying the determinants of virulence not only provides essential clues to the biology and evolution of an organism, it also paves the way for the development of novel vaccines and diagnostics. The complex progression of steps in the behaviour of a pathogen can be described in five stepsadherence, invasion, multiplication within the host, interference with host defenses, and damage to the host.

The corresponding categories of genes share many "themes," or mechanisms of action. These are *adhesins,* like pili/fimbrae; *invasins,* which allow bacterial cells to move to their preferred location within the host; *pabulins,* or genes that help the bacterium scavenge energy and essential nutrients like iron; *evasins,* or factors like capsule and surface proteins that provide mechanisms of surviving host defenses; and *toxins,* or factors that damage the host and the secretion systems that deliver these toxins into host tissues.

Genome sequencing not only provides an overview of all the virulence factors in an organism; it also gives insights into the relative genetic investment in each class of factor. This in turn provides ecological insights into the host-pathogen relationship and the relative adaptive strategies of ecologically different pathogens. The genome of every pathogen sequenced thus far has provided a significant list of newly discovered candidate genes ready for further empirical characterization.

The contribution of genomics to our overall understanding of pathogenicity and virulence is especially large when the target organism is a fastidious grower or an organism for which genetic tools have not been developed. For example, prior to the sequencing of the *T pallidum* genome, no virulence factors were known in this species.

In the annotation of this sequence, a total of 70 candidate virulence factors were identified. One interesting exception that highlights how little we know about certain molecular mechanisms of virulence came from the genome analysis of *Borrelia burgdorferi,* the spirochete agent of Lyme disease, which goes through a tick-rodent-human life cycle. In this case, the surprise was how few virulence factors could be found.

Instead of a wide range of recognizable genes involved in virulence or host-parasite interactions, the genome revealed a tremendous repertoire of lipoproteins unique to *Borrelia* species. Even more interestingly, 63 of the total 105 lipoproteins are carried on the 17 plasmids contained within the genome.

The loci encoding these largely unknown lipoproteins make up 8% of the genome, compared with only 1.3% in *Helicobacter pylori,* attesting to their biological importance. There are three methods of detecting novel virulence factors within genome sequences. One can search for genes homologous to those encoding known virulence factors in other species, detect virulence factors *de novo* based on their unusual DNA characteristics, or identify genes found only in pathogens.

The majority of virulence factors are found using homology-based methods. While this approach is extremely effective for well-studied taxonomic groups containing well-studied pathogens, it is obviously less useful for undersampled taxa, and can shed no light on virulence genes for which there are no known copies in public databases. Similarity analysis can also be used to characterize sequences (paralogs) within the same genome, including those that are not found in public databases. For example, completion of the *M. genitalium* genome allowed the identification of all copies of the *MgPa* operon, a key virulence determinant of this species.

Analysis of the *M. tuberculosis* genome led to the discovery of the large PE and PPE genes families, which comprise 10% of the total coding capacity of this species and are suspected to play a role in virulence and antigenic variation. Methods that are not reliant on the detection of homology include the detection of horizontally transferred DNA sequences, the identification of genes encoding surface-exposed peptides by sequence motifs related to their secretion or surface binding, and the detection of sequences potentially involved in generating antigenic variation by their repetitive nature or their membership of large gene families.

One of the most striking examples of the power of detection methods that are not homology based involves searches for simple sequence repeats (SSRs). Also known as microsatellites in eukaryotes, these hypervariable, short tandem repeats are prone to expansion and contraction through the gain or loss of repeat units via replication slippage. When found in promoters and coding regions, mutations in these repeats can cause "phase variation" of the genes with which they are associated.

The ability to vary phase, or undergo rapid switching between distinct phenotypic states, is a trait shared by a wide range of pathogenic bacteria, and such loci have been termed "contingency loci," as they generate random but useful genetic variation that plays an important role in adaptation. Phase variation through SSRs often produces antigenic variation, but it can also facilitate adaptation to different microniches within the host. A wide variety of molecular mechanisms drive the generation of mutations at contingency loci, but SSRs lend themselves extremely well to discovery by *in silico* analyses. Prior to the publication of the *H. influenzae* genome in 1995, only a few such SSR-containing contingency loci were known.

However, a computational survey of the fully sequenced genome revealed nine tetranucleotide repeats positioned in coding regions such that they had the potential to be translational molecular switches. A knockout mutant lacking one of these genes, *lgtC,* had reduced virulence.

All of these repeats, plus others not found in the index genome (Rd), have now been empirically demonstrated to cause phase variation of genes responsible for lipopolysaccharide biosynthesis, adhesion, and iron scavenging, and the power of this *in silico* method has been applied to other genomes.

Most notably, we now appreciate that the genomes of *H. pylori, Campylobacter jejuni, N. meningitidis,* and *Streptococcus pneumoniae* are each characterized by sets of between 10 and 35 contingency loci containing SSRs. The third method of identifying genes involved in virulence is perhaps the least specific, but offers a chance to find potential virulence determinants for which we lack any *a priori* knowledge upon which we can build detection methods.

This method requires two or more genome sequences and relies on using subtractive methods to detect unshared genes or noncoding regions between genomes. This approach can now be applied to very closely related genomes, since there are two or more genomes available for a total of eight pathogenic species. *E. coli* is one species for which we have the genomes of both pathogenic and nonpathogenic strains, and comparisons have revealed that as much as 25% of the genome of pathogenic strains is not shared by strain K 12.

It is interesting to note that these regions contain more genes for degradation than for biosynthetic capacity. The concept of these "electronic filtering experiments" can now be extended to the comparison of unsequenced genomes. There are already nine species of bacteria capable of causing disease in humans for which we have the genome sequence of more than one strain. We have both pathogenic and nonpathogenic strains from both *E. coli* and *S. pneumoniae.*

Microarrays containing large numbers of bacterial genes or complete genomes are now being developed for *genomotyping,* or the profiling the genetic similarities and dissimilarities between collections of strains. Such technologies promise to provide unprecedented levels of information

Table 6.3: Sequencing of Multiple Strains from a Single Species.

Species	Genomes	Nonpathogenic	Pathogenic
Chlamydia trachomatis	2	0	2
Chlamydia pneumoniae	3	0	3
Escherichia coli	3	1	2
Helicobacter pylori	2	0	2
Mycobacterium tuberculosis	3	0	3
Neisseria meningitidis	2	0	2
Staphylococcus aureus	2	0	2
Streptococcus pyogenes	2	0	2
Streptococcus pneumoniae	2	1	1

about the nature and significance of the extensive population-level genomic diversity that is known to characterize bacterial species.

Functional Behaviour

Complete genome sequences provide a catalog not only of genes, but also of pathways that, when combined into functional categories, explain the metabolism and biosynthetic capacities of an organism. We now know that the tight relationship between genome size and total numbers of genes also translates into relative functional complexity.

Since the publication of the first bacterial genome sequence, annotated genes have been assigned to functional categories, an approach originally pioneered by Riley in a seminal study of *E. coli*. Adherence to a standard classification schema allows a familiar schema to be produced for each genome.

The types and relative numbers of gene categories in a genome can be read to explain various biological attributes of a cell. Such analysis can be used both to confirm the molecular basis of observed traits as well as to discover new ones.

Even more importantly, this cellular blueprint can be compared across genomes to gain insight into how different bacteria are built to occupy their specific ecological niches. Knowledge of the unique metabolic and genetic features of pathogenic strains and species may make a large impact on human health by facilitating the development of cheaper, faster diagnostic methods. Analysis of pathways and metabolic capacity might also reveal clues to the culturing of fastidious pathogens.

Finally, an understanding of the genes required for essential metabolic tasks that are not shared with eukaryotes (especially humans) is the key to developing potential targets for novel antimicrobials. The stylized blueprint of the physiology and metabolism of a generalized cell is shown in Figure 6.1. This schema is vastly simplified compared to those reported in genome analysis papers, but is a composite of the essential and optional features of a bacterial cell.

Core features shared with even the smallest of genomes *(M. genitalium)* include DNA replication, transcription and translation, some form of energy production or energy scavenging, and the ability to obtain key metabolites either through biosynthesis or by scavenging through the use of transporters with various affinities.

Cells with larger genomes and more genes have increased functional diversity, both in terms of the overall number of unique pathways and with respect to genetic redundancy. Optional features include such things as genes for motility and chemotaxis, secretion systems, and drug efflux systems. The cellular blueprint of a bacterial cell reflects its lifestyle and, for comnsal and virulent bacteria, the nature of the relationship with their host.

Highly specialized, obligate intracellular pathogens have small genome sizes, and therefore low numbers of genes, and lack some or most biosynthetic pathways, having evolved over time in association with their hosts to scavenge rather than produce key metabolites. Fascinatingly, it is now known that there are many evolutionary routes to becoming an intracellular pathogen, and different pathways can be lost-in varying orders-during evolution of a reduced genome.

The number of pathways (e.g., the ability to synthesize amino acids, purines, or pyrimidines)

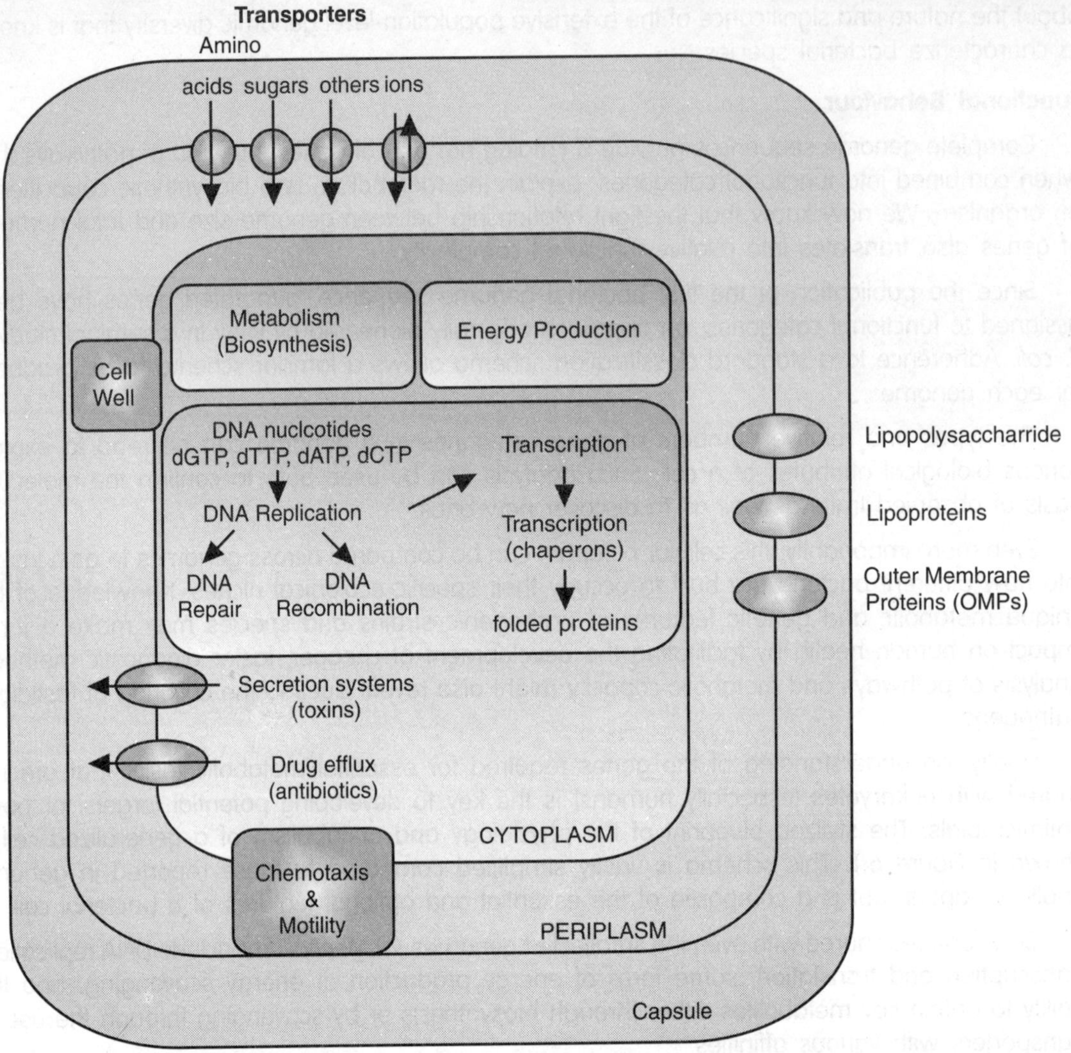

Figure 6.5: A stylized schema of a bacterial cell. This composite view of a hypothetical cell shows all the major features of bacterial cells, all of which can be reconstructed from the annotation of a complete genome sequence. Essential features include DNA replication, transcription, and translation, and energy production. Other features are optional. Secretion systems are found only in Gram-negative bacteria. Capsule is a virulence determinant only in certain species, and many of these species, like H. influenzae, have both capsulated and noncapsulated strains. Possession of different combinations and numbers of these components reflects different physiologies and, for pathogens, different relationships with the host.

lost reflects the degree of dependence a bacterium has on its host. Genome analysis shows in intricate detail the genetic trade-offs associated with adaptation to a given ecological niche-if a cell does not have a biosynthetic pathway for an essential molecule, it must have at least one transporter/scavenging mechanism in its place.

Moreover, the mechanism a bacterium evolves for such scavenging can determine the type and extent of damage done to host cells. In some cases, robbing from the host environment is not a passive but an aggressive activity that involves special enzymes (toxins) that dissolve host tissues to release component parts, as in the case of the flesheating bacterium *Clostridium perfringens.*

In combination with appropriate transporters, such systems can remove the need for entire biosynthetic pathways, which are then lost over evolutionary time, gene by gene. Larger genomes encode a more extensive genetic capacity that includes both larger numbers of regulatory sequences and more extensive and often redundant pathways for synthesizing and acquiring nutrients and other molecules.

The cosmopolitan, generalist, soil bacterium *P. aeruginosa* has the genetic capacity to become an opportunistic pathogen because the functional diversity and complex regulatory mechanisms in this large genome provide this species with resistance to *disinfectants* and *antimicrobials.* The analysis of the relative proportion of genes in each functional category is of special interest, especially when the lack of key metabolic capacities explains particular dependencies of pathogens on their host.

Also informative are the abundances and types of surface molecules that mediate interactions with the host. For example, the genome of *T. pallidum* is marked by a near absence of genes that encode outer-membrane proteins while the *M. tuberculosis* genome is unique for the unusually large portion of the coding capacity dedicated to the production of enzymes involved in lipogenesis and lipolysis.

Bacterial NIche

Genome sequences are shedding new light on how bacteria survive within the host. These insights include the identification of genes, pathways, and even global features of genome sequences that help to explain the ability of bacteria to invade and colonize specific locations in the host. For example, the ability of *H. pylori* to colonize the human stomach is due to the enzyme urease, but completion of the genome sequence also revealed that there are twice as many basic amino acids (lysine and arginine) in the proteins of *H. pylori* than in other bacteria.

This unusual amino acid composition provides an additional adaptation to life in an acidic environment. It is also possible to discover how species adapt to microniches within humans, and why certain species fail to colonize certain niches. Analysis of the genome of the bacterium *Fusobacterium nucleation,* which colonizes the oral cavity, revealed that the restricted ecological niche of this species is due to the lack of genes that encode cholinebinding proteins.

This is in contrast to other common commensals like *H. influenzae* or *S. pneumoniae,* which have these genes and can therefore colonize the nasopharynx. Genomes also provide insights into how pathogens modify their environments-a process that can contribute significantly to host-cell damage and death.

The genome sequence of the flesh-eating bacterium *C. perfringens* provides an extreme example of the ability of bacteria to change and utilize the resources in their chosen environments. This anaerobe uses its genetic capacity when colonizing wounds to create a preferred anaerobic environment by making carbon dioxide, the gas that leads to gangrene, through anaerobic fermentation.

The genome sequence revealed the limited biosynthetic capacities of *C. perfringens,* and its repertoire of more than 200 transporters. These features place the toxins and enzymes it uses to obtain nutrients by degrading host tissues into proper context.

Genetic Engineering

One of the most fascinating results to emerge from genome sequencing is an improved understanding of bacterial genome architecture, or the macro- and microlevel organization of genes, elements that fill noncoding regions, repetitive sequences, and compositional patterns within a genome.

A direct outcome of this knowledge is our growing appreciation of the fluidity of genome sequences. Genome analysis is revealing both the nature and extent of molecular mechanisms that both promote mutation (vertical inheritance) and gene transfer (horizontal inheritance). For example, the presence of SSRs near virulence determinant genes in mucosal pathogens suggests the potential for clones to generate quickly genetic variation of biological significance through mutation.

More striking, perhaps, is the extent to which certain genomes are "engineered" to rearrange the order and composition of their genes and noncoding regions by the presence of plasmids, bacteriophage, transposons, and other classes of repetitive DNAs, like uptake sequences (discovered as a result of the sequencing of the *H. influenza,* genome). In contrast, the genomes of other organisms, particularly those of intracellular pathogens and *symbionts* with very small genomes, are not.

Through genome sequencing, we now know that the small genomes of intracellular pathogens have lost most if not all of the mobile and self-propagating mediators of genome fluidity found in some other bacteria. In contrast, these elements can represent a significant proportion of the larger genome sequences of free-living, cosmopolitan bacteria, and represent a tremendous capacity for evolutionary change. The ability to acquire foreign DNA can impact on both the short- and longterm phenotypic evolution of pathogens. While the true frequency of horizontal transfer and the percentage of horizontally transferred sequences is debated, it is clear that genetic transfer can lead to biological innovation.

Traits such as antibiotic resistance, which protect bacteria against noxious chemicals they may encounter in certain environments, can circulate in bacterial populations and jump across evolutionarily distinct lineages to form new combinations of genes. In pathogenic strains, these new combinations can give rise to clones with changed pathogenic potential; and in nonpathogenic strains, the repeated uptake of foreign DNA can even give rise to new species of pathogens. Species of bacteria capable of causing disease in humans have repeatedly evolved in the tree of life.

Many species contain both pathogenic and nonpathogenic strains. Both these facts demonstrate the ease with which bacteria can apparently evolve strategies that result in virulence and damage to the host. In many cases, this ease can now be shown, through the comparison of appropriate sets of genome sequences, to be the direct result of the ability to acquire DNA horizontally.

The primary source of genetic diversity between the sequenced genomes of the pathogenic strains of *Streptococcus pyogenes,* and between genomes of virulent and avirulent *E. coli,* are

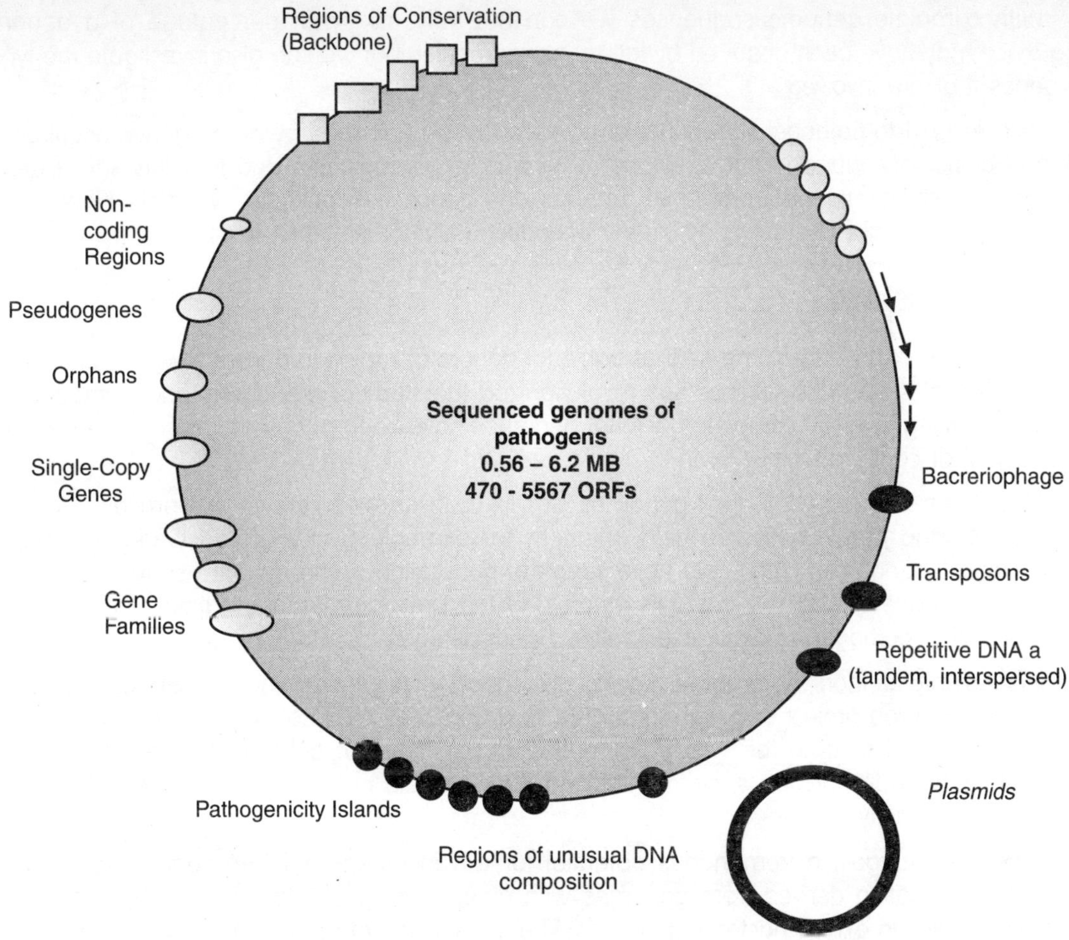

Figure 6.6: A stylized schema of a bacterial genome. This composite view of a hypothetical genome shows the core features of bacterial genomes. The abundances of these features vary among genomes, and some features are absent from particular genomes.A large portion of each genome (80% or more) will belong to the genetic backbone of a species. Features in black are those genetic elements and sequences most susceptible to genetic transfer between genomes.

phage and phage-like sequences that have inserted themselves and additional short segments of foreign DNA into the genetic backbone of these species. In contrast, the *Vibrio cholerae* genome provides an unprecedented opportunity to study how a free-living aquatic organism emerged to become a pathogen of humans through the lateral acquisition of a collection of virulence factors, such as the filamentous CTX bacteriophage that encodes the cholera toxin, and the TCP pathogenicity island.

Analysis of this genome supports the view that horizontal transfer has not only been a means for strains of this species to expand and reorganize micro-sections of their genetic capacity, but has also provided a mechanism for this species to acquire the genetic capacity contained on a second chromosome. The composition of genes on the second, smaller chromosome of *Vibrio cholerae* suggests that it is derived from a former captured megaplasmid.

With complete genome sequences we can not only ask what percentage of a genome sequence may have been acquired by lateral gene transfer, but we can also investigate the types of genes that are involved.

With respect to pathogens, genome analysis is revealing that many of the genes acquired by lateral transfer are found in only a subset of strains in a population, and that this set of genes contains a disproportionate number of orphans and genes with unknown function. However, of those genes that are characterized, many encode traditional virulence factors.

APPLICATION

The past 8 yr have seen the field of bacterial genomics rapidly progress toward becoming a mature field of study. This success has revolutionized the study of pathogens and pathogenesis. As of this writing, over 80 complete bacterial genomes have been published, half of which belong to strains that cause disease.

More than a hundred further genomes are in draft form, or are in the process of being sequenced, and many more are held privately, for example by Integrated Genomics (http://www.integratedgenomics.com/). We have seen the publication of the genome sequence of more than one pathogenic strain in a single report and the simultaneous publication of the genome sequence of a pathogen and a closely related nonpathogen.

Increasingly, authors of genome reports are choosing to present the combined results of a genomesequencing project and high-throughput experimental analyses (e.g., microarray studies). These facts attest to both the raised estimation of the value of genome sequencing and the reduced costs and improved experimental and bioinformatic technologies now available to the research community.

They also suggest a tremendous potential for advancement in future years. The ability to integrate information derived from genomes with other sources of information will be central to our future ability to exploit bacterial genomes. The sequencing of pathogen genomes has proved beyond doubt the value of using genomes to discover new determinants of virulence. It has also proved the power of using genomes to elucidate the molecular basis of physiology, adaptation to niche, and the mechanisms by which genomes change over both short and long evolutionary time spans.

In order to build on this knowledge, we must (1) find new ways to characterize the vast number of genes without known function, especially orphans; (2) continue to look for global patterns that help define unusual features of specific sequences, strains, and species; (3) more effectively translate the results of genome sequencing into hypotheses that can be tested by experiment; (4) explore the extent and significance of populationlevel genetic diversity; and (5) place each genome into its proper "organismal" context by understanding its specific ecology and evolution.

Perhaps most importantly, this heterogeneous and complex information must be captured in next-generation databases that will improve the scope, quality, and interconnectivity of our current invaluable electronic resources. With this integrative approach, based on a sound knowledge of genome sequences, we will be able to explore in more detail than ever before what makes a pathogen. This will occur through an increased understanding of the genetic capacity of a wider

Table 6.4: Internet Resources for Examining Genomic Resources Available for Pathogens.

ELECTRONIC RESOURCES
Get a full list of complete and in progress genomes at GOLD:
http://www.integratedgenomics.com/gold/
Access all complete bacterial genome sequences:
http://www.ncbi.nlin.nih.gov/PMGifs/Genomes/eub_g.html
Access the taxonomy of all sequenced pathogens:
http://www.ncbi.nlm.nih.gov/PMGifs/Genomes/new_micr.html *(also see http://www.cme.msu.edu/bergeys/)*
Find individual genes among all known public DNA or protein sequences:
http://www.ncbi.nlm.nih.gov/Entrez http://www.ebi.ac.uk/srs
Access curated lists of individual proteins in Swissprot:
http://www.ebi.ac.uk/swissprot/
Conduct similarity searches against finished and unfinished microbial genomes:
http://www.ncbi.nlm.nih.gov/cgi-bin/Entrez/genom table.cgi
See features shared by bacterial pathogen proteomes:
http://www.ebi.ac.uk/proteome/
See lists of known virulence factors in the Prints Database:
http://www.jener.ac.uk/BacBix3/Pprints.htm

Internet resources have grown quickly in number and complexity because the completion of the first bacterial genome sequence. For specific information on particular pathogens it is best to find a database dedicated to that pathogen. International genome sequencing centers like TIGR (http://www.tigr.org) and the Sanger Centre (http://www.sanger.ac.uk) have developed individual databases for each sequenced genome.

range of pathogens, a greater appreciation of what pathogens do with their genetic capacity, and a deeper knowledge of exactly how this capacity is engineered to change over time and how much of this change is biologically significant.

Chapter 7

Chemo Therapy

One of the challenges of medical research today is to find ways of bringing our genetic knowledge of cancer to clinical application and to develop improved *therapies* by exploiting gene-based strategies. By offering increased specificity and reduced toxicity, *gene-based* approaches promise alternatives when *conventional* treatments fail.

Gene therapy also offers improved responses to conventional treatments when used in combination with gene therapy. Chief among the cancers in urgent need of improved treatment options is prostate cancer—the most commonly *diagnosed* cancer in men. Advanced prostate cancer is resistant to most forms of hormone therapy, radiation therapy. and conventional chemotherapy. In fact, many of the most powerful *chemotherapeutic* drugs commonly used for other cancers are not effective against *prostate cancer.*

Among these agents is cisplatin (also known as *cis-diammincdichloroplatinum* [CDDP]. Platinol™. cis-platinum), which causes DNA damage in rapidly dividing cells by forming primarily bifunctional intrastrand adducts between adjacent guanines and *guanine-adenine dinucleotides. Cisplatin* is highly effective against ovarian cancer and bladder cancer and can achieve cures when used against testicular cancer.

Cisplatin has also been applied to colon and brain *tumors* but is not in routine use for these applications. A method for affecting, increased sensitivity to *cisplatin* and for *inhibiting* or reversing resistance to cisplatin would provide an extended application for this drug, an important agent with well-known properties. The cellular response to stress and DNA damage now appears to be central to the issue of resistance to DNA-damaging *chemotherapies* such as cisplatin, suggesting that genetic *approaches* that target these pathways may be effective in *reversing* resistance and *enhancing* the benefits of these agents.

In this chapter, we describe the potential of reversing drug resistance through gene therapies that target two major stress and DNA damage-response pathways: the Jun NH_2-terminal kinase/ stress-activated protein kinase (JNK/SAPK) pathway and the DNA damage-induced *apoptotic* pathway mediated by the p53 tumor suppressor. There are *suggestions* that inhibition of the Jun kinase pathway or restoration of the p53 pathway may offer new biologic approaches to therapy *sensitization* in *prostate* cancer.

These approaches would expand the potential of presently available treatments and offer *alternatives* when conventional treatment protocols fail. These approaches are illustrated by the use of a p53 adenoviral expression vector and by the use of highly efficient antisense JNK *oligonucleotides*.

Tumor Suppressor

The p53 tumor *suppressor* has attracted considerable attention, not only for its tumor suppressor properties but also for its ability to sensitize tumor cells to apoptosis following treatment with chemotherapy and radiation. Recent studies suggest that p53 gene therapy could have a broad application as a therapy sensitizer in cancer, a *possibility* that greatly expands the clinical application of p53-based approaches.

Because many cancers, including prostate cancer, acquire p53 mutations as the disease progresses and becomes resistant to therapy, p53 gene therapy in combination with *chemotherapy* could have *advantazeous* application for these advanced cancers.

Numerous *in vitro* studies now support the application of p53 gene therapy as a therapy

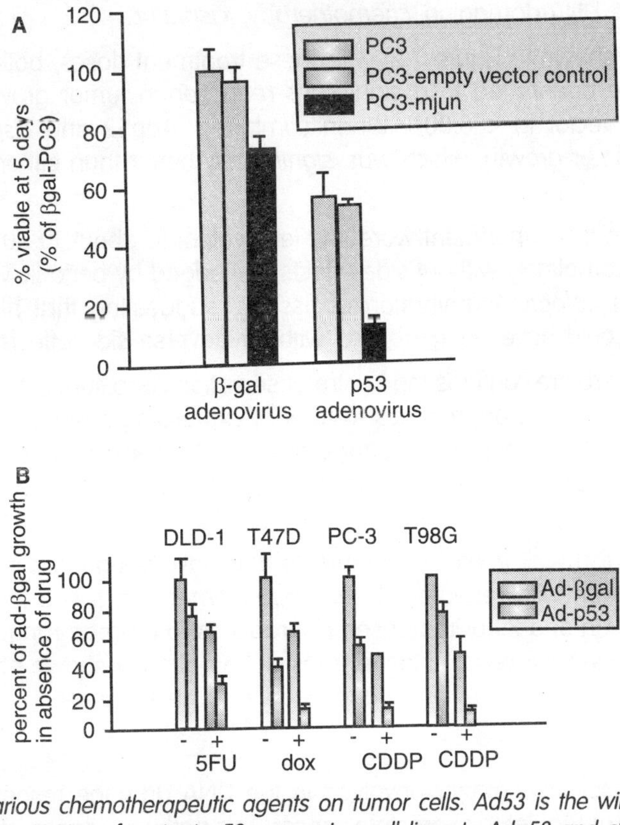

Figure 7.1: (A) Effect of various chemotherapeutic agents on tumor cells. Ad53 is the wild type whereas Ad-Bgal is the control vector.(B) Sensitivities of mutant p53-expressing cell lines to Adp53 and chemotera peutic drugs.

sensitizer for cancer. The results of screening tumor cells for sensitivity to various chemotherapeutic agents following exposure of cells to a *replication-defective adenovirus* encoding wild-type p53 (Adp53) compared to control vector (Ad-β-galactosidasc 1AD-βgal]) is summarized in Figure elsewhere in this chapter.

The cells were treated under conditions in which some 50-70% of them shows *transgene* expression, as evidenced by X-gal staining of parallel cultures infected with Ad-βgal. Viability is scored by the methanethiosulfonate (NITS) assay. Under these conditions, it was found that growth suppression by Adp53 alone is incomplete and that 7-d *viabilities* of Adp53-treated cultures were about 60-80% of control vector-treated cultures.

In addition, numerous tumor cell types lacking wild-type p53 expression, including the prostate cancer line PC-3, can be *sensitized* to *chernotherapeutic* drug treatment following *restoration* of wild-type p53 activity, consistent with the role of p53 in mediating DNA damage *recognition* and *apoptosis.*

Nude Mouse Model for Prostate Cancer

Recently it has been shown that p53 is highly effective at suppressing the growth of human prostate xenografts in nude mice but fails to achieve complete tumor eradication by itself. We have tested the possibility that more complete *suppression* can be achieved by combining p53 gene therapy with the DNA-damaging *chemotherapy, cisplatin.*

In the experiment shown in Figure 7.2, with these treatment doses, both Adp53 and cisplatin administered as single agents led to a significant *reduction* in tumor growth relative to tumors receiving only control vector ($p < 0.001$). When combined. Adp53 and *cisplatin* led to an even greater reduction in tumor growth, which was *significantly* better than either agent used alone ($p < 0.02$).

The vector doses in this experiment were low (equivalent to about 10-30 plaque-forming units [PFU]/cell) and derive completely without side effects, as judged by periodic weight measurements of the animals and *histologic* examination of tissues, suggesting that higher doses or more prolonged *treatment* could have been *applied* without adverse side effects.

This in vivo study therefore confirms the in vitro observations and supports the clinical *application* of p53 adenovirus combination approaches to tumors expressing mutant p53. *Adenovirus*-based approaches are among the most promising for clinical applications of gene therapy. *Adenoviruses* are relatively easy to prepare, are stable, can be obtained in high titer, and provide high *gene-transfer efficiencies.*

Adenoviruses are also well *tolerated* in patients and have been used in phase I clinical trials with few side effects even with repeated doses. *Improvements* in tissue targeting of adenovirus will extend the application of adenovirus-based approaches to a broad range of clinical situations. Nevertheless, *adenovirus-mediated* p53 gene transfer has already shown efficacy in *clinical* trials for lung cancer, *hepatocellular carcinoma,* and head and neck cancer.

JNK Pathway

Another important cellular pathway involved in the DNA damage response is the JNK/SAPK pathway. which is often upregulated in prostate cancer. This pathway represents one of the *mitogen-*

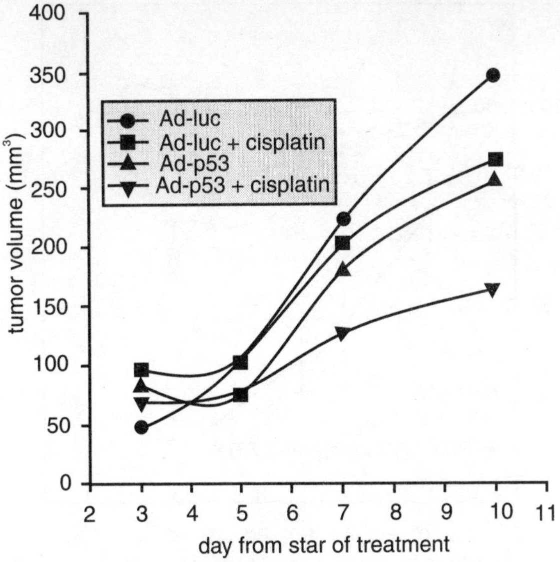

Figure 7.2: Suppression of PC-3 prostate tumor growth in nude mice by Adp53 + cisplatin.

activated protein kinase pathways and plays a role in growth factor signaling, oncogene expression, and cellular transformation, as well as the DNA damage response. *Numerous* reports document the activation of the JNK pathway in cells following treatment with a variety of DNA-damaging agents, including ultraviolet (UV) radiation, ionizing radiation, methylethane sulfate, N-nitroN'-nitroso-guanidine, l-β-d-arabinofuranosylcytosine), and *cisplatin*.

A major substrate of JNK is the transcription factor c-Jun, which is greatly activated on *phosphorylation* of serine residuals 63 and 73. We have demonstrated that tumor cells that stably express a *nonphosphorylatable* dominant negative mutant Jun (c-Jun [863A, 573A]) arc defective in repair of *cisplatin* adducts. *consistent* with their increased sensitivity to cisplatin.

The DNA repair assay that we used was based on the observations that cisplatin adducts will inhibit the *Taq* polymerase such that the decrease in yield of the polymerase chain reaction (PCR) product is directly *proportional* to the degree of *platination* of the template. The density of adducts can then he calculated based on a Poisson distribution predicted from a random process of *platination*.

Thus, (*P*) the relative PCR signal strength (damaged template relative to undamaged template) is related to the platination level (n), the number of adducts per PCR amplicon-sized fragment by the formula $P = e^{-n}$. In our case, we amplified a 2.7-kb fragment of the HPRTase gene, a fragment providing a sufficiently large target size to enable us to detect significant decreases in PCR signals from templates from cisplatin-treated cells. *Amplification* of a smaller fragment of 150 by in length, a fragment too small to register significant levels of *platination* under our conditions. served as a control for the efficiency of the PCR amplification.

Table 7.1 summarizes the frequency of adduct formation calculated after quantitating the PCR signal strengths (results of two experiments performed in triplicate). As shown in Table 7.1, cisplatin adducts form at a greater frequency on genomic DNA in mutant Jun-expressing cells than in

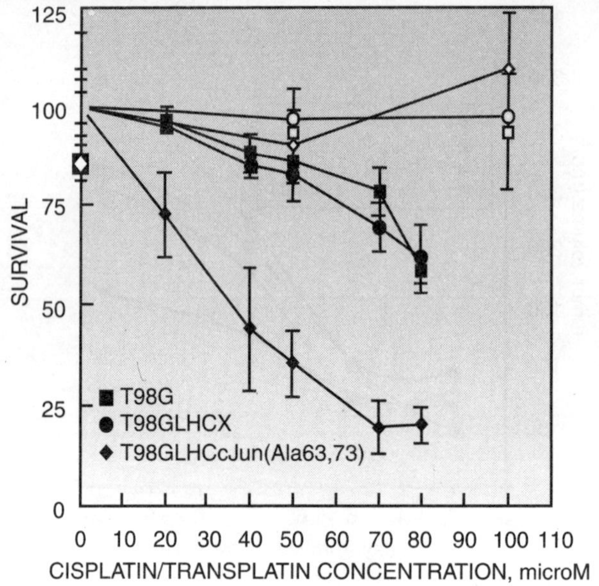

Figure 7.3: Cisplatin and Transplatin sensitivity of T98G cells and clones. Cell survival (viability) assay showing that expression of mutant Jun (dnJun) sensitizes clonal T98G cells (") to cisplatin. As controls, similar viability results are shown for wild-type c-dun overexpressing clonal T98G cells (•), and empty vector expressing clonal T98G cells (•). Open symbols represent the corresponding experiments with the inactive transplatin control indicating that decreased survival by inhibition of the JNK pathway is specific to cisplatin-generated DNA damage.

parental cells. In addition, mutant Jun-expressing cells show no repair of adducts in a 6-h recovery period following a 1-h treatment with cisplatin.

Whereas parental cells repair about half of the adducts in that time period. These results have been confirmed using a plasmid recovery DNA repair *assay*. In this assay. cells are transfected with chloramphenicol acetyltransferase (CAT) reporter plasmids that have been damaged ex vivo by *previous* treatment with 25 μM cisplatin for 3 h. Reporter gene expression 24 and 48 h *posttransfection* is taken as an indication of DNA repair.

Table 7.1: Cisplatin-DNA Adducts per 2.7 kb.

Treatment	T98G Parental cells	T98G Mutant Jun cells
200 μ*M* cisplatin for 1 h	0.48 ± 0.08	0.63 ± 0.2
200 μM cisplatin for 1 h. 6 h recovery	0.20 ± 0.05	0.55 ± 0.2

As shown Figure 7.3, cells that express mutant Jun are suppressed relative to parental cells in their ability to repair the damaged plasmid. consistent with the PCR-based DNA repair assays.

Although the mechanism by which the JNK pathway affects DNA repair is not known, there are several *candidate* target genes whose roles in DNA synthesis and repair could account for observations that we have made. The *promoters* of DNA *synthesis* and repair genes contain the

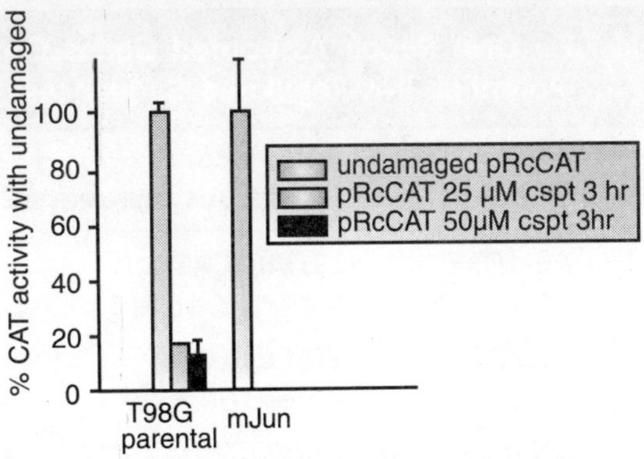

Figure 7.4: Plasmid reactivation assay using CAT reporter plasmid damaged ex vivo by treatment with cisplatin (espt).

transcript ion factors AP- 1 or ATF2/CREB regulatory elements. These include DNA polymerase B, *topoisomerase I. topoisomerase II, uracyl glycosylase, proliferating cell nuclear antigen,* and *metallothionine.*

In several cases, such as DNA polymerase β, exposure of cells to UV radiation is known to induce the gene, and ATF2/CREB sites are required for this induction. Thus, a number of genes involved in DNA excision repair may be coordinately *upregulated* on activation of the JNK pathway following *cisplatin* damage. Some or all of these could serve as potential targets for gene therapy approaches to drug sensitization of cancer.

Jun Kinase Pathway

In a parallel arm of the in vivo study shown Figure elsewhere in this chapter, we tested the antitumor efficacy of downregulation of JNK combined with cisplatin. In this study, we *downregulated* JNK with *antisense oligonucleotides targeting* either the JNK1 or JNK2 family of JNK isoforms.

These compounds were previously developed and characterized as described. An important *methodological* consideration is the use of high *affinity antisense oligonucleotides* that completely suppress target mRNA and protein at low *concentration.*

In collaborative studies with Isis Pharmaceuticals. we have developed a systematic method that now has been adapted to a *multiwell* procedure. The *salient* features are that a large series of phosphorothioate oligonucleotides *complementary* to 20 nucleotide stretches of JNK1 and JNK2 spaced approximately every 50 by along the *transcribed* portion of the gene are prepared on a small scale. Thus, the target gene sequence must he known.

These trial *oligonucleotides* are then tested for the ability to suppress *steadystate* transcript levels in culture cells 24 h after a 4-h lipid-mediated transfection. Most of these trial oligonucleotides we tested were not efficient at promoting suppression of target mRNA at low (0.2-0.4 µM) *concentration.*

However, it was readily possible to identify antisense compounds that were >90% efficient at eliminating target mRNA at these low concentrations after a single 4-h *transfection.* This step is key in avoiding any temptation to use less efficacious antisense compounds by increasing the

Table 7.2 : DNA Repair Associated Genes Containing Potential JNK-Regulated Sequences.

Repair Associated Gene	Element	Sequence (consensus sequence)2	LLG Position	Score
DNA polymerase B	AP-1/TRE	CTGACTCA (- t g a c t c a)	337	2.0
	ATF/CREB	TTACGTAA (t t a c g t c a)	282	2.0
DNA polymerase α	ATF/CREB	CACGTCA (t/g a c g t c a)	-82	
		GGGGTCA (t g a g t c a)	-149	
Topoisomerase I	AP-I	GGGGGCGG (t g a g t c a)	753	2.0
	AP-1	TGACCCA (t g a c t c a)	21.7	2.0
	ATF/CREB	TGACGTCA (t g a c g t c a)	792	2.0
Topoisomerase IIα	ATFJCREB	TGACGCCG (t g a c g t c a)	286	2.0
Topoisomerase IIβ	AP- I	TGATTGG (t g a g t c a)	337	
	AP-I	TGACTCA (t g a c t c a)	3	
	AP-1	AGAGTCA (t g a g t c a)	65	
ATF-3	AP-1	TGAGTAA (t g a c g t c a)	-1600	
	AP-1	ATAGTCA (l g a c g t c a)	-1353	
	AP-1	AGACTAA (t g a c g t c a)	-605	
	ÅP-1	GAGTCA (t g a c g t c a)	-380	
	ATF/CRE	TTACGTCA (t t a c g t c a)	-92	

(Table Contd.)

Repair Associated Gene	Element	Sequence (consensus sequence)2	LLG Position	Score
c-Jun	ATF/CREB	TTACCTCA (t t a c g c a)		2.0
Uracil glycosylase	AP-1	TGGGTCA (tgagtca)	141	2.0
PCNA	AP-I	TGACTCA (t g a c t c a)	489	2.0
PCNA	ATF/CREB	TGAGGTCAGGG (t g a c g t c a- - -)	209	1.64
	ATF/CREB	GTGACGTCAC (-t t a c g t c a- - -)	1253	1.60
GADD 153	ATF/CREB	ACTCCTGACCTT (t g/t a c g t c a - - - -)	207	1.63
Growth arrest and DNA-damage-inducible gene	AP- I	TGACTCA (t g a c t e a)	710	2.0
XRCC I	ATF/CREB	ACGTCA (a c g t c a)	1815	2.0
X-ray damage repair cross complementing gene product.	ATF/CREB	GGACGTCAA (t a c g t c a)	1814	2.0
	ATF/CREB	CCTGACCTCA (- - t g a c g t c a)	2029	1.64
	ATF/CREB	GCTGACGTCAG (- -t g a c g t c a-)	466	1.60
		CCAATCA (t g/t a c/g t c a)	93	2.0
MGMT	ATF/CREB	TGCGTCA (t g a c g t c a)	1661	2.0
06-Methylguanine-DNA-methyl-transferase	ATF/CREB	GTGACATCAT (- t g a c t c a -)	1195	
	AP-1	TGAGTCA (t g a g t c a)	734	2.0
	AP-1	TTACTCA (t t a c t c a)	285	1.73
MSH2	ATF/CREB	"I'GGCGTCA	108	1.62

(Table Contd.)

Repair Associated Gene	Element	Sequence (consensus sequence)2	LLG Position	Score
		(t g a c t c a)		
	AP-1	TGAATCA	569	2.0
		(t g a c/g t c a)		
	AP-I	TGATGAAA	884	1.62
		(t g a c/g t c a)		
Metal loth ioneinITA	AP-1	GAGCCGCAAGT	188	2.0
		(g a g t c a- - - -t		
		GACTTCTAGCG		
		g a c t c a a g l c		
		CGGGGCGTG		
		a - - - - - - - - t g)		

[1]Repair-associated protein for which only partial promoter sequences are known ti.c., in Genbank) without recognizable AP-I-regulated sites include ADP ribose polymerase, tif2/ref I, SSRP, Ercc1. and thymidylate synthetase.

[a]AP-1: activator protein-1 complex, a Jun and a Fos family member; ATF/CREB, cAMP: response element binding proteins: LLG: log likelyhood score, which is 2 for a perfect match of the candidate response, element with the consensus sequence (TESS criteria) and all ambiguous matches yielding a score of 0. MMS: methyl methanesulfonate: TESS: transcription element search system; PCNA. proliferating cell nuclear antigen; PKA. protein kinase A: PKC, protein kinase C.

[b]AP-1 consensus; TC;/TAC/GTCA; CREB/ATF-2 consensus: TG/TACGTCA. Positions are based on TESS numbering of promoter sequences unless preceded by (-).

concentration to high levels (>1 μM) at which many nonspecific interactions leading to aberrant cellular localization and weak *membrane-protein* complex formation have been observed. Moreover, compounds that require >1 μM cannot be considered as potential drugs owing to the generation of nonspecific interactions and to the many *drawbacks* in attempting to achieve local concentrations on the order of ~1 M in vivo.

These considerations are likely important in understanding the many difficulties some researchers have had in "getting antisense to work." For the studies summarized here, it was possible to select highly efficient antisense oligonucleotides that are complementary to a sequence that is invariant in all isoforms of JNK1 or JNK2, thereby making it possible to reduce or eliminate all isoforms of JNK1 or JUNK with a single *antisense* oligonucletide.

Next, the elapsed time of suppression of target mRNA and target protein was determined. These studies showed that a single antisense treatment suppressed target mRNA and protein from approximately 24 h to 72 h, thereby defining a 2-d window in which anti sense-mediated effects could be observed.

However, for growth studies in which cell numbers are counted or tumor volume is measured, any loss of growth owing to inhibition of a growth-promoting target protein will remain apparent as decreased total growth for at least 3 wk. Because a variety of human tumor cells, including PC-

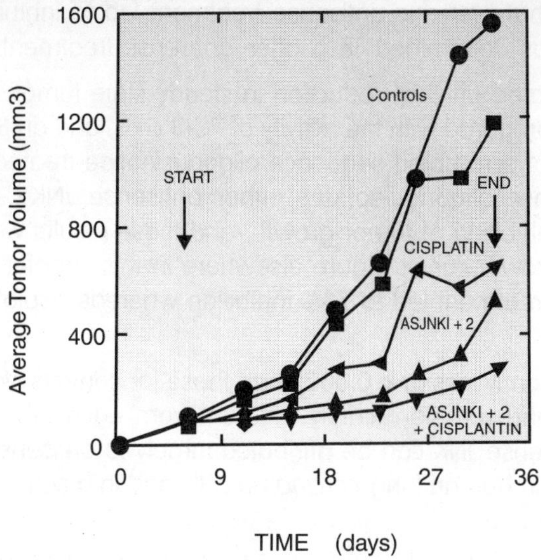

Figure 7.5: Application of antisense JNK and cisplatin chemotherapy to established xenografts of PC-3 human prostate carcinoma cells. Groups of 10-11 or 15 athymic female mice that had been inoculated with PC-3 cells and allowed to develop visible tumors were started on treatment ("start" arrow) of either one of the indicated antisense oligonucleotide solutions by daily IP injection or cisplatin weekly or a control consisting of vehicle alone or a scrambled sequence oligonucleotide. The average size of the tumors for each treatment group is plotted from the time of inoculation to the end of treatment ("end" arrow). The percentage of inhibition of growth was calculated by integrating each curve and expressing the result as 100 x (1-(growth/growth of control) j. These experiments were carried out in parallel with the p53 adenovirus-treated PC-3 tumors.

3 cells, were sensitized to the *cytotoxic* effects of DNA-damaging cisplatin in vitro upon inhibition of the JNK pathway, it was decided to test whether xenografts of PC-3 cells could be sensitized to cisplatin.

Moreover, because many of the potential JNK-regulated genes encode gene products that facilitate DNA synthesis, it appeared logical to expect that inhibition of JNK in vivo may affect tumor growth even in the absence of cisplatin. This experiment was carried out as a separate arm of the same in vivo study as for the evaluation of Adp53 Figure 7.2. Following *inoculation* of the mice with the PC-3 cells, the animals were monitored until visible and palpable tumors developed. At this point and for all subsequent observations, *tumor* volumes were estimated from the length and width of the tumors. Treatment consisted of *intraperitoneal* injection of *oligonucleotide* solution daily for 6 of 7 d/wk.

On the d 7, oligonucleotide treatment was omitted and the mice received either vehicle or cisplatin. Treatment (systemic antisense oligonucleotide. 25 mg/kg of an equimolar mixture of antisense JNK1 and antisense JNK2 termed combined-antisense JNK1 + JNK2, or a scrambled-sequence oligonucleotide or vehicle phosphate-buffered saline [PBS] alone) was initiated on development of readily visible and palpable tenors.

Every seventh day, antisense treatment was omitted and a subgroup of 15 animals receiving the combined-antisense treatment also received cisplatin, i.e., the same dose of cisplatin and timing as for the p53 treatment regimen. As a control, a separate group of animals received cisplatin alone. To confirm that systemic antisense treatment led to inhibition of JNK activity. JNK activity of tumor extracts was determined 18 h after *antisense* treatment.

These studies demonstrated an *89%* reduction in steady-state tumor JNK activity in tumors of antisense-treated animals compared with the activity of PC-3 cells, and an 80% reduction compared with the tumor JNK activity of scrambled sequence oligonucleotide-treated animals. We observed that treatment with antisense oligonucleotides either antisense JNK2 or combined-antisense treatment—led to marked inhibition of tumor growth, and these results were superior to those for cisplatin alone. When the growth curves Figure elsewhere in this chapter were integrated, it was found. e.g., that combined treatment led to 78% inhibition whereas cisplatin treatment led to 47% tumor growth inhibition.

These results are significantly less ($p < 0.002$) than those for controls (vehicle alone, scrambled-sequence oligonucleotide) and are significantly different from each other ($p < 0.02$, analysis of variance). The effect of antisense JNK can be attributed largely to antisense JNK2. The dominance of JNK2 has been observed in human lung *carcinoma* cells and in a series of nine human prostate *carcinoma* cell lines.

When antisense cisplatin treatments were combined, growth inhibition was further enhanced to 89% of maximum growth inhibition. Thus, these in vivo studies suggest that it may be possible both to inhibit growth and to sensitize solid tumors to chemotherapy by eliminating the JNK pathway.

Technique

The approaches described here have implications for therapeutic approaches targeting the DNA damage response in tumor cells. Upregulation of the DNA repair machinery may accompany tumor *progression* and the development of drug resistance. This process may involve, in part, AP-1 and ATF2/CREB regulated DNA synthesis and repair genes.

In fact, earlier studies by Scanlon et al. demonstrated that downregulation of AP-1 activity using a c-fos ribozyme was effective at sensitizing ovarian carcinoma cells to cisplatin, and this correlated with *downregulation* of DNA polymerase β, topoisomerase I, and thymidylate synthetase, all known to have AP-1 sites in their promoter regions. Independent of JNK DNA damage response, loss of the p53 tumor suppressor contributes to resistance to DNA-damaging *therapies* by removing

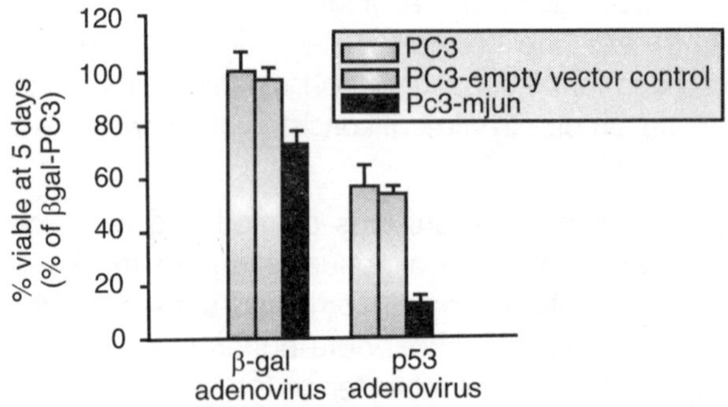

Figure 7.6: Cells modified to express mutant jun are sensitized to p53. Relative 4-d growth following Ad p53 treatment or Ad-13gal treatment (control) of PC-3 prostate carcinoma cells (parental), PC-3 empty vector transduced cells, and PC-3 mutant Jun-expressing cells (PC3-mjun). Growth was assayed using the MTT assay.

an important component of the DNA damage recognition machinery. Because p53 induces apoptosis in response to the level of DNA damage, it is likely that the success or failure of DNA repair contributes to the suppressive effects of p53. We therefore hypothesized that cells in which the JNK pathway was inhibited would show greater growth inhibition following exposure to p53 adenovirus.

As shown in Figure elsewhere in this chapter, this is indeed the case. In this experiment, PC-3 prostate cancer cells expressing mutant Jun were much more growth suppressed following treatment with p53 adenovirus in 7-d 96-well growth assays than were parental cells or cells Modified with empty vector.

These results support the combined application of p53 gene therapy along with antisense therapy to inhibit the JNK pathway, or some downstream target of the pathway. When used in combination with conventional *chemotherapies* such as cisplatin, the combined approach, which can be foreseen today, may extend the application of *conventional* treatments to prostate cancer and provide a new *strategy* for treatment of therapy-resistance advanced disease.

BASIC REQUIREMENTS

1. Adenoviral vectors (Deborah Wilson, Introgen Therapeutics. Houston, TX).
2. Antisense phosphorothiates dexyoligonucleotides:
   a. JNK1:5'-CTCTCTGTAGGCCCGCTTGG-3'.
   b. JNK2:5'-GTCCGGGCCAGGCCAAAGTC-3'.
3. Delbecco's modified minimum medium supplemented with 10% fetal calf serum (Irvine Scientific, Irvine, CA). PC-3 human prostate carcinoma cells, as well as other cell lines used in in vitro assays (T47D breast cancer cells, DLD-1 colon cancer cells, and T98G glioblastoma cells) were grown at 37°C: in an environment of 10% CO_2.
4. Animals: 5- to 6-wk athymic female postweaning I-larlan Sprague-Dawley mice (Harlan, Indianapolis, IN).
5. 10X PBS: 2 g of anhydrous KH,PO_4, 11.4 g of anhydrous Na_2HPO_4, 2 g of KCl, 80 g of NaCl. Make up the volume to I L with H_2O. Dilute L 10 before use.
6. X-gal staining solution: 1mg/ml of 5-bromo-4-chloro-3-indoyl-β-D-galactopyrano-side(X-gal), 5 mM potassium ferricyanate, 5 mM potassium ferrocyanate. 2 mM $MgCl_2$ in PBS.
7. *Tay* polymerase (Qiagen).
8. 2501nM dNTPs (Pharmacia Biotech.).
9. PCR Reaction mix: 25-μL. reactions contain 0.03-0.25 μg of DNA, 25 pmol each of forward and reverse primer, 250μM dNTPs (Pharmacia Biotech.), 1.25 *U Taq* polymerase (Qiagen). 1X buffer (Qiagen), Solution Q (Qiagen). The deoxyribonucleotide pruners are 5'-TGGGATTACACGTGTGAACCAACC-3' and 5'-GATCCACAGTCTGCCTGAGTCACT-3', respectively, with a 5' nested primer of 5'-CCTAGAAAGCACATGGAGAGCTAG-3'.
10. CAT assay reaction mix (per reaction): 70 μL of cell lysate, 30 μL of 5 mM chloramphenicol, 0.4 μL of ^{3}H-acetylCoA (200 mCi/mmol; NEN®Life Science), 0.6 μl.. (4.4 μg/ImL) of nonradioactivc acetylCoA (Pharmacia Biotech.).

11. Whole-cell extract (WCE) buffer: 20 m,14 HEPES. pH 8.0; 75 rnM NaCl-, 2.5 mM $MgCl_2$, 0.05% (v/v) Triton X-100. 0.5 mM dithiothreitol (DIT); 20 mM β-glycero-phosphate: 0.1 mM Na_3-$V_{04;}$ 2 μg/mL leupeptine, 100 μg/ml., paramethyl-sulfonyl fluoride.
12. JNK assay buffer: 20 *mM* HEPES, pH 7.7; 20 *mM* $MgCl_2$ 20 mM β-glycero-phosphate; 20 mM p-nitrophenyl phosphate; 0.1 mM Na_3VO_4, 2 mM DTT; 20 μM adenosine triphosphate (ATP); 5 μCi of [y*³²P] ATP.
13. Oligonucleotide transfection (lipofection) solution: Mix 10 μg/mL of lipofectin (Gibco-BRL, Gaithersburg, MD) reagent in MEM (Gihco-BRI., Gaithersburg. MD) with an equal volume of oligonucleotide solution, incubating this mixture at room temperature for 15 min and diluting it with lipofectin solution to a final oligonucleotide concentration of 0.4 μM.
14. X-gal staining solution: 1 mg/mL 5-bromo-4-chloro-3-indoyl-β-D-galacto-pyranoside, 5 m.M *potassium ferricyanate,* 5 mM *potassium ferrocyanate,* 2 mM MgCt in PBS.
15. Lipofection solution for plasmid transfection: Prepare a 1 μg/μL stock of DOTAP *lipofection* reagent (Boehringer Mannheim, *Indianapolis,* IN). Prepare the mix (per well) of a 24-well tissue culture plate as follows:

    a. Plasmid/HEPES mix: 1.25 μL (1 (μg/μL) of plasmid + 11.25 μL 5 *mM* HEPES, pH 7.8.

    b. DOTAP/HEPES mix: 7.5 μL of DOTAP (μg/μL) + 17.5 μL of 5 mM HEPES, pH 7.8.

    Mix the entire contents of (a) and (b) together. and incubate for 30 min at room temperature.
16. Cisplatin (diaminodichloro cis-platinum): Aqueous Platinol (Bristol Myers Squibb).

MECHANISM

96-Well Growth Assay

1. Plate cells in complete medium 1 day prior to vector treatment in 24-well tissue culture plates so that their density at the time of treatment is about 70-80% of confluence.
2. Treat the cells with Adp53 or Ad Egal for 2 to 3 h (100 PFU/cell).
3. Incubate the cells for additional 2 d in one of the Ad-βgal-treated wells.
4. Remove the medium and wash the wells two times with PBS.
5. Fix the cells by overlaying with a solution containing 3.7% parafomaldehyde (v/v) in PBS for min.
6. Wash the wells with PBS and overlay with X-gal staining solution.
7. Incubate the cells at 37°C overnight to allow development of blue stain in β-galactosidase-expressing cells for estimating the infection efficiency.
8. Following treatment with vector, plate the cells at low density (1000 cells/well) in 96-well plates. Treat (in triplicate) with drug (e.g., cisplatin) or antisense for 1-4 h depending on the drug, followed by incubation for an additional 5-7 d.
9. Incubate the cells for an additional 6 d and score viability by adding the tetrazoliutn dye MTT for I h and determining the A260 of the formazan product as described by the manufacturer (Prornega).

Subcutaneous Tumor Model

1. Inoculate 5 $\times 10^6$ PC-3 prostate cancer cells (mutant p53) subcutaneously on the back of nude mice by injecting 100 μL of a cell suspension in PBS at 5 $\times 10^7$ cells/ml.
2. Allow tumors develop 5 d to a size of 50-100 nun^3.
3. Estimate tumor volume by measuring the length and width of the tumor with a caliper and calculate using the following formula: volume = n*/6 (length × width^2).
4. Initiate treatment (designated day 1) and monitor tumor volume every 2 to 3 d.
5. For oligonucleotide treatment. inject mice intraperitoneally at 25 mg/kg and based on the average weight of all mice by daily ip-injection with a solution at a concentration yielding the total dose in 0.2 mL of PBS. Treat control animals with either vehicle alone or the *scrambled*-sequence control *oligonucleotide*.
6. For *cisplatin* treatment, inject trice *intraperitoneally* at 0.4 mg/kg once per week. For combined antisense JNK cisplatin treatment, inject antisense JNK *oligodoxynucleotides* for 6 d and cisplatin on the day 7.

DNA Repair by PCR Stop Assay

1. Prepare genomic DNA from about 5 × 10^5 cultured cells using the QIAmp blood kit (Qiagen) essentially following the manufacturer's protocol, except lyse the cells directly on the plate in the presence of PBS, Qiagen protease. and lysis buffer supplied in the kit.
2. Following purification, adjust the DNA concentration to 0.25 mg/mL in sterile water and store at -20°C until use.
3. Carry out quantitative PCR as described for the tneasurement of cisplanin adduct formation on specific regions of DNA. For each primer pair. verify that product formation is directly proportional to input template by performing a pilot experiment with serial twofold dilutions of template, followed by electrophoresis on a 1 % agarose gel containing 0.5 μg/mL of ethidium bromide.
4. Quantify hands using a Kodak digital camera and analysis software or an equivalent apparatus for the integration of band intensity from *photographic* film and determine *adduct* numbers using $P = e^{-1}$
5. For the PCR reaction, use primers at a *concentration* of 0.03-0.25 μg per 25-μL reaction. depending on the amount of template used in the PCR reaction.
6. Carry out the PCR reaction using the following amplification program: 1 cycle of 94°C for 1 min 30 s; 25 cycles of 94°C for 1 min. 57°C, for 1 min, and 70°C for 2 min 30 s: 1 cycle of 94°C for 1 min; 57°C for I min, and 70°C for 7 min.
7. Use two independent templates for each treatment *condition* and set up each for analysis in triplicate. For such analysis carry out an internal PCR control using primers to generate a 150-bp fragment of the *dihydrofolate* reductase gene.

DNA Repair Using CAT Reporter Assay

1. Treat 100μg of CAT reporter plasmid (in which the CAT gene is expressed from the RSV promoter), with 25 μM cisplatin in 300 μL total of PBS on ice for 3 h to create cisplatin-

damage reporter DNA.

2. Precipitate the DNA by adding 150µL of 7.5 M NH_4Ac and 450 µL of isopropanol.
3. Leave the mixture at -20°C for 1 h.
4. Centrifuge the DNA at 14,000g for 30 min.
5. Perform two washes with 70% ethanol, followed by a final wash in absolute ethanol.
6. Resuspend the DNA in sterile H_2O and use for transfections.
7. Process untreated DNA the same way but without the addition of cisplatin, and use this plasmid preparation as a control.
8. Plate the cells in 24-well plates the day prior to transfection so that their density at the time of transfection is 50% (about 10^5 cells/well).
9. Set up transfections in triplicate wells with either 1-1.25 µg of undamaged plasmid or damaged plasmid for the 24-h time point together with a second set of triplicate wells transfected for the 48-h time point, using the DOTAP liposomal transfection reagent.
10. Add 500 µL of medium to the entire contents of step 9 and mix well.
11. Wash the cells in the well of a 24-well plate twice with 1× PBS.
12. Add 537.5 µL of transfection to the well.
13. Incubate the reaction overnight in the 37°C, 10% CO_2 incubator.
14. Aspirate off the transfection mix and wash the cells twice with 1X PBC.
15. Aspirate off the PBS and replace with I rnL of fresh complete culture medium.
16. Incubate for 24-48 h at 37°C. 10% CO_2.
17. Remove the medium, wash the cells twice with PBS, and store the plate dry at-70°C until required.

CAT Assay

1. Thaw the plates from step 17 in Subheading 3.4. for 5 min at 37°C.
2. Add 100 µL of 1 X Reporter Lysis Buffer (Promega) to each well of t 24-well plate.
3. Shake the plate vigorously for 15 mnin.
4. Transfer the entire 100 µL of lysate to microcentrifuge tubes.
5. Spin the lysates for 2 min at 14,000g (4°C).
6. Transfer the supernatants to fresh tubes.
7. Heat the supernatants for 10-15 min at 65°C to destroy deacetylase activity. If particulate material is present after this heating step, centrifuge again and collect the supernatants.
8. Measure the protein concentration of each sample using the Bio-Rad Protein Assay (Bio-Rad), following the manufacturer's instructions.
9. Equalize the protein concentration of the samples using 1 X Reporter Lysis Buffer (Promega). Add 31 µL of CAT reaction mix to 70 µL of the samples.
10. Incubate the mix at 37°C for 2 h.

11. Add the entire sample to 1 mL of 7 M urea.
12. Add 1 mL of toluene:PPO (8 g of PPO (2.5-diphenyloxazolel/mL toluene). Shake well and count using a scintillation counter. The resulting counts are proportional to accumulated CAT expression and, therefore. total CAT activity.

Antisense Oligonucleotide Transfection

1. Wash 70% confluent cultures (24 h after plating) in tissue culture plates twice with PBS.
2. Incubate the cells with lipofectin-0.4 μM oligonucleotide solution at 37°C in 10% CO_2 from 4 h to overnight.
3. Following lipofection, wash the cells once with serum-free MEM and continue the culture in complete medium for 24 h prior to JNK assay or 4 d prior to growth assay.

JNK Assay

1. Transfect the cells with oligonucleotides as described in Subheading elsewher in this chapter
2. Wash the cells with ice-cold PBS and suspend in WCE buffer.
3. Determine the protein concentration of the cell extracts by the Bradford dye method (Bio-Rad).
4. Carry out the kinase assay as follows:

   a. Mix 50 μg of WCE with 10 μg of GST-c-Jun (1-223) for 3 h at 4°C.

   b. Wash four times and incubate the beads with 30 μL of kinase reaction buffer for 20 min at 30°C.
5. Stop the reaction by adding of 20 μL of Laemmli sample buffer.
6. Elute the phosphorylated GST-c-Jun protein by boiling the sample for 5 min.
7. Resolve the components by 10% sodium dodecyl sulfate-polyacrylamide gel electrophoresis.
8. Quantify the ^{32}P-phosphorylated GST c Jun by digitization and integration of the respective "band" values of the autoradiograph *of* the: dried gel.

POINTS OF REMEMBER

1. These vectors are replication-defective adenoviral recombinants in which the early region genes EIA and E1B required for viral replication have been deleted and replaced with an expression cassette containing the *transgene* of interest. In the case of the p53 adenovirus (Adp53), the expression *cassette* consisted of the human wild-type p53 coding sequence, flanked by the cytomegalovirus (CMV) promoter, and the simian virus 40 (SV40) *polyadenylation* signal. In the case of the control adenovirus (Adβgal), the expression cassette consisted of the bacterial β-galactosidase coding sequence *flanked* by the CMV promoter and SV40 polyadenylation signals. In the case of the control adenovirus (AdLuc). the expression cassette consisted of the firefly luciferase coding sequence flanked by the Rous sarcoma virus promoter and SV40 *polyadenylation* sequences. Viral stocks were stored at -70° C and repeated freezing and thawing were avoided. The concentrations of tile stocks, expressed in plaque-forming units per milliliter were 1 to 2 × 10^{11}.

2. The sequences were determined in preliminary studies utilizing a messenger walk procedure as described. Since in vivo studies require large amounts ofoligonucleotide. a single *"scrambled sequence"* control sequence was chosen consisting of a scrambled 20-nt sequence *previously* used as a control in the analysis of antisense protein kinase Ca: 5' TCGCATCGACCCGCCCACTA-3'. Both ASJNK sequences contain a single CpG sequence thought to have potential *immunostimulatory* properties that, therefore, may influence xenograft growth by *immunologic* mechanisms. The control *oligonucleotide* used here has a nucleotide composition closely approximating the average of antisense JNK1 and JNK2 but contains three CpG dinucleotide sequences thereby providing a control for the potential influence of CpG sequences. All three oligonucleotides were prepared and purified as previously described.
3. The 0.15-kb segment of genomic DNA amplified by use of the nested primer sustains undetectable levels of DNA damage under our conditions and serves as an internal PCR control and the basis for *normalization* of the amount of amplification of the 2.7-kb fragment. The *nested* amplification product varies by <5-10% among the various templates that we have used.
4. Wells that do not receive drug serve as a control to which drug-treated wells are compared. Example drug treatments are as follows: 5-fluorouracil (Adrucil™, Pharmacia Biotech.), 10 mM for 1 h: doxorubicin (*doxorubicin hydrochloride,* Aldrich), 3.7 μM for 1 h, cisplatin (PlatinolTnr, Bristol-Myers Squibb). 30 μM foi l h for PC-3 and 20 μM for 1 h for T98G.
5. Because *Taq* polymerise is blocked at cisplatin adducts. the relative efficiency of PCR amplification of genomic DNA from cisplatin-treated vs control cells drops in proportion to platination levels. The relative PCR efficiency is equal to the frequency of undamaged strands, *P,* within a population. *P* is related to the average number of cisplatin adducts, n, per fragment, by the Poisson formula $P = e^{-n}$, or – (1nP) = *n.* A drop in the PCR signal to 0.6 of control would therefore reflect an average cisplatin adduct density of –(in 0.6) = 0.51 adducts per fragment. PCR signals ranging from 0.9 to 1.0 of control are generally indistinguishable from the control, owing to standard deviations in the range of ±0.1. Since a PCR signal equal to 0.9 of control reflects an adduct density of 0.1 adduct per fragment. we consider that the assay is not sensitive to adduct densities of <0.1 adduct per fragment. In most cases with genomic DNA from cisplatin-treated cells, the assay requires PCR amplifying a fragment of about 2 to 3 kb long.
6. An example treatment consisted of ip administration if cisplatin (Platinol) on d 1 and 8 at a dose of 4 mg/kg (LD 10), i.e., 88 μL/22 g mouse. The dose corresponds to approx 20% of the IC50 of cisplatin for mice and is the maximum tolerable dose for a repeated weekly treatment. Vector (Adp53 or AdLuc) was administered intratumorally at a dose of 10^8 PFU/ injection on d 3, 5, and 7 and again on d 10, 12, and 14. Vector was diluted into sterile cold PBS prior to injection so that injection volumes were 100 μL.
7. Generally a 4-h exposure is required for the reduction in target JNK rnRNA by >80%. This is observed after 24 h of lipofection as judged by western analysis. Resistant cell types are treated for longer periods, and this approach is favoured over increasing the oligonucleotide concentration or altering the *oligonucleotide:lipofectamine* ratio.
8. Typically all *preparations* yield quite similar protein concentrations, but volumes should be

used so as to provide equal amounts of total cellular protein in all samples prior to subsequent analysis.

9. Prepare in advance GST-c-Jun fusion proteins by expression and purification (Qiagen plasmid DNA purification system kit) from *Escherichia coli* and by the addition of Glutathione Sepharose® 4B beads (*Pharmacia Biotech*).
10. For *digitization* software that provides interactive *designation* of the bands by, e.g., drawing boxes around the region to be digitized. If not *automated,* subtract background by using half-sized boxes placed exactly above and below the hand in question or by subtracting one half of the sum of same-sized boxes placed above and below the band in question. Subtract background values in all cases, thereby yielding *"net"* hand values. Determine the relative JNK activity by *normalizing* the resulting net band values by division by similar results for control cases such as the *scrambled-sequence* or mock transfected cell case.

Chapter 8

Meiotic Chromosomal (22q11.2) Aberration

Meiotic unequal crossover events between blocks of low-copy repeats (LCRs) may lead to gene dosage *imbalance* resulting in genomic disorders. *Genomic* disorders are frequently associated with mental *retardation* or learning disabilities and mild to severe congenital anomalies. Chromosome 22q11.2 is particularly susceptible to chromosome *rearrangements* leading to several genomic disorders including *velocardiofacial* syndrome/DiGeorge syndrome (VCFS/DGS), der(22) syndrome, and cat-eye syndrome (CES), associated with a *monosomy, trisomy*, and *tetrasomy* of 22q11.2, respectively.

Most VCFS/DGS patients have a similar *hemizygous* 3-Mb deletion mediated by *meiotic interchromosomal homologous* recombination events between LCRs termed LCR22s. The reciprocal *duplication* of the same interval, predicted on expected products of unequal crossover events, results in a more mild condition termed dup(22)(q11.2;q11.2) syndrome. In contrast to VCFS/DGS, dup(22)(q11.2;q11.2) and CES, der(22) syndrome is caused by a different *molecular* mechanism.

Der(22) disorder arises in offspring of normal carriers of the constitutional t(11;22) (q23.3; q11.2) *translocation* by recombination between AT-rich (high AT sequence composition) *palindromic* sequences on 11q23.3 and 22q11.2. The *palindromic* sequence on 22q11.2 is within one of the LCR22s.

Interestingly, both recurrent and novel breakpoints occur most often in LCR22s, making them an important *architectural* feature associated with *susceptibility* to genome rearrangements. To gain further insight into the mechanisms of how the LCR22s are involved in chromosome rearrangements, efforts are underway to determine the molecular evolution, structure, size, orientation, and their level of *variability* in humans.

MEIOTIC 22Q11.2 REARRANGEMENTS

Diseases resulting from rearrangements of the human genome are referred to as genomic disorders and most are mediated by region-specific LCRs. The 22q11.2 region is particularly susceptible to meiotic *chromosome* rearrangements associated with genomic disorders including VCFS/DGS (OMIM192430/OMIM188400); the reciprocal *duplication* syndrome, dup(22)(q11.2;q11.2), CES(MIM 115470); 13, and der(22) syndrome. All of these are congenital *anomaly* disorders with varying levels of learning disabilities and mental *retardation*.

22q11.2 Deletion Syndrome

In 1965, a disorder in a group of infants with thymus gland aplasia and neonatal hypocalcemia was described and termed DiGeorge syndrome. In the 1970s other sets of patients were *ascertained* with related, but more mild sets of clinical findings and were termed VCFS and separately, *conotruncal anomaly* face *syndrome*. In early 1990s, when molecular cytogenetic technologies became available, it was realized that all these clinically similar disorders shared a hemizygous deletion on chromosome 22q11.2.

These disorders occur quite frequently in different populations, in approx 1/4000 live births. The main clinical findings are learning and behavioural *abnormalities, craniofacial anomalies, thymus gland hypoplasia*, and *cardiovascular malformations*. Many other clinical findings occur in association with the disorders but with variable *expressivity*.

The great majority of the cases are sporadic owing to *de novo* deletions. When inherited, it is in an *autosomal* dominant manner and the penetrance is complete. Integrated physical and genetic maps were constructed and it was found that most VCFS/DGS patients have a similar 3-Mb deletion, whereas some have nested distal deletion *breakpoints*, the most common of which is a 1.5-Mb deletion.

The severity of the phenotype did not depend on the size of the deletion, suggesting that gene(s) within the smaller 1.5-Mb interval are responsible for its etiology. To understand the mechanism responsible for the recurrent deletion, the chromosome breakpoints for common features were *analyzed*. Three blocks of LCRs termed LCR22 were found to be associated with the breakpoints.

Cat-Eye Syndrome and Der(22) Syndrome

CES is a more rare disorder that the 22q11.2 deletion syndrome and it is associated with ocular colobomata, anal atresia, congenital heart defects, *renal malformations, craniofacial anomalies, male genital anomalies, skeletal defects*, and *mild mental retardation*. CES patients carry a *supernumerary bisatellited* chromosome 22, resulting in a *tetrasomy* of 22pter-q11.2.

There are two common breakpoints in CES, type I and type II, and they occur in the same regions as for the common 3-Mb VCFS/DGS deletion breakpoints. Der(22) syndrome is also a rare congenital anomaly disorder associated with similar features to CES, although there are some minor differences such as the absence of ocular colobomata and more severe mental *retardation* than in CES patients.

Der(22) syndrome arises in offspring of normal carriers of the constitutional t(11;22) translocation, the most common recurrent non-Robertsonian constitutional *translocation* in humans. Patients with this disorder have a partial trisomy of the 11q23.3qter and 22pter-q11.2 regions. The recurrent t(11;22) translocation breakpoint occurs in the same region as for the 1.5 Mb distal deletion breakpoint in VCFS/DGS patients, suggesting that this region is particularly labile.

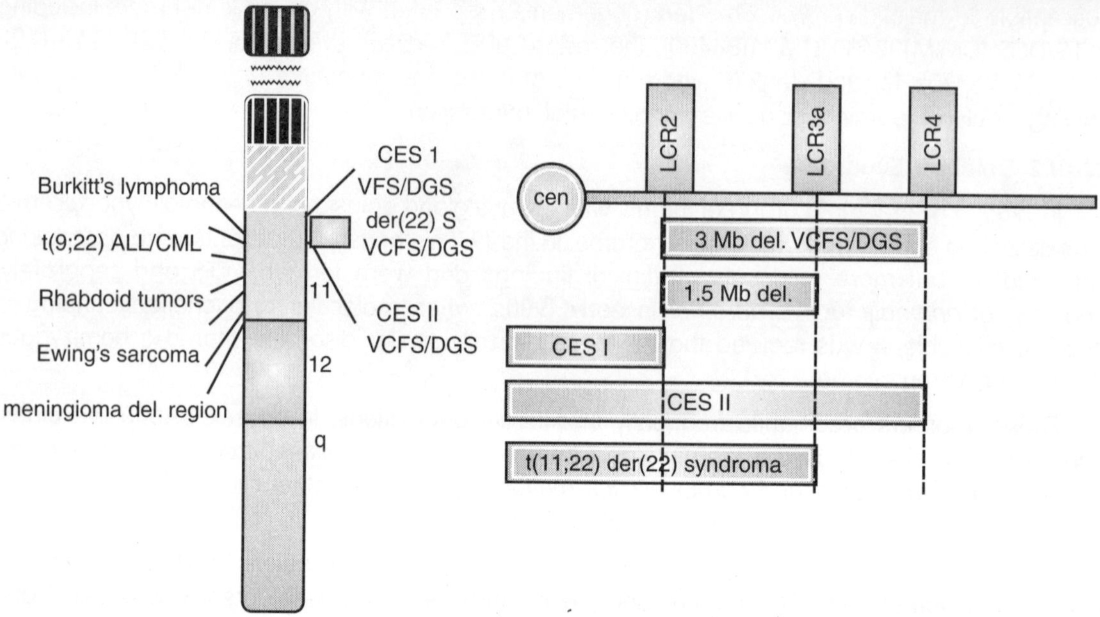

Figure 8.1: Chromosome rearrangements on 22q11.2. To the left of the ideogram of chromosome 22, the list of human cancers breakpoint mapping in the 22q11q12 interval. On the right side, the list of the recurrent congenital anomaly disorders (dup(22)(q11.2;q11.2) disorder maps to the same interval as the 3 Mb VCFS/DGS deletion). Other duplication breakpoints occur more telomerically (not shown). The blue box represents the 3-Mb region associated with all three disorders. On the right, the horizontal line represents chromosome 22q11.2, with the centromere to the left. The recurrent chromosome rearrangement breakpoints are shown.

Malignant Disorders 22q11.2

The most common recurrent rearrangement on 22q11.2 associated with malignant tumors is the balanced t(9;22) translocation in patients with acute lymphocytic leukemia and chronic myeloid leukemia. A characteristic *Philadelphia* chromosome der(22)t(9;22)(q34;q11) is present in tumor cells.

A balanced t(8;22) translocation is associated with Burkitt's lymphoma and it disrupts the immunoglobulin light chain locus on 22q11.2. Other rearrangements in *malignant* tumors include the t(11;22) translocation in Ewing sarcoma and the deletions or translocations in malignant rhabdoid tumors and *meningiomas*. Thus, it is likely that sequences in the 22q11.2 region are prone to breakage also during mitosis. Detailed molecular characterization of the breakpoints remains to be performed.

DISEASES DUE TO 22Q11.2 REARRANGEMENT

The establishment of physical maps of chromosome 22 made it possible to determine the position of the common recurrent breakpoints associated with chromosome 22q11.2 rearrangement disorders. Probes used to screen filters containing large genomic insert bacterial artificial chromosome (BAC) *clones* identified clones mapping to several locations on 22q11.2 including clones in the *vicinity* of the breakpoints.

LCRs have been noted already on 22q11.2 over 16 years ago (Heisterkamp and Groffen) and in the general vicinity of VCFS/DGS locus over 10 years ago, but the connection between them was not made until *integrated* physical maps were generated. *Interchromosomal* recombination events lead to the 22q11.2 deletion and a reciprocal duplication, both associated with congenital *anomalies.* Using *microsatellite* genetic markers and screening more than 150 VCFS/DGS patients and their unaffected family members, it was possible to narrow the region containing the common recurrent 1.5- and 3-Mb deletions.

Further analysis of the breakpoints occurring in VCFS/DGS and CES patients suggested that they occurred in the same regions. More recently, the constitutional t(11;22) breakpoint on 22q11.2 was placed in the vicinity of the distal deletion breakpoint in VCFS/DGS patients with the nested 1.5 Mb deletion. Blocks of region-specific *segmental* duplications or LCR22-2, LCR22-3a and LCR22-4; also known as LCR22-A, -B, and -D, respectively were found to map to the three regions of breakage, associated with VCFS/DGS, CES, and der(22) syndrome, all 1.5 Mb apart, implicating them in their etiology.

The LCR22-2 and LCR22-4 are more than 240 kb in size and more than 99% identical in sequence. This suggested that misalignment of nonallelic copies of the homologs could occur during meiosis and through the mechanism of *nonallelic* homologous recombination (NAHR) resulting in a deletion. Both intrachromosomal and *interchromosomal* unequal crossover events were found to be responsible for the 3-Mb deletion.

This was determined by performing haplotype analysis with microsatellite markers, of three generations, the patient, parents, and grandparents. For interchromosomal events, which are more predominant, there should be individuals with the reciprocal duplication, dup(22)(q11.2q11.2). An analogous mechanism has been described for chromosome l7p 11.2 for Smith-Magenis syndrome (SMS).

Indeed, interchromosomal unequal recombination events between two misaligned chromos-

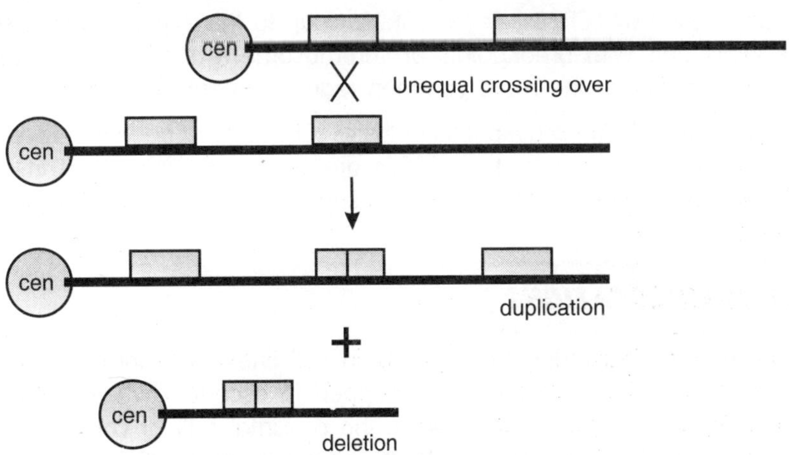

Figure 8.2: Molecular mechanism for 22q11.2 rearrangements. Unequal crossover between two homologs can generate a duplication or a deletion. The filled box represents LCR22-2 and the open box represents LCR22-4. Both LCR22s are more than 225 kb in size and are 3 Mb apart.

omes 22 during meiosis resulted in a reciprocal duplication, dup(22)(q11.2)(11.2). *Unrelated* dup(22)(q11.211.2) patients had been ascertained based on *suspicion* that they had VCFS/DGS.

Dup(22)(q11.2q11.2) disorder is associated with some clinical findings occurring in VCFS/DGS, but also with other features which are distinct including characteristic facial features with mild *craniofacial anomalies, hearing loss,* and *urogenital malformations.* Thus, based on mechanisms proposed, a newly recognized genetic syndrome was identified. Interestingly, unlike for VCFS/DGS which are characterized by deletions between LCR22-2 and LCR22 (LCR22-A-D), the size of the duplications vary in size, extending to more distal LCR22s suggesting that larger duplications are more tolerated than larger deletions or that the mechanism for deletion and duplication may involve other sequences within LCR22s.

CONSTITUTIONAL TRANSLAOCATION T(11;22)

In contrast to VCFS/DGS, CES and dup(q11.2q11.2), mediated by NAHR events between LCR22s, a different mechanism is responsible for the balanced recurrent t(11;22) translocation leading to the der(22) syndrome. The site of breakage for the constitutional t(l 1;22) translocation is within the center of AT-rich palindromes on chromosomes 11823 and 22q11.2.

The AT-rich palindrome on 22811.2 is in LCR22-3a, which is equidistant between LCR22-2 and LCR22-4 that mediate the 3 Mb VCFS/ DGS deletion. Two cases of t(17;22)(q11.2;q11.2) with AT-rich palindromes on chromosome 17 at the breakpoints were reported. Because the same sequences are present at the translocation breakpoints, a similar molecular mechanism is likely responsible for the t(11;22) and t(17;22) translocations based on one shown in experimental systems. The model proposes that palindromic sequences can form unstable cruciform structures susceptible to double-stranded breaks and repair by *nonhomologous* end-joining (NHEJ) mechanisms.

As part of the repair, a translocation may ensue, thereby stabilizing the DNA. Therefore, both homologous and nonhomologous (NAHR and NHEJ) mechanisms are responsible for recurrent *genomic* disorders caused by LCR22s. Seven LCR22s map to the 22q11.2 region, suggesting that there might be individuals with breakpoints or chromosome rearrangements within LCR22s in addition to the ones *harboring* the more common rearrangement endpoints.

Rare individuals with deletion breakpoints in these LCR22s have been identified. Based on the analysis of chromosome breakpoints, the LCR22s are sequences that result in *susceptibility* to recurrent rearrangement mutations and understanding the molecular mechanisms should shed light on this process.

22Q11.2 TRANSLOCATIONS

Chromosome regions harboring breakpoints in cell lines from individuals carrying novel *translocations* within the 22q11.2 region were mapped to ascertain whether there could share common features. Most breakpoints occurred in the proximal half of chromosome 22q11.2. Of these, half map to the LCR22s. Among the LCR22s, most were in LCR22-3a, the LCR22 associated with the t(11;22) translocation. In two reports of translocations disrupting LCR22-3a (LCR22-B), one showed the breakpoint occurred between AT-rich palindromes in LCR22-3a and on the partner chromosome 4q35 but the other did not (chromosome X), *implicating* different mechanisms.

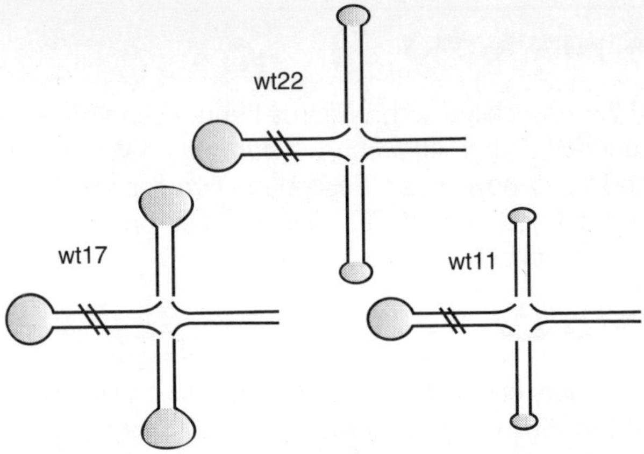

Figure 8.3: AT-rich palindromes mediate the t(11;22) and the t(17;22) translocation. AT-rich palindromes on chromosomes 11q23.3, 17q11.2, and 22q11.7 can form cruciform structures as shown. The regions on chromosomes 11 and 22 form perfect palindromes, whereas the region on chromosome 17 contains non-palindromic sequences in the center of the cruciform structure, depicted as a look. Although the sequences in all three chromosomes are AT-rich, they are not homologous to each other.

In our study, half of the translocations disrupting 22q11.2 occurred distal to the most telomeric unique sequence probe. The breakpoints on the partner chromosomes occurred more frequently in telomeric bands, supporting the *epidemiological* data that break points in telomeric bands are more common events than other regions within chromosomes. Previous reports have noted a higher rate of telomeric rearrangements than other chromosomal bands. Telomeric regions are very dynamic and have shuffled during evolution between nonhomologous chromosomes. An *ascertainment* bias toward imbalance of smaller segments as opposed to larger segments could explain the occurrence of 22q11.2 and telomeric rearrangements, but it is also possible that two *susceptible* regions misalign and pair following chromosome *breakage*.

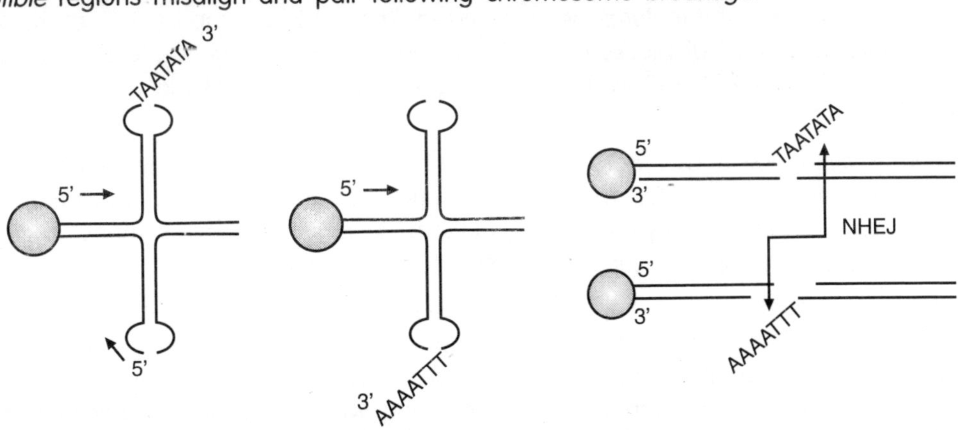

Figure 8.4: Model of mechanism for translocations mediated by AT-rich palindromes. Palindromes are prone to double strand breaks which lead to translocations repaired by non-homologous end-joining (NHEJ). Although the sequences in all three chromosomes are AT-rich, they are not homologous to each other. When a breakpoint occurs in the labile loop created at the tips of the cruciform structure, a translocation may occur, resulting in a more stable non-palindromic sequence. The arrows depict the orientation of DNA sequence with respect to the centromere.

INVERSION POLYMORPHISMS

In chromosome 7q11.23, associated with Williams-Beuren syndrome and chromosome 15q11-q13 associated with Prader-Willi and *Angelman syndromes* (PWS/AS), *polymorphic* inversions between flanking inverted LCRs have been described. Two reports have shown that such an inversion *polymorphism* do not occur on 22q11.2. This is not surprising as 80% of the LCR22-2 and LCR22-4 sequences are in the direct, not indirect, orientation.

PHYLOGENY OF LCR22S

To determine the evolutionary origin of LCR22s, high- resolution fluorescence *in situ* hybridization (FISH) mapping was performed. Utilizing human chromosomal paints, orthologs to human chromosome 22 were identified were not rearranged in Great Apes, Old and New World monkeys. Using LCR22-specific probes derived from large insert genomic clones, multiple signals were detected in interphase nuclei from the different primate species including apes *Pan troglodytes* (common chimpanzee), *Pan paniscus,* (pygmy chimpanzee or bonobo), *Gorilla gorilla* (gorilla), and Old World monkeys (rhesus monkey [Maraca mulatto]).

These data are similar to what was ascertained for other genomic disorders mediated by LCRs, such as PWS/AS, and SMS, showing that multiple copies of LCRs predate the divergence of Great Apes from Old World monkeys, at least 20-25 million years ago. The LCRs that are associated with human *genomic* disorders are not present in the mouse. This suggests that they have evolved after the divergence of the common ancestor between mouse and *primates.*

MOLECULAR STRUCTURE OF LCR22S

To better understand *mechanisms* involved in the *evolutionary* origin of LCR22s, their sequence composition and *architectural* features were analyzed. The LCR22s comprise 12% of the 22q11.2 interval and are composed of blocks or modules *harboring* known genes, *pseudogenes,* and *predicted genes.* They are 10-250 kb in size, contain a variety of blocks in direct and inverted *orientations.*

The genes within (*USP18, GGT, BCR, GGTLA*) have become duplicated during evolution, thereby shaping the architecture of LCR22s. The full-length functional genes each lie in different LCR22s, whereas duplicated copies are present in many of them. Because the genes comprise a significant part of the LCR22s, they can be used to understand their evolution.

Although important for the evolution of LCR22s, they are not likely candidates for the malformations in VCFS/DGS patients as the copies in the deleted region are non-functional *pseudogenes*. The three proximal LCR22s *mediate* the recurrent chromosome 22q11.2 rearrangement disorders. Low stringency FISH mapping with LCR22 probes, revealed multiple signals on 22q11.2 and a few other chromosomes suggesting that they have been unstable during evolution.

As previously stated, the genes are present in a single copy in the mouse, on different *chromosomes,* and have therefore evolved more recently in primates. The sequences of the LCR22

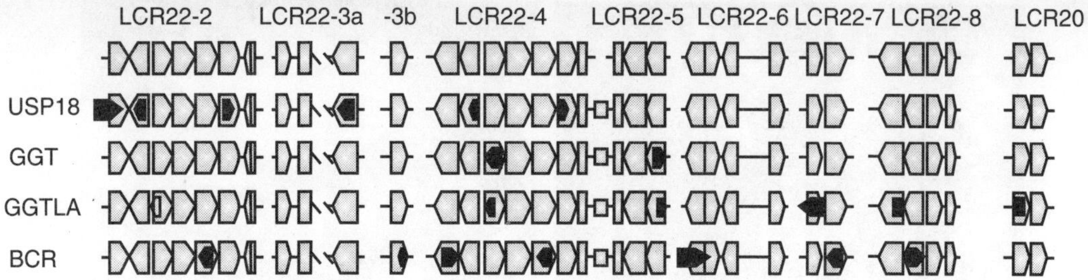

Figure 8.5: Block structure of the LCR22s and genes within. The known genes, USP18 *(68a),* GGT *(77),* GGTLA *(78), and* BCR *(79) and their pseudogene copies, comprise part of the LCR22s. We performed a comparative computational analysis of all the clones spanning the 22q11.2 region to identify similarities. We also took into account a recent annotation of the LCRs in chromosome 22 (67). The different regions of similarity are depicted by different colours. The genes and predicted genes, in their normal genomic organization with respect to the centromeres are shown. LCR22-3a has an uncloned region. The LCR22s are drawn to scale.*

genes and copies therein were compared by examining the 22q11.2 region to determine if there was a specific mechanism responsible for gene duplication.

ALU-MEDIATED RECOMBINATION EVENTS

The results of a genome-wide effort to identify sequence elements at the junctions between LCRs and unique sequences were reported. A total of 9464 junctions and 2366 duplication alignments were examined throughout the genome and a significant enrichment of *Alu* elements in the vicinity of the junctions was found (27% of all LCRs had an *Alu* at least one end).

This suggested that the duplication process forming the LCRs coincided with a primatespecific burst of *Alu* retroposition activity. Because the *Alu* sequences were identical in sequence, it was possible that they could undergo Alu *Alu* based unequal recombination events forming mosaic *Alu* products.

In the case of 22q11.2, the breakpoints in some of the *Alu* elements involved in the rearrangements occurred by unequal crossover mechanisms, whereas others occurred by different mechanisms, resulting in breakpoints at the end of *Alu* elements as opposed to the middle of *Alu* for unequal crossover events. The *IGSF3* gene maps to chromosome 1p13.1. Part of this gene became copied onto 22q11.2, to LCR22-2 and LCR22-4. It is adjacent to the *GGT* locus in these LCR22s.

The rearrangement responsible for this *"trams"* rearrangement involving non-homologous chromosomes appeared to have been created in meiosis by a replication-dependent mechanism similar to the model proposed by *Richardson* et al. In this model, recombination occurs by Alu-mediated *misalignment* of two different *chromosomes*.

A gene conversion then takes place, thereby avoiding crossovers, which would lead to aberrant translocations. These results suggest that mammalian genomes have a mechanism scanning the entire genome to find stretches of homology.

GENES OF PSEUDOGENES

Besides genes and pseudogenes, there are other architectural features in the LCR22s that are

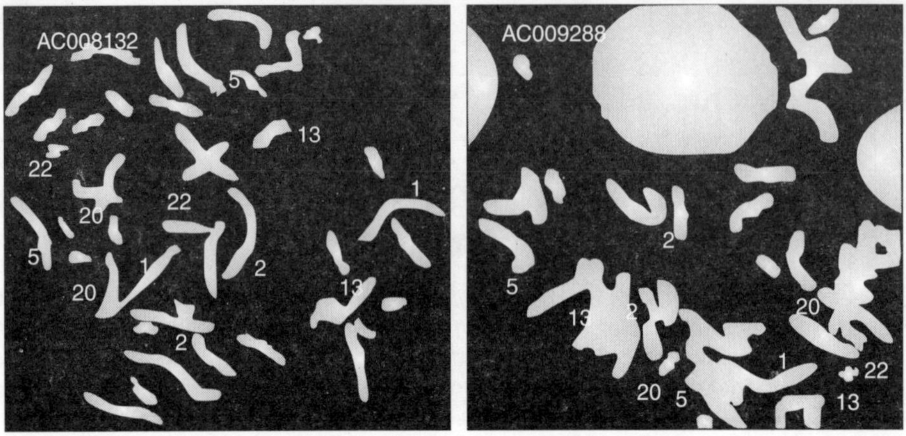

Figure 8.6: Low-stringency FISH mapping. Probes from LCR22-2 (GenBank A0008132) and LCR22-4 (GenBank A0009288) were used for low-stringency FISH mapping. Hybridization signals were detected on chromosomes 1p13,2p11,5pl3,13p11, and 20p12. The strongest signals were detected on chromosome 22q11 owing to the presence of multiple copies of sequences contained within the LCR22s.

noteworthy. As mentioned, LCR22-3a harbors an AT-rich palindrome that mediates the recurrent t(11;22) translocation and perhaps the 1.5-Mb deletion in VCFS/DGS patients. The AT-rich sequence and surrounding sequences have remained uncloned and are missing from the human genome *libraries*.

It is likely that they are absent because of the palindromic nature of the AT-rich stretch of sequence in humans. Interestingly, a similar pattern of AT-rich sequences, although non-palindromic, is present in LCR22-2 and LCR22-4. It is possible that in some humans, they could be *palindromic* in nature but have rearranged in bacteria during library formation, thereby deleting sequences.

The presence of potential palindromic sequences in LCR22-2 and LCR22-4 is particularly interesting as recently palindromic sequences have been found near the hotspot for the common 17p11.2 deletion associated with SMS and its reciprocal duplication dup(17)(p11.2p11.2) syndrome.

TRANSFORMATION BETWEEN THE LCR22S

The LCR22s are more than 98% identical in sequence. Of all LCR22s, LCR22-2 and LCR224 are the most similar, and they are more than 99% identical over large regions. The high level of identity between the two may be the reason why unequal homologous recombination events cause 22q11.2 rearrangements.

Another type of recombination that is possible in regions of high sequence similarity is gene conversion, where one sequence invades and replaces another. *Paralogs* are defined as sequence elements, which are present more than once per genome, arising by duplication processes during evolution. The different LCR22s are examples of paralogs.

A nucleotide variation in one vs another LCR is termed a paralogous sequence variant (PSV; such as an A in LCR22-2 and a T in LCR22-4). A series of contiguous PSVs would represent a signature for a particular LCR22. A single *nucleotide* alteration in LCR22-2 (A/T) and in LCR22-4 (A/T), suggests that gene conversion could have been taking place.

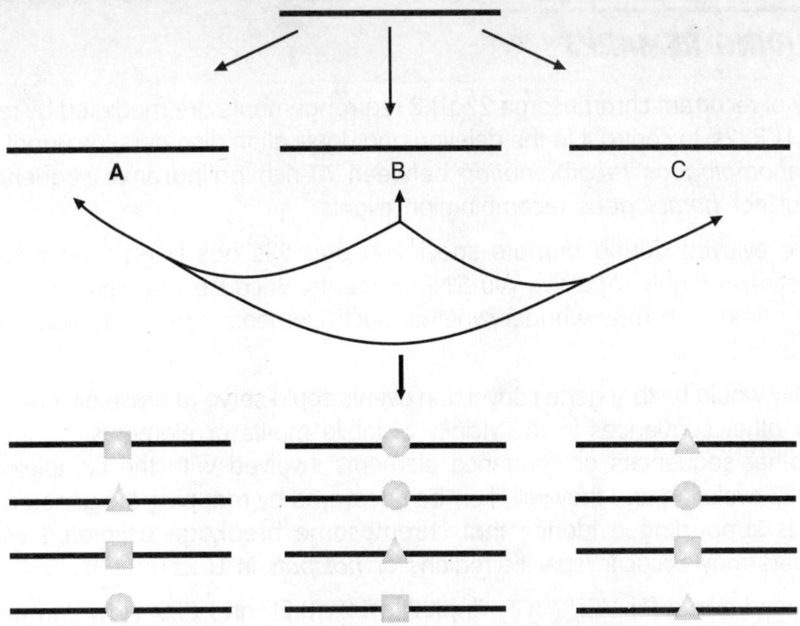

Figure 8.7: Model of gene conversion. An original locus duplicated to copies 1, 2, and 3 during primate evolution. The coloured shapes represent individual SNPs, which developed over time. The gene conversion events between copies 1, 2, and 3 result in a homogenization process whereby variation occurs in different individuals.

This is especially significant if there is a consecutive series of such nucleotide variants in *cis.* DNA sequence analyses support the occurrence of gene conversion events in LCR22-2 and LCR22-4 based on our analysis of overlapping BAC clones from different genomic libraries spanning the interval (Morrow et al., unpublished data). The evidence is quite similar to that for the large olfactory receptor gene family *segmental* duplication and for telomere associated repeats.

A model has been proposed to explain how segmental duplications can become *homogenized* during evolution. This process was hypothesized from data generated in studies of the HERV sequences involved in the *AZFa microdeletion* on chromosome Y and the CMT1 A rearrangement on chromosome 17p12.

Recently, it has been suggested that regions harboring higher rates of gene conversion might be hotspots for common rearrangements in LCRs. A gene conversion event, detected as a simultaneous loss and gain of a single *nucleotide* variation, would confirm that the LCR22-2 and LCR22-4 are more *dynamic* than *originally* believed.

Gene conversion could cause stretches of homology to become larger in length and could more strongly stimulate unequal crossover events, thus leading to a deletion or duplication. It is therefore important to define the rate of gene conversion and identify *subregions* within LCR22s that have altered rates of gene conversion. Long stretches of sequence identity, perhaps created by gene conversion events, between the two LCR22s might serve as a point for unequal crossover.

Based on recent studies of the SMS deletion and reciprocal duplication, there is a potential association between the positions of the unequal crossover and an increased rate of gene conversion.

CONCLUDING REMARKS

The majority of recurrent chromosome 22q11.2 rearrangements are mediated by recombination events between LCR22s. In contrast to the deletion and duplication disorders, recurrent transloctions result from *nonhomologous recombination* between AT-rich palindromic sequences or other mechanisms, but not *homologous* recombination events.

LCR22s have evloved during primate speciation and this has been mediated, in part, by *misalignment* between highly repetitive Alu SINE elements. Because Alu elements have diverged in sequence over time, one may wonder whether such elements could sitimulate recombination events in meiosis in humans.

One possibility would be that gene conversion events could serve to erase divergences between Alu elements or other sequences in the vicinity of labile motifs or elements. Besides Alu SINEs, there may be other sequences or sequence elements involved with the complex evolutionary pathway for LCR22 evolution and they will likely be uncovered by mapping the genomes of promate species. There is supporting evidence that chromosome breakage assoiated with recurrent gemomic disorders may occur in specific regions or *hotspots* in LCRs.

The same may be true for VCFS/DGS, dup(22)(q11.2;q11.2), and CES. How can one find these hotspots? It may be possible that hotspots for *chromosome* rearrangements have higher gene conversion rates. In this way, premeiotic germ cells from some individuals will be more or less *susceptible* to rearrangements than others.

To determine this, it would be necessary to compare sequences of parental DNAs in a large numberof trios, with the child harboring a *de novo* deletion. In addition to the seven LCR22s, there are other LCR blocks, mostly smaller in size as compared to the ones described here, containing different genes or pseudogenes, in the same interval of 22q11.2 but they have not been associated yet with any particular human disorder. It would also be of interest to determine their *evolutionary* origin in primates to compare this with the larger, disease associated LCR22s.

Chapter 9

Matrix-assisted Laser Disorption

Matrix-assisted laser *desorption/ionization* (MALDI) mass spectrometry (MS) is widely used in proteomics research as a high-throughput, *high-sensitivity* technique to identify proteins and their *post-translational* modifications. In addition, MALDI-MS produces single charges on most *peptide/protein ions*, thus simplifying the *interpretation* of the resulting data.

In this chapter, we address issues related to the analysis of MALDI-MS data, aiming at developing a robust and efficient biomarker discovery *methodology*. Generally, we follow the framework outlined in Figure elsewhere in this chatper for *biomarker* analysis of MALDI-MS spectra. First, baselines are corrected to reduce the influence of *background* noise.

Then, the data are smoothed to remove high-frequency noise. After that, biologically significant peaks are identified and aligned together across all sample spectra. These aligned peaks serve as the feature points for subsequent statistical analysis. Among the previously mentioned steps, *baseline correction* and smoothing can generally be handled using well-established methods (e.g., during baseline correction we usually use *curve-fitting* techniques to eliminate the influence of background intensity variation before detecting and analyzing peaks, whereas approaches for the peak identification and alignment steps that occur during data *preprocessing* are less well established.

Finally, feature selection and classification are the critical last steps that occur during biomarker discovery. In the following, we will address these issues in more detail. Before our general discussion, the critical need for developing statistical as well as visualization tools for biomarker studies is demonstrated in Figure elsewhere in this chapter.

The five m/z ratios indicated in this figure by vertical lines were identified by computational analysis in as biomarkers of ovarian cancer. However, a closer look at representative mass spectra around these biomarkers strongly suggests they do not arise from the ionization of biologically *meaningful* peptides that are differentially expressed in sera from normal vs ovarian cancer patients. Rather, the positions of these biomarkers seem to mark regions with differing baseline and/or noise *intensities*.

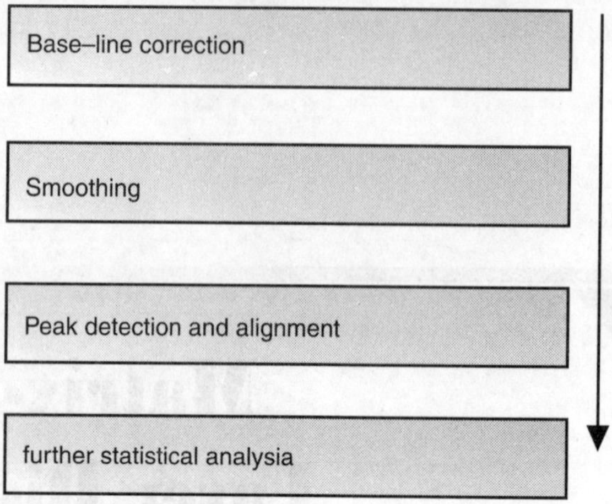

Figure 9.1: General framework of biomarker analysis of mass spectrometry data.

MATERIALS

The MALDI-MS spectra used in our research were acquired on desalted human sera samples by a Micromass M@LDI-L/R instrument. First, 2 µL of each serum sample is deposited into a 96-well plate (ABgene Thermo-Fast 96, Skirted) and acidified by the addition of 10 µL 0.1 % *trifluoroacetic acid* (TFA; Pierce). A Packard MassPrep robot then desalts the samples in groups of four.

The C-18 ZipTips (P10 size, 0.6 µL, Millipore) are washed with 50% acetonitrile (Baker, HPLC grade), 0.05% TFA, and are then *equilibrated* with 3×20 µL 0.1% TFA. Each sample is aspirated up and then expelled back into the original sample well eight times, and the ZipTip is then washed five times with 20 µL 0.1 % TFA.

Peptides/proteins are eluted from the ZipTip with 10 µL of 50% acetonitrile, 0.1 % formic acid, and deposited into a new 96-well plate. A 2-µL aliquot of each sample eluent is manually removed (which was found to be more reliable and efficient than *robotic spotting* on the Packard robot), mixed with 1.0 µL of a saturated (approx 4.5 mg/mL) solution of alpha-cyano-4 *hydroxycinnamic acid* matrix (Waters) in 50% acetonitrile, 0.05% TFA, containing an internal standard of 25 fmol/2 µL bradykinin (Sigma) (M+H C12 *mono-isotopic* mass: 1060.569), as a marker for the *reflectron* analysis, and 100 fmol/2 µL of insulin chain B, oxidized (Sigma) (M+H average mass: 3496.95) as a linear marker.

This sample solution was then spotted manually onto the MALDI-MS target and subjected to automated MALDI-MS on a Micromass M@LDI-L/R mass spectrometer. The M@LDI-L/R mass spectrometer automatically acquires two sets of data in positive ion detection mode. The mass range acquired is dependent on the mass *analyzer* being used, with 700-3500 Da for *reflectron* and 3450-28,000 Da for linear.

Although the mass range is adjustable, meaningful data are not acquired below approx 700 Da, because of interference from the matrix. The mass of 28,000 Da is about the upper limit for the α-cyano-4-hydroxy *cinnamic acid* matrix. For higher masses, the sample would have to be respotted using a different matrix. The linear analyzer is used for the high-mass region because

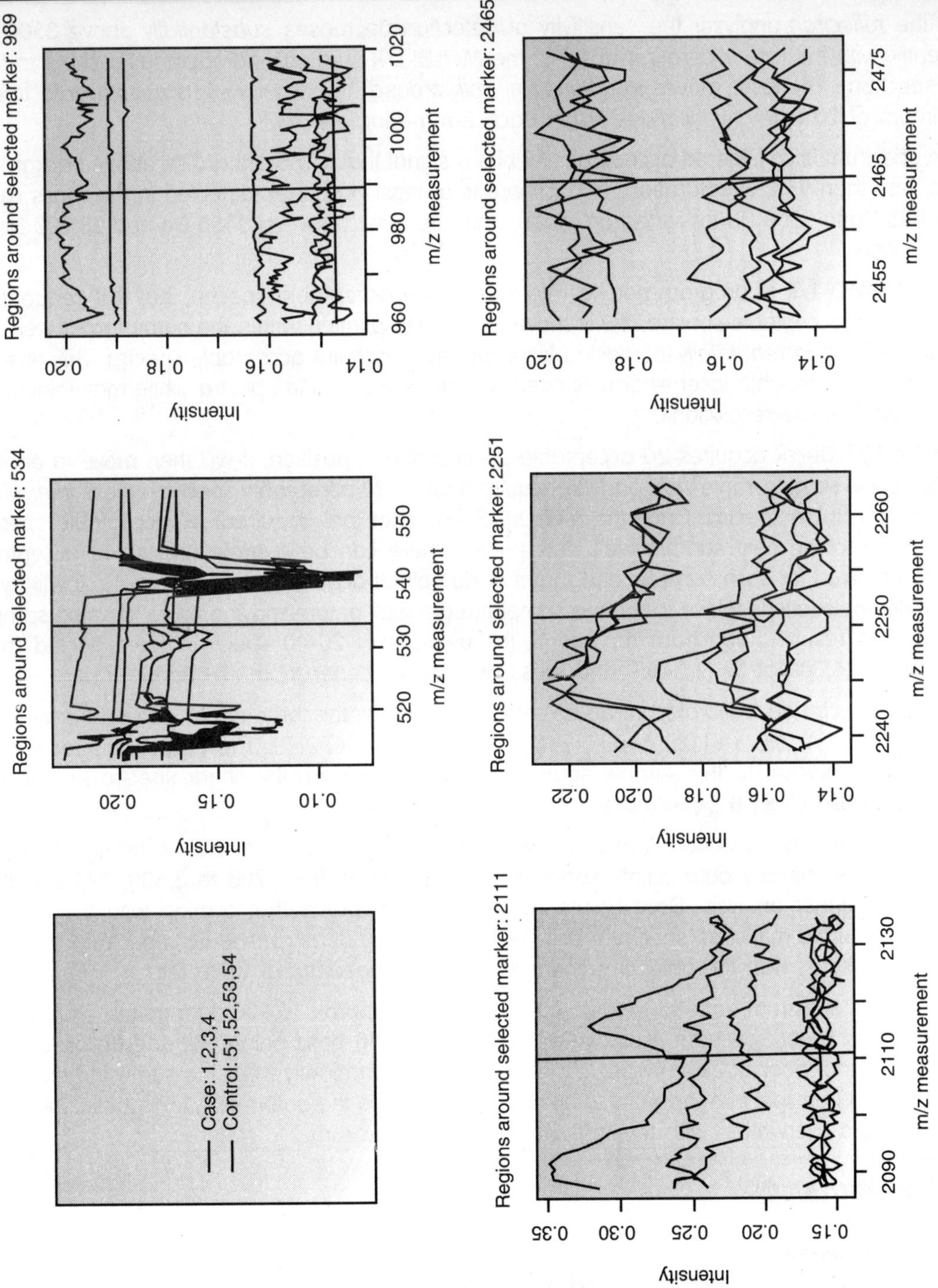

Figure 9.2: Five vertical lines denote the location of five biomarkers that were identified in the ovarian cancer.

with the *reflectron* analyzer the sensitivity of detection decreases *substantially* above 3500 Da. Currently, with the laser operating at 5 Hz, the M@LDI-L/R sums 10 individual laser shots to form one spectrum. The laser moves in a *random walk* around the target well, acquiring data from a maximum of 20 different *locations* within each 2-mm-diameter well.

A spectrum is considered acceptable if it has a signal that is more than 2% above background noise, less than 95% of saturation, and if there is at least one peak detected in the mass range from 1125 Da to 3500 Da for *reflectron mode*, and one peak between 3450 Da and 28,000 Da for linear mode.

The M@LDI-L/R is programmed to retain up to 40 acceptable spectra, but if it sequentially acquires 4 unacceptable spectra, it will move to another location within the same target well. The instrument uses *incrementally* increasing laser energy to acquire acceptable spectra. This ensures that the lowest possible laser energy is used to obtain acceptable spectra while maintaining the best possible mass resolution.

If the M@LDI-L/R acquires 20 acceptable *spectra* at one position, it will then move to another position in the same sample well, and will acquire another 20 acceptable spectra, unless *interrupted* by 4 unacceptable spectra. Once the M@LDI-L/R has shot (not acquired) 40 acceptable spectra, it will move to the next sample well. This means there can be a maximum of 40 acceptable spectra acquired for each sample, and that if at no point it acquires acceptable data, it will try up to 10 different locations within the same sample target well before moving on to the next sample. Typically, the resulting spectrum represents the average of 20-40 spectra. The expected mass resolution is 14,000 at M+H 2465, and mass accuracy is better than +70 ppm.

Linear spectra are calibrated using as external calibrants the three peaks derived from equine cytochrome C (M+2H, 6181.05; M+H, 12361.1; and 2M+H, 24721.2). The *calibrant* is located on target spots adjacent to the sample spots. For data processing, the linear spectra are aligned using the insulin chain B peak/mass.

Each (averaged) *reflectron* MALDI-MS spectrum is converted to a text-file listing of approx 94,900 *m/z* vs intensity data points spanning the *m/z* range from 700 to 3,500, which is then suitable for further analysis. Data points in the linear MALDI-MS that overlap with the 3450 to 3500 region of the reflectron spectrum are deleted, resulting in a "*continuous*" spectrum from 700 to 28,000 daltons, with the reflectron/linear breakpoint *appearing* at 3500 Da.

The linear spectrum from 3500 to 28,000 Da contains approx 36,900 data points. The merged reflectron plus linear spectrum thus contains approx 131,800 data points. All *spectra* are visually screened so that poor quality spectra can then be reshot manually from the same targets. If the manually acquired spectrum appears to be visually superior to the *automatically* acquired spectrum, then the file is overwritten with the manually acquired spectrum.

METHODOLOGY

Data Preprocessing

1. Global alignment: To circumvent the slight variation in m/z values for the same relative data points in different spectra, reflectron data points are numbered consecutively by assigning the observed m/z value that is closest to the expected MH+ for the C12 isotope

of bradykinin, which is 1060.569, as data point zero. Linear data points are *numbered* consecutively by assigning the observed *m/z* value that is closest to the expected average MH+ of oxidized insulin chain B, which is 3496.95, as data point 100,000.

2. Exclusion of data around *bradykinin* internal standard: To eliminate the bradykinin peak from consideration as a possible marker and to prevent its intensity from impacting on normalization of reflectron mass spectra (discussed later), m/z values that range from 1060.4 to 1063, which include the C 12 and the first two of three apparently observable isotope peaks, are excluded from consideration. The third isotope peak potentially could be erroneously determined to be a marker, which is a possibility that needs to be kept in mind. However, by excluding only up to 1063 we leave open the possibility of finding a real marker in the 1063 to 1064 region. Assuming that we exclude the 1060.4 to 1063 region and that data points are numbered as indicated above, then data points numbered from approx -6 to +112 will be excluded from consideration as possible markers.
3. Initially, all isotope peaks from each peptide ion are included in the search for biomarkers. This approach will better allow peptide biomarkers to be identified whose isotopic envelopes overlap with envelopes resulting from *nonbiomarker* peptides. Hence, if we want the program to search for 10 peptide markers, we need to allow 40 data point markers, with one for each of the four isotopes that might *commonly* be observed from reflectron data.
4. Baseline correction is carried out separately for linear and reflectron data as follows:

   a.Take the natural log of all *intensities.*

   b. Use a sliding window of 1000 data points; determine the least-squares, robust local *polynomial* fit curve that best represents the baseline, and for each data point subtract the *corresponding* baseline intensity.
5. Link together each reflectron spectrum with its corresponding linear spectrum at 3500 Da to produce a spectrum from 700 to 28,000 Da, and exclude the linear internal standard of oxidized B-chain of insulin.
6. Linearly normalize each spectrum and delete *unacceptable* spectra.
   a. Determine an overall, baseline-corrected median spectrum.
   b. Determine a linear normalization factor for each spectrum that will minimize the summed difference between all intensities observed within the specified mass range from 710 Da to 27,990 Da and the *corresponding* intensities in the over all, baseline-corrected median spectrum.
   c. Because different spectra may have slightly different mass ranges, which would change the dimensions of the dataset, we normalize with respect to only those points in the mass range 710 to 27,990 Da, even if the actual data extend beyond this range.
   d. The normalization factor, f_n, is calculated by minimizing the difference between I_n and I_{median}, which requires finding a f_n to make $f_n \times I_n = I_{median}$. This involves a least-square regression analysis of all observed intensities in the specified mass range in the median curve vs those in each *individual* spectrum, with the slope of this curve = f_n.
   e. Re-normalize each spectrum's raw *normalization* value (as determined previously) by

setting the median overall raw normalization value equal to 1.0, then delete from the dataset all spectra with re-normalized factors F below 0.5 or above 2.0 (the latter spectra should be *flagged,* individually examined, and if needed, be repeated, as they are likely to include poor spectra).

f. If any spectra are eliminated via the above test, then re-normalize the remaining spectra.

g. Multiply each background-corrected, data point intensity in each spectrum by this re-normalization factor.

7. Remove background noise that otherwise would lead to instability in the classification program and limit peak consideration to one mass-vs-intensity data point for each observed peptide ion isotope. Hence, each data point must pass the following *criteria* in order to be considered as a possible marker:

   a. Minimum intensity level test: rank the normalized *intensities* for each spectrum and take only data points whose intensities are within the top 20%.

   b. Peak test: assume that only data points in completely or partially resolved peaks result from peptide ions and are likely to be meaningful. Data points in partially *resolved* peaks may represent the intensity sum of a useful marker *superimposed* on an unrelated, *nonmarker peptide ion.*

      i. At least three of four successive data point intensities before or after each candidate marker data point must show a progressive increase or decrease in background corrected, normalized peak intensity.

      ii. Extend each peak by an additional four data points before/after the last data point that passed the peak test. Hence at this point we include a band width around each peak.

      iii. At least 10% of the cases or controls need to pass the peak test for this peak to be considered as a useful marker peak.

8. Unique peptide origin test: because there is no benefit in having multiple markers that all originate from different isotopes of the same peptide ion, all potential markers must have *m/z* values that differ from each other by at least 3.1 Da. Hence, if multiple potential markers are found that differ from each other by less than 3.1 Da, only the best marker (as judged by its individual ability to discriminate all cases from control samples in the training set) is considered. It is possible (e.g., if there are incompletely resolved, unrelated peptide ions that overlap with the C 12 isotope peak of a marker peptide ion) that the best *isotopic* representative of a marker ion is not the C12 isotope.

   Without carrying out the unique peptide ion origin test, the variable importance measure determined by the RF algorithm would be unstable. That is, markers arising from different isotopes of the same peptide should be surrogates for each other. If one marker (i.e., from the C12 isotope) from a peptide is selected in a tree construction, then *randomly* permuting the values of the corresponding C 13 isotope peak from this same peptide would not be expected to significantly change our classification ability. Hence, both these markers would appear to be unimportant. Instead, for each peptide we select only one *mass-vs-intensity* data point from whichever isotope (assuming the marker arises from a reflectron spectrum; note that this will not be a problem with linear spectra where peptide/ protein isotopes are not resolved), which can best *discriminate* all cases from controls in the training set based on some criterion. For this purpose, we carry out an initial rough filtering, using the normalized difference between case/control group for all candidate markers within the 3.1 -Da mass range to identify the best marker.

Sampling

The data points obtained from the data preprocessing step then represent potential biomarkers. The following is a step-by-step outline of the basic features of a random forest algorithm we have implemented to find markers that can best *discriminate* cases from control samples.

1. Prespecify the number of trees to grow (currently, we are using 5000 trees), which will constitute our forest. Each tree will be grown as follows:

   Randomly divide the training sample into training/testing set-approx 67% as the training set, 33% as the testing set. Our tree will be constructed based on the training set according to the following *procedure*. Keep in mind that each training set is a resample with replacement from the original training sample; on average it contains approx 67% of the original training samples, and the remaining "*testing*" setcontains approx 33% of the spectra in the training set.

   a. Start from all samples as one root node. Randomly select m_{try} variables without replacement from all *variables* (say 60 mz vs intensity values that passed the peak test and other tests described above); for each variable, find the best cutoff intensity value to best split the root node, which means to find the intensity cutoff value that will divide the maximum number of cases vs control samples correctly. This will then provide a measure of the decrease of impurity (i.e., two child nodes should contain a more *homogeneous* mix than the root node) from root node to two child nodes for each selected variable. Among these m_{try} variables, select the best one, based on the decrease of impurity, as our final split variable for the root node. As for measure of *impurity*, we use the Gini index, $1-(P_1\wedge 2 + P_2\wedge 2)$, where P1 is the proportion of case samples, P2 is the proportion of normal samples, and $P_1 + P_2 = 1$. When P1 = 0 or 1, Gini is 0, which is least impure. When P1 = P_2 = 0.5, Gini is 0.5, which corresponds to a maximum impure sample. Our overall goal is to reduce this *Gini index*, and the overall decrease of Gini index due to one *variable* can reflect the importance of this variable. After we select a split variable, we can split the root node into two child nodes, and the decrease of impurity is measured as: $P_1\wedge 2 + P_2\wedge 2 - (R_1\wedge 2 + R_2\wedge 2) - (L_1\wedge 2 + L_2\wedge 2)$, where P_1 and P_2 are the proportions of cancer/control for all samples in the root node, R1 /R2 and L_1 /L_2 are the proportions in the right/left child nodes, $P_1 + P_2 = R_1 + R_2 = L_1 + L_2 = 1$. This function is always positive, and we want to maximize it by choosing different cutoff values for selected split variables.

   b. For each child node, repeat the previous growing step as for the root node. We keep on growing until the tree perfectly fits the data, i.e., all terminal nodes contain only one class of samples.

   c. Repeat the above two steps 5000 times to get 5000 trees, with each tree having an associated training and testing set.

2. Use these 5000 trees to get a prediction for each sample and a measure of importance for each variable, as described in the following:
   a. For each sample S, find those trees which are based on a training set not containing this sample, label this set as Pred, and suppose Pred has m trees. Put S down Pred to get m predictions for S. The final prediction for S will be the majority vote of these l>m predictions. For these in *predictions*, we will get a margin for S: margin = proportion of correct prediction - proportion of incorrect prediction (this assumes there are only two classes of samples; for >2 classes, we will minus the maximum of incorrect class *predictions*). Ideally we want margin to be as large as possible, and one of our importance *measures* will be based on this margin.
   b. The importance measure for the mth variable is calculated as follows: For each tree, we have a testing set. Randomly permute the value of the mth variable in the testing set, and we can get a new prediction for testing samples. Proceed as previously for all 5000 trees. Finally, we can get a new prediction error and also a new margin for each sample. The decrease of margin after permutation will reflect the importance of the mth variable. The increase of prediction error can also serve as an importance measure, but intuitively this measure will be very crude, because essentially it loses some information by summarizing the continuous margin with a discrete value (0/1).
   c. Select mk (e.g., 28) biomarkers based on the importance measure, and construct a random forest for these ink variables as previously. As a new sample comes in, use this forest to predict the class of this sample. Our confidence in the prediction can be measured with the margin.
3. Export a list of potential biomarkers. For each marker, plot the following:
   a. m/z vs intensity for all background corrected, normalized case vs all control samples for the m/z region extending for 3 Da on either side of the marker.
   b. Make a similar plot as described previously for every fifth case vs control sample.
   c. Overall median case vs control intensities around each marker, extending for 3 Da on either side of the marker.
4. Delete all potential markers that cannot reasonably contribute to a partially or completely resolved *peptide isotopic cluster.*
5. Use the resulting markers to develop, as discussed later, the quantitative *criteria* for the diagnosis of unknown samples, with the first step being normalizing the unknown sample with respect to the training set.
   a. For these selected markers, construct binary classification trees for future prediction. We grow 5000 trees instead of one tree. Basically, the more trees we have, the more stable the final prediction error is. We can run several times with 1000 to 10,000 trees to evaluate the effects. Typically, 5000 trees will give a very stable result. We use random selections in the tree construction process. Consequently, we can have different results, although the markers used are exactly the same for different trees.
   b. The following is one typical run of our construction process:
      i. Sample with replacement from all samples to form a new training set.

ii. For this new training set, construct one tree. At each node split, we randomly select one m/z marker value to split.

iii. Repeat steps i and ii until desired number of nodes and trees is reached.

iv. For future samples, we do a majority vote using all of the trees.

Method of Analysis

This can be addressed by estimating a "sample size" vs "prediction accuracy" curve. This basically requires the following procedure:

1. Suppose we have a very large population S, which contains all potential samples. 2. Given a sample size N.
3. Randomly take N samples from S to form a training set T_N, and construct a classifier $C(T_N, .)$ using our procedure.
4. Use $C(T_N, .)$ to classify a random sample from S, and calculate its error rate

$$E_{a \in S} C(T_N, a)$$

Here the expectation is taken over all potential samples from S.

5. Take the average over the previous error to get the error estimation for sample size N:

$$E_{IT} N_I = {}_N E_{a \in S} C(T_{N,a})$$

Generally, we do not have the full population S. Instead, we only have a finite sample, S_n (n = 187 for our ovarian cancer dataset described later), which will be used as an estimator of S. In practice, we use the following estimation procedure, based on a finite sample S_n:

a. Given sample size N (<n), we take N = 40, 80, 120 to get a rough estimate for the general trend of error vs sample size curve.

b. Randomly take N samples (without replacement) from S_n to form a training set TN. Then construct a classifier $C(T_N, .)$.

c. Use $C(T_N, .)$ to classify S_n and get an error estimate

$$E = \frac{\sum_{i=1}^{n} C(T_N, a_i)}{n} = \frac{N}{n} E_0 + \frac{n-N}{n} E_1$$

where

$$E_0 = \frac{\sum_{a_j \in T_N} C(T_N, a_j)}{N}$$

is the total error rate we reported using our RF program, and

$$E_1 = \frac{\sum_{a_j \notin T_N} C(T_N, a_j)}{n-N}$$

is similar to an out-of-bag error estimation (error for samples not in training set). d. Repeat steps b and c M times, e.g., M = 100, to get M error estimations $E^1, \ldots, E^M$

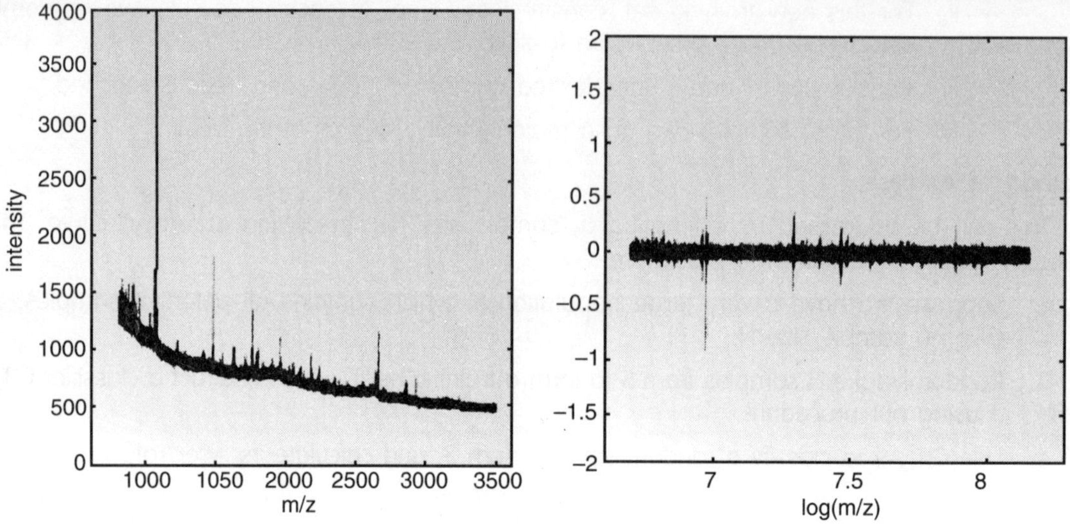

Figure 9.3: Data preprocessing example. Left: original raw data. Right: mass spectrometry data after baseline correction.

e. Use $(E^1 + \ldots + E^M)/M$ to estimate the final error rate for sample N.

Significance

In this section, we apply the methods described in the previous section to an ovarian cancer dataset to identify cancer-related biomarkers. The MALDI-MS spectra were acquired as described previously on sera from 77 control cases and 93 disease cases, using a Micromass M@LDI-L/R instrument, with 700-3500 Da for reflectron and 3450-28,000 Da for linear, and this resulted in a merged dataset consisting of MS spectra that extend from 700 to 28,000 Da. Figure 9.3 shows how our workflow preprocesses the input raw data. On the left is an example of a MALDI-MS raw spectrum obtained on serum, the decreasing tendency is obvious.

After taking a logarithm transformation on both the m/z and intensity axis, we estimate the baseline of the curve and subtract the baseline from the data. The resulting curve on the right plot clearly shows that the decreasing tendency has been removed.

After data preprocessing, we apply the random-forest method on the peaks to identify biomarkers that are most informative in terms of distinguishing disease cases from control cases.

Figure 9.4 gives a two-dimensional visualization of the samples based on the proximity matrix generated when we run the random-forest algorithm. In the graph, most samples are well separated, but some of them mix together.

Figure 9.5 shows the impact of training size on the classification rate, as well as the additional information offered by the data from the linear range.

We can clearly see the trend that larger training sets result in smaller classification errors. And for a fixed training set, classification error drops dramatically from 5 to 20 biomarkers and then it levels off at approx 20-40 biomarkers for the combined reflectron + linear data. With 136 samples in the training set, we can achieve approx 20% classification error.

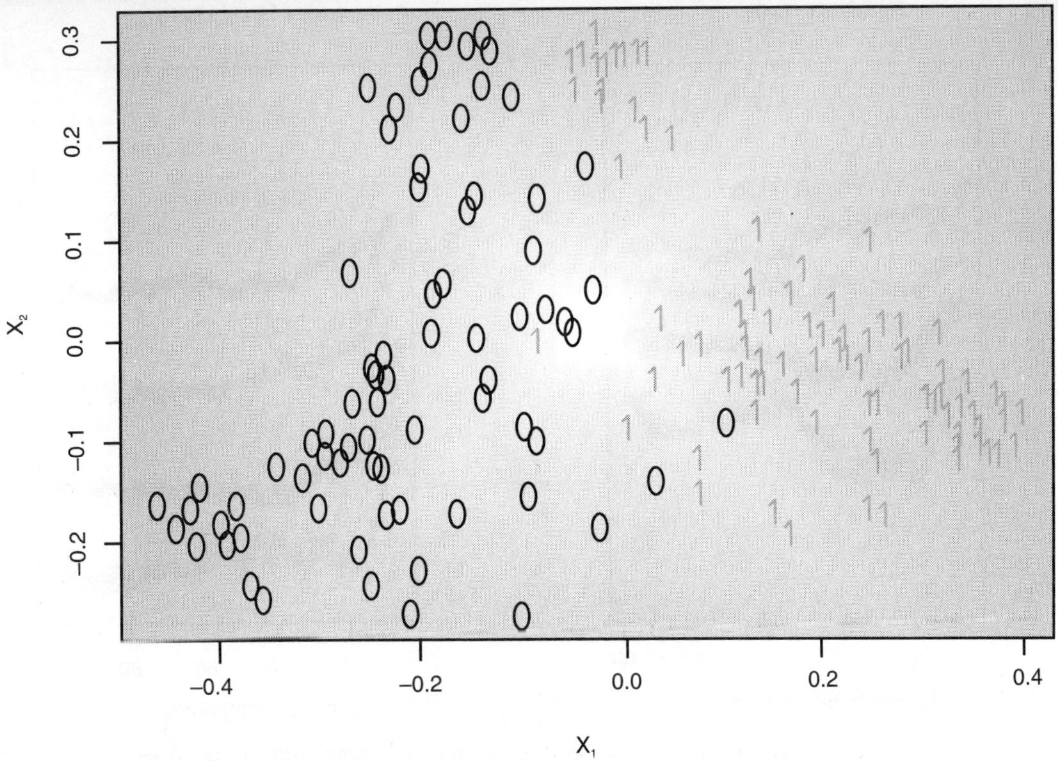

Figure 9.4: The symbol "0" represents the samples from control group and "1" represents cancer group.

POINTS TO REMEMBER

1. During peak alignment, simply using a linear shift may not work all the time, as previous studies have shown that the variation of peak locations in different datasets is nonlinear, and this variation still exists even for technical replicates. The reasons that underlie data variation are extremely complicated, including differences in sample preparation, chemical noise, co-crystallization and deposition of the matrix-sample onto the MALDI-MS target, and laser position on the target. Here we are interested in reducing the variation and aligning these peaks together. As for peak alignment, we are exploring a superset idea that is very similar to the clustering approach in. Based on the assumption that the peak variation range is smaller than the minimal distance dmin, we construct a superset of all peaks as follows:
   a. Set a distance threshold as the smallest distance between neighbouring peaks in all peak sets.
   b. Initialize the superset as the first peak set.
   c. For each point in a new peak set, check its counterpart in the superset that has the smallest distance:

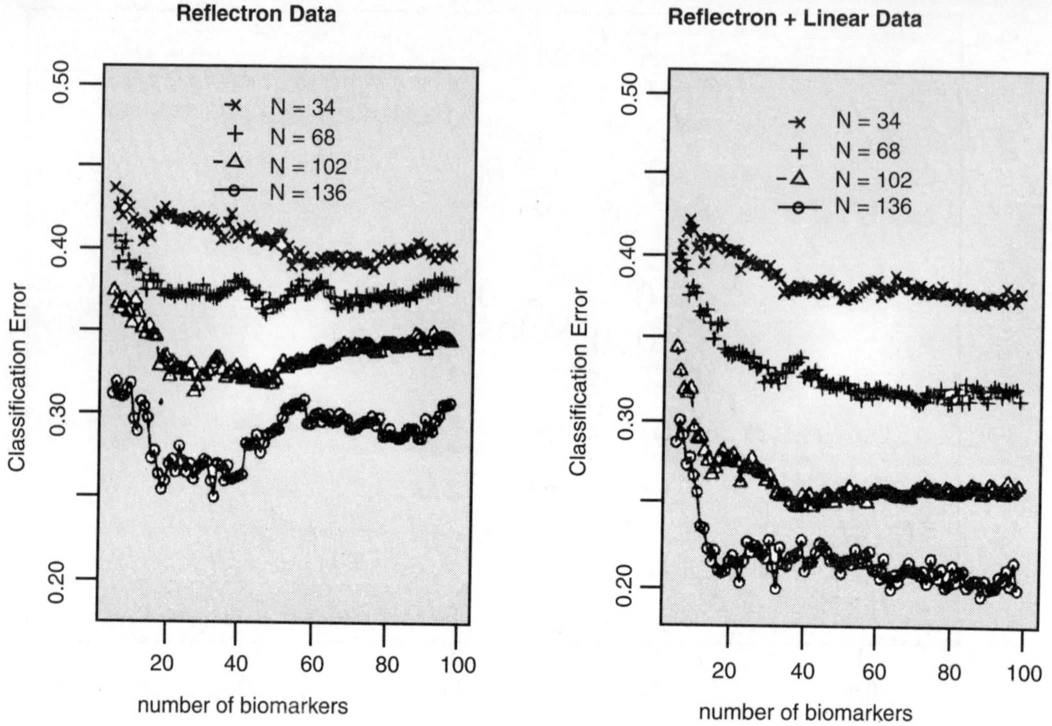

Figure 9.5: The fivefold cross validation classification error estimations. After adding the linear data, the best classification error achieved drops from approx 25% to 20% and the classification error estimation is also more stable.

i. Over the threshold: new peak, add to the super set;

ii. Below the threshold: already in the superset, ignore.

Repeat step 1c until all new peak sets are processed. After the superset is constructed, we can align every peak in each peak set to the superset. It is guaranteed that each peak point in a peak set has at least one possible counterpart in the superset.

2. There are related approaches to address the peak alignment problem. Coombes et al. pooled the list of detected peaks that differed in location by three clock ticks or by 0.05% of the mass. Yasui et al. believed that the *m/z* axis shift of peaks are approx ±0.1 % to ±0.2% of the m/z value. Thus, they expanded each peak to its local neighbourhood, with the width equal to 0.4% of the m/z value of the middle point. This method certainly oversimplifies the problem. In another study, they first calculated the number of peaks in all samples, allowing certain shifts, and selected m/z values with the largest number of peaks. This set of peaks is removed from all spectra, and the procedure is iterated until all peaks are exhausted from all the samples. In a similar spirit, Tibshirani et al. proposed to use complete linkage hierarchical clustering in one dimension to cluster peaks, and the resulting dendrogram is cut off at a given height. All the peaks in the same cluster are considered as the same peak in further analysis. Randolph and Yasui used wavelets to represent the MS data in a multi-scale framework. They used a coarse-to-fine method to

first align peaks at a dominant scale and then refine the alignment of other peaks at a finer scale. From the signal representation point of view, this approach is very interesting. But it remains to be determined whether the multiscale representation is biologically reasonable. Johnson et al. assumed that the peak variation is less than the typical distance between peaks, and they used a closest point matching method for peak alignment. The same idea was also used in Yu et al. to address the alignment of multiple peak sets. Certainly, this method is limited by the data quality and it cannot handle large peak variation. Dynamic programming (DP) based approaches have also been proposed. DP has been used in gene expression analysis to warp one gene expression time series to another similar series obtained from a different biological replicate, where the correspondence between the two gene expression time series is guaranteed. In MS data analysis, however, the situation is more complicated, because a one-to-one correspondence between two datasets does not always exist. Although it is still possible to apply DP to deal with the short-of-correspondence problem, some modifications are necessary (such as adding an additional distance penalty term in the estimation of correspondence matrix). It also remains unclear how DP can identify and ignore outliers during the matching. Eilers proposed a parametric warping model with polynomial functions or spline functions to align chromatograms. In order to fix warping parameters, he added calibration example sequences into chromatograms. Although the idea of using a parametric model is interesting, it is difficult to repeat the same parameter estimation method in MS data, because we cannot add many calibrator compounds into the MS samples. Also, it is unclear whether a secondorder polynomial would be enough to describe the nonlinear shift of MS peaks.

3. Peak detection is an essential step in data preprocessing. The results in this chapter are obtained using a simple peak identification method. Right now, we are applying the amplitude modulation (AM) technique to form a peak envelope signal for each peptide. The AM technique is widely used in telecommunication and broadcasting, where a carrier with a fixed frequency f0 moves the original signal up to a transmitting frequency range. After receiving the transmitted signal, we use an envelope detection technique to reconstruct the original signal. Our idea here is to apply the envelope detection on the MS data before searching for the local maxima. In this way, the multiple peaks corresponding to the same peptide can be considered as a sampled version of a much smoother envelope signal. Envelope detection makes it much easier to group these successive peaks together, thus to reduce the complexity of the spectrum such that each peptide will be represented by only a single peak. It should be noted that applying a simple low-pass filter (like a Gaussian filter) will not guarantee the extraction of the envelope signal, because the carrier frequency f0 may move the envelope signal beyond the low-pass region. Consequently, the lowpass filtering will over-blur the signal. Here we apply a Gabor filter centered at fn on the data and take the amplitude of the filter response as output. The Gabor filter consists of an even real part (i.e., Gaussian weighted cosine function) and an odd imaginary part (i.e., Gaussian weighted sine function). Because the amplitude response is not sensitive to the phase variation, this process is also called quadrature filtering. We can then search for local maxima with a discrete differentiation method, followed by a peak width check, in which we keep only those local maxima whose left and right neighbouring local minima have distances larger than a threshold value. In this algorithm, several parameters need

to be determined, and they are different based on specific applications.

4. As we have mentioned, most ions produced in MALDI-MS have only single charges. Thus, the isotopically resolved peaks are assumed to have a constant distance of 1 Da. The equal spacing property of the MALDI data has been used in several approaches to identify peptides from isotopically resolved peaks. Breen et al. used morphology and watershed algorithms to first detect a set of individual peaks and then use a Poisson distribution model (based on equal spacing between isotopic peaks) to detect peptides. They also demonstrated that their algorithm can separate two overlapping peptides using the Poisson isotopic distribution model. Similarly, Gras et al. used a template-matching algorithm to identify peptides from a set of detected peaks, while they used the equal spacing property in the construction of their parameterized template. The same equal spacing property is the basis for extracting the envelope signal in this approach.
5. Envelope detection does not preserve the total intensity of a single peak, which is very important for quantitative studies. The point here, however, is to reduce the redundancy among multiple peaks that correspond to the same peptide. Guided by the envelope signal-based peak detection results, we still can go back to the data before envelope detection and detect different isotope peaks for the same peptide. Roughly speaking, this is a kind of top-down strategy, whereas previous approaches in used a bottom-up strategy.
6. Normally, spectral peaks are local maxima in MS intensity data. In most published algorithms, peaks are mostly defined with respect to their nearby points. All these methods are based on similar intuitions and heuristics. Several parameters need to be specified beforehand in these algorithms, e.g., the number of neighbouring points and the intensity threshold value. In fact, the parameter settings are related to our understanding/modeling of the underlying noise. Coombes et al. defined noise as the median absolute value of intensity. Satten et al. used the negative part of the normalized MS data to estimate the variance of noise. Wavelets-based approaches have also been proposed to de-noise the MS data before peak detection. Based on the observation that there are substantial measurement errors in the intensity value, Yasui and his coworkers argued that binary peak/nonpeak data is more useful than the absolute values of intensity, while they still used a local maximum search method to detect peaks. Clearly, the success of noise-estimationplus-threshold methodology largely depends on the correctness of the noise model, which remains to be validated.
7. Another issue in peak detection is to avoid false-positive peaks. This is often done by adding an additional constraint (e.g., the peak width constraint) or by choosing a specific scale level after wavelets decomposition of the original MS data. In the case of high-resolution spectra, it has been proposed that more than one isotopic variant of a peptide peak should be present before a spectral peak is considered to result from peptide ionization. It may also be possible to use prior information about the approximate expected peak intensity distribution of different isotopes arising from the same peptide during peak detection (i.e., the theoretical relative abundance of the first peptide isotope peak may range from 60.1% for polyGly (n = 23, MW 1329.5 Da) to 90.2% for poly Trp (n = 7, MW 1320.5 Da). Certainly, we also have to consider the issue of limited resolution and the consequent overlapping effect of neighbouring peaks.
8. There are many classification methods, e.g., linear and quadratic discriminate analysis,

and k-nearest neighbour. Most of these methods were developed in the pregenome era, where the sample size was usually very large whereas the number of features was very small. One can intuit that the sample classification error rate will increase with more noise in the data. Therefore, directly applying these methods to proteomic datasets may not work. Instead, feature selection methods are usually applied to select some useful features at first, and then the selected features are used to carry out sample classification based on traditional classification methods. In this context, feature selection serves two purposes: for biological interpretation and for reducing the impact of noise. Two sample t-test statistics or variants thereof are often used to quantify the difference between two groups in the analysis of gene expression data. However, determining the right dimension of features in feature selection still remains a difficult problem: if we only select a small number of features, we may miss many useful features. If we select a large number of features, we may over-fit the data with noise. Another problem with a tstatistic is robustness: the univariate feature selection based on a t-statistic is very sensitive to noise.

9. Instead of using univariate feature selection methods, it may be useful to consider multivariate feature selection methods. Lai et al. analyzed the co-expression pattern of different genes to select genes that have differential gene-gene coexpression patterns with a target gene. Some interesting genes have been found to be significant and have been reported to be associated with prostate cancer, yet none of them showed marginal significant differential gene expressions. Generally, the multivariate feature selection is a combinatorial approach. To analyze two genes at a time, we need to consider n2 possibilities instead of n for the univariate feature selection. For analyzing the interaction of K genes we need to consider nK possibilities, this quickly bccomes intractable.

10. Machine learning-based approaches have been proposed to couple feature selection and sample classification. They implicitly approach the feature selection problem from a multivariate perspective. The significance of a feature may highly depend on other features. Isabelle et al. have reported using support vector machines (SVM) to select genes for cancer classification from microarray data. Qu et al. applied a boosting tree algorithm to classify prostate cancer samples and to select important peptides using MS analysis of sera. Wu et al. reported using random forest to select important biomarkers from an ovarian cancer dataset based on MALDI-MS analysis of patient sera, and showed good performance of randomforest feature selection compared to univariate t-statistic selection.

11. For the random-forest algorithm from Breiman, randomness is introduced at each node split. Specifically, at each node split, a fixed number of features is randomly selected from all the features, and the best split is chosen among these selected features. On the other hand, for the random subspace method developed by Ho, a fixed number of features is selected at first and is used for the same original data to produce a tree classifier. Thus, both models have the effect of randomly using a fixed subset of features to produce a classifier, but differ in the underlying tree-building method.

Chapter 10

Genes and Smelling Ability

The completion of the human genome sequencing has opened new *opportunities* to better understand complex *biological systems*. In this realm, the human sense of smell is an excellent example of how genome analysis provides new information on genome organization and on *deficits*. Before the advent of genomic tools, the understanding of this highly *sophisticated* sensory neuronal pathway has been rather *sketchy*.

In this chapter we summarize the relevant progress made in the last decade, and highlight the initial *elucidation* of two classes of olfactory deficits and their possible *underlying* genetic mechanisms. Olfaction, the sense of smell, is a versatile and sensitive mechanism for detecting volatile odorous molecules (*odorants*). Many organisms rely on olfactory cues for a wide range of activities such as food *acquisition*, *reproduction*, *migration*, and predator alarming. For that, the olfactory system is characterized by a remarkable ability to detect and *discriminate* thousands of low-molecular-mass compounds.

This *sophisticated* chemical detection apparatus, which has evolved over approx 1 billion years, has long intrigued scientists attempting to understand its molecular facets.

MECHANISM OF PERCEPTION

Olfactory perception is a result of a cascade of biochemical and electrophysiological processes, in which the intrinsic information in the molecular structure of an odorant is converted into the perception of a characteristic odor quality. Odorants of various chemical configurations are inhaled and reach the *olfactory epithelium*, situated at the posterior region of the *nasal* cavity.

This is the interface where odorants are bound and recognized, and where from *axonal* electrical messages are transmitted to the olfactory bulb of the brain. The human *olfactory epithelium* accommodates approx 10 million olfactory sensory neurons, each extending a single dendrite to the surface of the epithelium.

The mucus-bathed *dendritic* ends bear specialized cilia, which expand the receptive membrane surface. *Odorants,* carried by a stream of inhaled air, dissolve in mucus, and then interact with receptor proteins within the *ciliary* membrane.

The binding of odorants to one or more of a repertoire of hundreds of specific receptor proteins initiates a cascade of signal transduction events, involving the G proteindependent elevation of cyclic adenosine monophosphate (cAMP), opening of cation channels, and membrane *depolarization.* This process triggers action potential in the *unmyelinated* axons of the olfactory sensory *neurons,* leading to the olfactory bulb.

There, the axons form synapses with apical dendrites of mitral cells within structures called glomeruli. Mitral axons leaving the olfactory bulb project widely to other brain structures, such as the olfactory cortex, where further information processing occurs. It is generally accepted that the *olfactory* system employs a combinatorial strategy to *discriminate* among the millions of odorous compounds and their mixtures.

Odor perception appears to be a *multidimensional* task, whereby every olfactory receptor (OR) binds numerous types of odorants with different *affinities* and vice versa. Consequently, a unique combinatorial code is generated for every odor *stimulus,* suggesting that olfactory *discrimination* is a *polygenic trait.*

However, based on the receptor affinity distribution model, most of the odorant binding affinities of an OR are weak, and only few have biological significance. Of these, the strongest affinity receptor determines the detection and recognition *thresholds,* which may, therefore, be treated as a single gene trait.

SENSE ORGANS

OR proteins, of which the Nobel-awarded discovery took place 13 years ago, belong to the G protein-coupled receptors (GPCRs) *superfamily, transducers* of a wide array of extracellular signaling molecules that constitute important targets in the *pharmaceutical* industry.

Members of this protein superfamily share features of sequence and structure, including seven hydrophobic transmembrane helices, as well as three intracellular and three extracellular loops that link these helices. ORs belong to GPCR family A, or rhodopsin-like GPCRs, which is the largest and most well-studied. Although the overall sequence similarity between the various members of this family is very low, with mean amino acid *pairwise* identity of 17%, they share several *highly* conserved sequence motifs that are thought to play an essential role in either maintaining the structure and functional conformational transitions of these proteins, or in *interacting* with upstream and downstream partners. In most mammals, the olfactory repertoire has 1000-1400 OR genes, constituting the largest gene family within their genomes.

The OR coding region spans approx I kb, almost always without introns, a property that facilitates their identification and cloning from genomic DNA. OR genes are divided into two classes containing 17 families, which further divided into subfamilies, all based on sequence similarity scores. ORs with more than 40% protein sequence identity are considered within the same family and if they share more than 60%-as belonging to the same subfamily.

It was suggested that ORs of the same subfamily might recognize molecules with similar chemical and/or physical characteristics. However, more experimental evidences are needed to

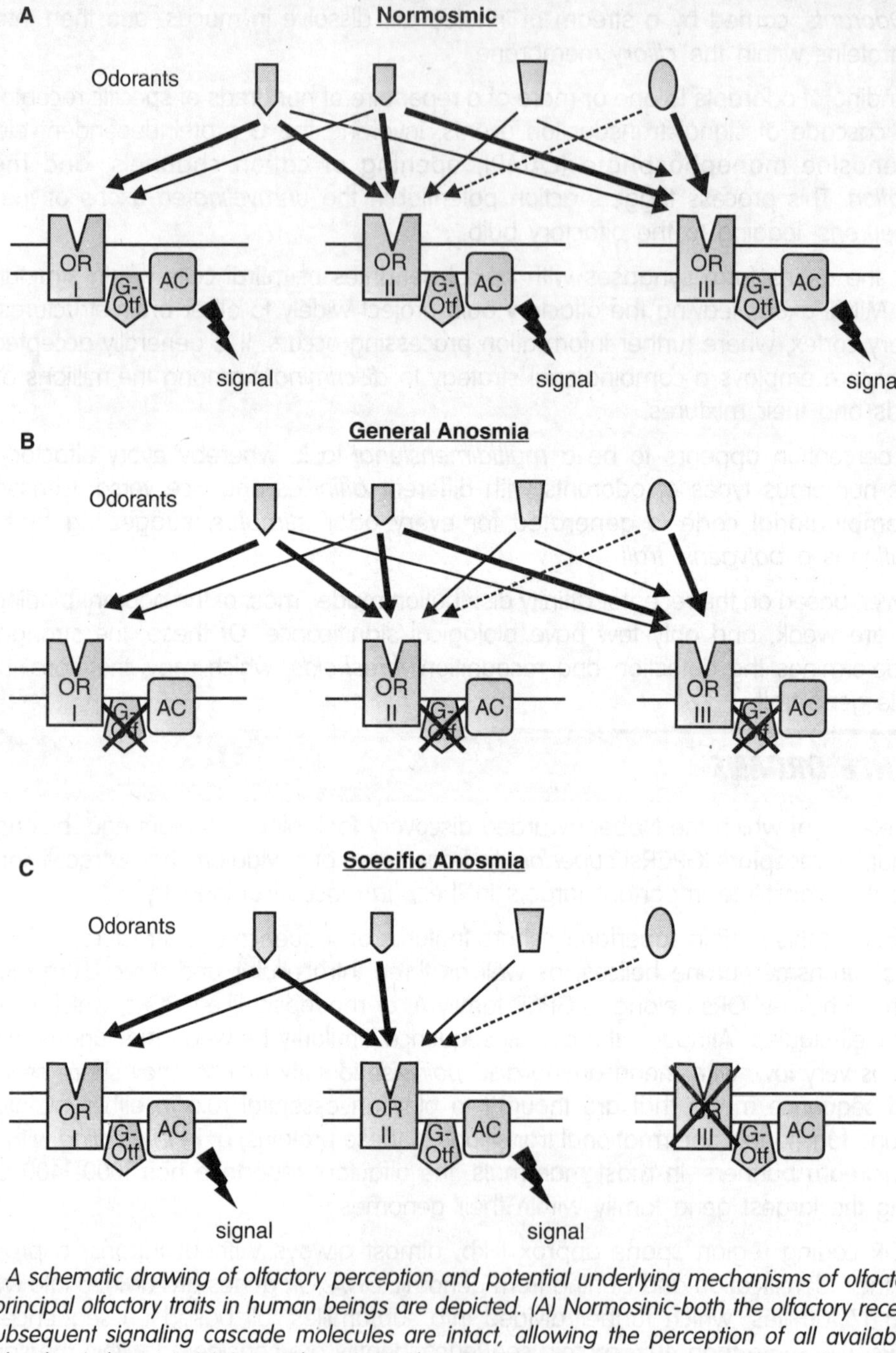

Figure 10.1: A schematic drawing of olfactory perception and potential underlying mechanisms of olfactory deficits. The three principal olfactory traits in human beings are depicted. (A) Normosinic-both the olfactory receptors (ORs) and the subsequent signaling cascade molecules are intact, allowing the perception of all available odorous volatile molecules (odorants). (B) General anosmia-damaging mutation in one or more of the olfactory signal transduction proteins would extinguish the olfactory signal stemming from all OR types. (C) Specific anosmia—inactivation of one OR would eliminate the response or significant decrease the threshold toward one or a few particular odorants. Such specific OR inactivation would likely not affect perception of other odorants.

fully confirm this assumption. Vertebrate OR genes are organized in genomic clusters and are distributed on almost all chromosomes. As an example, in humans they are absent only on chromosomes 20 and Y. This wide genomic distribution is believed to have evolved from a single OR gene cluster on human chromosome 11 (11 at 4.96) that contains only OR genes of class I.

An intrachromosomal duplication may have originated the first class II OR gene cluster. This cluster is inferred to have further duplicated to the q-telomeric region of human chromosome I and from there to have expanded to many other genomic loci. This evolutionary process of OR genes migration was favoured by strong selective pressure toward *expanding* and *diversifying* the OR gene repertoire, as increased OR count likely enhances both sensitivity and selectivity.

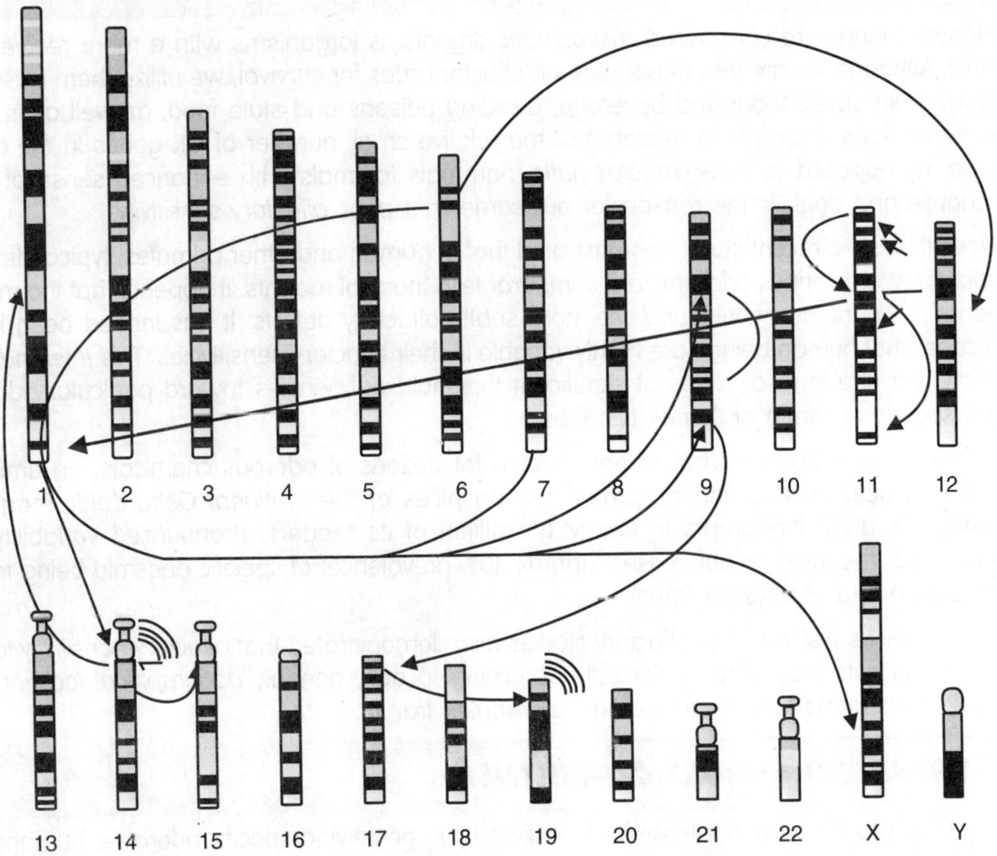

Figure 10.2: A tentative schematic view of olfactory receptor (OR) genes genomic expansion. The proposed steps, in rough chronological order, are: (1) a duplication on chromosome II that resulted in the formation of the first class II cluster out of an original class I cluster (thick cyan). (2) A duplication that led to the formation of a cluster on chromosome 1, most likely in the framework of a whole genomc diploidization (thick magenta). (3) Internal duplications within chromosome II (thin green), and expansion from chromosome 1 onto a number of other chromosomes (thick green). (4) Additional isolated duplications (thin black), gene scattering of family 4 (red "radio waves") and family 7 (blue "radio waves"). Only the generation of clusters larger than five members is explicitly shown. The small red histograms to the right of the chromosomal bands indicate OR gene cluster locations, with the area proportional to cluster size.

Interestingly, this repertoire augmentation process appears to have been reversed in primates in the last 20 million years, probably because they became less dependent on olfactory cues. This is manifested in the observation that in such species OR genes underwent a massive accumulation of *pseudogenizing* mutations, generating in-frame stop codons. This process of OR gene loss has remarkably accelerated in the human lineage leaving less than half of the OR genes intact.

The high prevalence of defective human OR coding regions is a wide evolutionary deterioration, and its phenotypic impact awaits elucidation. Each such *pseudogene* may be regarded as a natural *knockout,* potentially affecting the human ability to detect and *discriminate odorants.*

HUMAN OLFACTORY ANOMALIES

Human beings are considered *microsmatic* organisms (organisms with a more *feeble* sense of smell). Although we are less dependent on olfactory cues for survival, we utilize them extensively in enjoying perfumes, food, and beverage, avoiding poisons and stale food, as well as in subtle social interactions. It stands to reason that the relative small number of OR genes in the human genome, as opposed to other *macrosmatic* mammals (animals with enhanced sense of smell, e.g., mouse and dog), is the reason for our somewhat poor *olfactory* sensitivity.

Nevertheless, a recent study demonstrated that in human and other primates, typical detection *thresholds* toward various odorants are *comparable* to those of rodents. It appears that the massive OR gene loss in human could *underlie* more subtle olfactory deficits. It has indeed been known for *decades* that human beings are highly variable in their olfactory sensitivities. This *inter-individual* variability is in the form of cases of significant threshold deficiencies toward particular odorants, termed specific anosmia or "*smell blindness*".

Such human deficiencies have been studied for dozens of odorous chemicals. An ambitious study in this field was carried out under the auspices of the *National Geographic* magazine, examining six distinct odorants in nearly 1.5 million of its readers. Pronounced variability was observed, with the most notable cases (approx 30% prevalence) of specific anosmia being toward *Androstenone* and *Galaxolide* (musk).

This survey as well as other olfactory studies also demonstrated that olfactory *sensitivity* toward particular odorants may vary significantly according to age, gender, *geographical location,* and various *environmental* factors suggesting a complex trait.

VARIABLE OLFACTORY CAPABILITIES

In addition to the human inter-individual diversity in perceiving specific odorants, humans also vary in their general olfactory capabilities. This covers a wide range of phenomena, from general *anosmia* through hyposmia (diminished sensitivity to smell).

At the other end of the scale is general hyperosmia (enhanced smell sensitivity). Moreover, during aging. a general olfactory loss occurs, which is also a feature of several neurodegenerative diseases such as *Alzheimer's* and *Parkinson's* diseases. It is estimated that approx 1% of the Western world population suffers from *chemosensory* disorders. Most of the people suffering from smell disorders have an acquired condition, which develops during life, owing to allergy, viral upper respiratory tract infection, nasal *sinus* diseases, *head trauma, inhalation* of *noxious* chemicals, or *medicinal* drug intake. A much smaller minority is born without a sense of smell, an affliction referred to here as congenital general anosmia (CGA). Two broad categories of CGA can be considered: CGA occurring with other *anomalies* (syndromic) and CGA seen as an isolated condition.

Prevalence for isolated CGA is roughly estimated to be between 1:5000 and 1:20,000. Several past reports have described the condition, including a few familial cases. One of them shows 22 patients with 8 being familial. On *examination* by biopsy of the olfactory region in several anosmic patients, their *respiratory epithelium* was found to be normal, but not their olfactory epithelium, some of them lacking it totally.

In all cases, *axonal abnormalities* and the absence of mature olfactory sensory neurons were observed. It has been proposed that the olfactory epithelium may degenerate because of functional failure in olfaction. Another study found a large four-generation family with 27 CGA-affected individuals, and recently, two Iranian families with isolated CGA have been reported with a suggested 46-cm linkage interval on chromosome 18.

A parallel study has collected 27 individuals in 10 families and 56 sporadic cases of CGA. The three major candidate genes *(ADCY3, GNCA2, GNAL)* were screened in these individuals, but no causative mutations for CGA could be found. Although the number of affected families and individuals is small, all familial cases described are consistent with an autosomal dominant mode of inheritance with partial penetrance. The most well studied group of syndromic CGA is related to the Kallmann syndrome. Such patients exhibit hypogonadotropic hypogonadism and anosmia, secondary to failure of gonadotropin-releasing hormone (GnRH)-producing neurons to migrate from the olfactory placode to the brain, and to agenesis of the olfactory bulbs.

The prevalence of the disease has been estimated at 1 in 10,000 in males and five to seven times lower in females. Three different modes of inheritance have been reported in familial cases of Kallmann Syndrome, X chromosome-linked, autosomal dominant, and autosomal recessive. The X-linked form of the Kallmann Syndrome has been well characterized, being caused by mutations in the gene *KAL I* (chromosome Xp22.3).

The *KAI1* protein, anosmin-1, is a locally restricted component of basement membranes and/ or extracellular matrices during the organogenesis period. It has been found to enhance axonal branching from olfactory bulb output neurons and to affect the migratory activity of GnRH-producing neurons. To date, no causative genes have been described for isolated human CGA. However in mouse, three transduction genes have revealed *behavioural* phenotypes consistent with general anosmia when they are inactivated.

Mice generated by homologous recombination, lacking the functional olfactory cyclic nucleotide-gated channel (Cnga2), the stimulatory olfactory G-protein (Gnal), or enzyme adenylyl cyclase III (Adcy3), display profound reductions or even absence of physiological responses to odorants. Most of the *homozygously* deficient mice die within a few days after birth owing to an apparent inability to locate their mother's nipple and suckle. *Congenital* and *progressive* blindness have been studied much more extensively. Inherited forms of blindness can be caused by anomalies in the central nervous system or in different parts of the eye: cornea, iris, lens, retina, and the optic nerve.

About 140 genomic loci have been associated with retinal dystrophies and more than 60 genes have been identified. The partial similarity in the transduction processes between olfaction and vision may facilitate the future identification of some of the causative genes of CGA and its mode of genetic transmission.

LOSS OF SPECIFIC SENSE OF SMELL

Specific anosmia closs of sense of smell is defined as incapacity of an individual with an otherwise normal smell sensitivity to detect particular odorants. Specific anosmia is seldom absolute, and most often a person has a 10- to 100-fold diminished sensitivity to a given odorant, hence, the more exact term is specific *hyposmia*.

In contrast, specific hyperosmia, is described as enhanced sensitivity toward a specific *odorant*. Both hyper- and *hyposensitivity* toward specific odorants are likely driven by the same molecular mechanism, yet despite the wide variety of reports of specific *anosmia*, no comparable evidence has been reported for specific hyperosmia. Considerable evidence indicates that specific anosmia is genetically determined.

For example, various studies showed that anosmia to Androstenone (16-Androsten-3-one), a steroid of gonadal origin that serves as a boar pheromone, is highly concordant in monozygotic twins. Whissell-Buechy et al. demonstrated that anosmia to the odorant Pentadecalactone behaves as a recessive trait in human beings. A similar *Mendelian* recessive inheritance was observed in mice with specific deficiency to detect *Isovaleric acid*.

A subsequent linkage analysis study has associated this phenotype with two distinct genornic loci, on mouse chromosomes 4 and 6. Interestingly, a small fraction in the human population is also incapable to specifically detect Isovaleric acid. A comparable linkage analysis in human, potentially focusing on the human *syntenic* regions would be advisable. Although specific anosmia has a strong genetic element, other confounding factors contribute to the overall olfactory variability of a human individual. For example, olfactory faculties are age dependent, reaching a maximum in the late *teenage* and then declining gradually.

This deterioration accelerates and becomes significant during the sixth decade. Similarly olfactory sensitivity may be influenced by the gender, whereby women have been reported to perform better than men in certain olfactory tests. Other *environmental* and behavioural factors were also suggested to affect olfactory performance, whereby subjects allegedly anosmic to *Androstenone* became capable of smelling it following repeated exposure.

This phenomenon of "*olfactory plasticity*" could explain part of the variability in *Androstenone perception*, and in general, incomplete penetrance has to be taken into account for this odorant and its functional homologs. Overall, this evidence illustrates that human response to odorous molecules is a complex trait, regulated by genetic, developmental, and environmental factors. The molecular basis of "*odor blindness*" (*specific anosmia*) is believed to be analogous to colour blindness and specific taste deficits.

Colour vision is controlled by three genes encoding cone opsin receptors for long (red) and medium (green) wavelength (on chromosome X) and for short (blue) wavelengths (autosomal). Mutations that inactivate one of these genes abolishes significantly our colour *discrimination* ability and resulting in a different form of colour blindness. A similar *phenomenon* was recently observed in the more complex system of taste perception.

A member of the taste receptors class II (T2R38) gene was associated with the *gustatory* capability to detect *phenylthiocarbamide* (PTC). Like many cases of specific anosmia, the ability to taste the bitter taste of PTC was subject to various psychophysical studies that indicated a *Mendelian* recessive trait with possible genetic and environmental confounding factors. Indeed, the two haplotypes of the *T2R38* gene that segregated significantly between tasters and the nontasters of PTC could not explain the entire diversity in the study.

Thus, it was suggested that PTC detection is a complex trait with a major quantitative trait locus, with relatively high phenotypic effect. In this aspect, smell perception is more closely related to gustation than to colour vision, because it also employs a group of *chemoreceptors* to distinguish between large varieties of chemicals. Still, as all these three sensory pathways utilize the same operating principles that are based on GPCRs it may be suggested that their *inter-individual phenotypic* variability is driven by broadly similar genetic mechanisms.

HEREDITARY MECHANISM

The potential underlying is hereditory/ genetic mechanisms of human olfactory deficits may be deduced from the biochemical pathways of olfactory perception. The binding of an odorant to an OR initiates a cascade of events leading to olfactory *signaling*, and damage in different steps of the transduction chain may lead to a deficient phenotype. The first group of candidate genes includes genes that participating in the olfactory signal *transduction* and its termination.

Among these, the three main components of the olfactory transduction chain are the human orthologs of the mouse transduction genes previously described. Olfactory marker protein (OMP), a wellestablished marker of olfactory tissues is another candidate gene, as it has been also shown to modulate the *kinetics* of olfactory *electrophysiological* responses. In addition, there are two kinds of *phosphodiesterases* (PDEs): calmodulin-PDE (PDEI C) and cAMP-PDE (PDE4E), and various protein kinases that were implicated in the termination of the olfactory signal transduction.

A second group of candidate genes for CGA are biotransformation enzymes proposed to play a role in the post signaling processing of the odorant molecules themselves, thus eliminating them from the vicinity of ORs. These include cytochrome P-450 (CYP2G1) and uridine diphospho (UDP) glucoronosyl transferase (UGT2A1). The third group consists of transcription factors that are crucial for the development and functionality of the olfactory system.

These proteins are members of the Olf/Early B-cell factor and nuclear factor I transcription factor families, which mainly expressed in postmitotic olfactory neurons and control the expression of several genes *(ADCY3, CNGA2,* and *OMP)*. Finally, odorant-binding proteins *(OBP2A)*, lipocalins that may mediate the binding of odorants to the olfactory receptors or prevent saturation of olfactory receptors by excessive odorant concentration, have a potential to underlie CGA.

It should be noted that some of these candidate genes are functionally expressed in other non-olfactory tissues, and mutations in these components may result in more compound disorders or syndromic anosmia. In contrast to the diverse genetic mechanisms that could underlie CGA, the main obvious candidate genetic determinants to be associated with specific anosmia are OR genes.

Moreover, odorant-specific threshold deficits are significantly more prevalent than general anosmia and, hence, are likely caused by more frequent genetic variations. Single nucleotide polymorphisms (SNPs) sites at which two alternative bases occur at appreciable frequency, are the most common genetic variation in the human genome.

Consequently, they are believed to constitute the genetic component of most *multifactorial* human traits. In the case of specific anosmia, SNPs that damage a particular OR function could lead to significant threshold sensitivity differences toward a particular odorant. This could be rationalized by the "*threshold hypothesis*". According to this premise, the highest-affinity receptor toward a certain odorant is the one that determines the odorant's threshold sensitivity.

If this OR is being inactivated, the threshold will be defined by the next highest-affinity receptor. In the case of a larger OR repertoire, as in rodents, a higher level of functional redundancy will obtain, and the loss of the highest-affinity *receptor* will have a diminished probability to generate a recognizable olfactory deficit. In contrast, in humans, where the OR repertoire has diminished *significantly*, affinity values would tend to be more widely spaced, and *threshold* variations could become more *prevalent*.

In these cases, an inactivating polymorphism in an OR encodes the best receptor for a certain odorant, is expected to cause significant threshold difference between individuals who carry the functional OR to those in whom it is deleted. Various types of SNPs in OR genes might underlie *odorant-specific* olfactory deficits. The most obvious ones are the approx 600 nonsynonymous SNPs that may change a crucial residue to the protein function. For example, changing the arginine (R) in the highly conserved MAYDRY motif that is believed to participate in the coupling of the G protein to the receptor was demonstrated to terminate the protein function.

Despite the relatively high conservation of this residue, it has been found to display polymorphisms in 15 different OR genes, which is significantly more than the average of the polymorphism count per residue along the protein sequence. Other types of candidate polymorphisms are those that change amino acids in the complementarity-determining region for odorant recognition in a functional OR.

These might not *inactivate* the receptor but rather change the affinity spectrum of its corresponding odorants. In addition, polymorphisms in the *promoter* or other regulatory regions of OR genes might also cause OR *inactivation*. The present partial knowledge about the genomic disposition and control mechanisms of ORs renders the detection of such important deleterious polymorphisms less straightforward. The most promising candidates for underlying odorant-specific olfactory deficits are recently discovered SNPs of a highly unusual *disposition*.

These generate a premature stop codon in the OR gene sequence and consequently segregate between an intact gene and an inactive one (pseudogene). Several *dozens* such *segregating pseudogenes* have been discovered in the human OR subgenome. These define a remarkable genetic variability, whereby almost every human being possesses a unique assortment of intact and disrupted ORs. Furthermore. significant differences were observed between various ethnic groups in the degree to which certain *polymorphic* pseudogenes are conserved in their intact forms.

Although these ethnic differences in functional OR repertoires could be explained in terms of geographic isolation and *bottleneck* events, an alternative evolutionary mechanism is related selection, suggesting that different intact ORs tend to be conserved more effectively depending on geography or lifestyle. These OR segregating pseudogenes introduce a remarkable genetic diversity to the human genome, whose phenotypic correlates awaits elucidation. Another type of potential causative genetic polymorphisms in human OR genes are the observed variations in copy number of several OR genes, mainly on chromosomal telomeric regions.

These polymorphisms are believed to have emerged via meiotic nonallelic homologous recombination (NAHR) events that are associated with genomic segmental duplication in the human genome. In rare meioses, these segmental duplications are mistaken for *allelic* sequences, with the result that chromosomes are incorrectly spliced.

Consequently, NAHR could account for various types of *genomic* rearrangement in OR clusters, such as deletions, duplication, and inversions. Similarly, it may account for gene conversion events that may underlie disruptive mutations, by introducing nonfunctional sequences from a pseudogene into a functional gene. Although ORs are the best candidates, the potential involvement of nonreceptor genes in specific anosmia cannot be ruled out. A well-known example is the involvement of Acj6, a POU-domain transcription factor, in specific anosmia to a subset of odorants in the fruit fly *Drosophila melanogaster.*

SIGNIFICANCE OF OLFACTION

The significance of olfaction to human's life quality is often underestimated. This sensory pathway plays a key role in food and beverage recognition and enjoyment, it enables people to avoid dangers such as smoke, spoiled food, and poison, and it affects human social interactions. The realization of the underlying genetic mechanisms of human olfactory diversity would open new opportunities to both the food and fragrance industry.

For example, human panels for the testing of new products would be selected such that their genetic profiles will fit those of the targeted consumers in an optimal way. Similarly, products could be directed to certain segments of the population according to their predictable olfactory capabilities. Alternatively, specific anosmia or hyperosmia could be *artificially* induced to eliminate or enhance specific odors.

This could be done both in the genomic level where specific OR gene would be silenced or overexpressed, and at the protein level, using agonists and antagonist. Thus, much of the future of olfactory research and development might depend on deciphering the genetic basis of human-specific deficiencies. *Pseudogenes,* nonfunctional copies of genes, have been considered always as molecular relics with no effect on human traits.

Therefore, they have been attributed always to the neglected genomic majority of non-coding "*junk*" DNA. With the completion of the human genome sequencing, it became clear that pseudogenes are comparably distributed in our genome as coding genes. Consequently, there is a higher interest in improving the *pseudogene* annotation in the human genome and in studying their functional roles.

In this realm, the special evolutionary state of the human OR gene family where many members exist at the border between functional genes and nonfunctional pseudogenes provides an unusual opportunity to explore the effect of pseudogene accumulation to human fitness. Finding the phenotypic correlates of segregating pseudogenes in the human olfactory system could shed new light on the function of pseudogenes in the human genome. Finally, in the post-genome era, considerable efforts are devoted to decipher the genetic basis of human multi-factorial traits, attempting to correlate between phenotypes and genotypes.

In this realm, genetic research of olfactory deficits could be an excellent model system for understanding the genetics of human variation in gene families in general. As in olfaction, there are many biological systems where genes families subserve a particular functional role in identification and processing various ligand libraries. We propose, therefore, that a better understanding, through genetic studies, of odorant-OR relations would significantly help to form a better picture of similar complex systems. .

CONCLUDING REMARKS

The sense of smell, a discriminator of millions of volatile compounds, is crucial for survival and well being. Mutations in the olfactory G protein transduction pathway result in complete olfactory deficiency in mouse. Owing to the rarity of the human counterpart deficit, CGA, only a few affected families have been described, and no causative mutations have been identified as of yet.

In contrast, genetic polymorphisms in OR genes are prevalent and amply documented. Some of these genetic variants exchange between a functional and nonfunctional forms of the coded OR protein (*segregating pseudogenes*), whereas others segregate between gene existence and absence.

These genetic polymorphisms are promising candidate to underlie the abundant human olfactory threshold variations-the specific anosmia phenotypes. Future studies of human olfactory deficits would shed further light on the molecular mechanism of odor perception, as well as on the genetics of analogous biological systems.

Chapter 11

MicroRNAs in Cancer Cells

MicroRNAs (miRNAs) are *noncoding*, single-stranded RNAs of 21-25 nucleotides that negatively regulate the gene expression. They *down-regulate* protein expression by either blocking translation of proteins from mRNA or mediating the degradation of target mRNAs. These tiny RNAs are encoded by hundreds of genes as long primary transcripts termed *"pri-miRNAs,"* with hairpin structures that are processed in the nucleus by the double-strand-specific ribonuclease III Drosha into miRNA *precursors* of 60-80 nucleotides.

These precursors are transported to the *cytoplasm* by Exportin 5 and further processed by the *enzyme Dicer* into partially double-stranded RNAs in which one strand is the mature miRNA. Subsequently, mature miRNAs are incorporated into the RNA-induced silencing complex (RISC), leading to RNA interference.

In *mammalian cells*, the bound mRNA typically remains intact but is not translated, resulting in reduced expression of the corresponding gene. Presently, 474 human miRNAs have been experimentally verified. (miR-Base, version 9.1, http://microrna.sanger.ac.uk). In addition, computational predictions have estimated the existence of at least 1,000 human miRNAs. Current estimates suggest that about one third of human mRNAs appear to be miRNA targets.

This makes the miRNA genes one of the most abundant classes of regulatory genes in mammals. The picture gets even more *complicated* when one miRNA may have several mRNA targets and several miRNAs may regulate a single target mRNA. Despite the success in identification of genes *encoding* miRNAs, knowledge about their function and specific mRNA targets is limited, and only a handful miRNAs have been fully characterized.

In mammals functions for specific miRNAs have been described in the processes of B-cell differentiation, adipocyte differentiation, *insulin secretion*, *embryonic brain* development, and *embryonic cardiogenesis*. Therefore, one major function is to control processes that underlie cell *proliferation* and differentiation in *diverse* organisms during normal development. The essential roles of miRNAs in vertebrate development have been strongly demonstrated by removing the *Dicer enzyme* in zebra fish and mice.

In both cases, removal of Dicer resulted in severe developmental defects and *embryonic* death. Further, it has been demonstrated that an individual miRNA can trigger large-scale changes in development. Moreover, some miRNAs are differentially expressed in stem cells, and recently it was found that miRNAs are required for stem cell division. It is speculated that *abnormalities* in miRNA expression might similarly contribute to the generation or maintenance of "cancer stem cells," tumor initiating cells.

One of the arguments connecting several miRNAs to cancer was the finding that more than half the Mapped miRNA genes are localized to fragile sites and genomic regions involved in cancer. And, the first evidence for direct involvement of miRNAs in cancer was the finding that miR-15 and miR-16 were deleted or underexpressed in most cases of *chronic lymphocytic leukemia.*

Reduced accumulation of specific miRNAs has further been reported for let-7 in lung cancer; miR26a and miR-99a in lung cancer cell lines; miR-143 and miR-145 in colon cancer; and miR-125b, miR-145, miR-21, and miR-155 in human breast cancer. Recently, a general down-regulation of miRNAs in *tumors* compared with normal tissues was observed.

An accumulation of specific miRNAs in cancer has also been observed: miR-155 in lymphomas and overexpression of seven miRNAs encoded by the gene cl3orf25 amplified in certain *lymphomas* nominates this miRNA cluster the first noncoding oncogene. Another example of a miRNA-oncogene overexpressed in brain tumor is the antiapoptotic factor miR-21. On the other hand, *down-regulation* of the RAS oncogene by let-7 implicates let-7 as a tumor suppressor in lung tissue.

Thus, depending on what genes the miRNAs regulate, the miRNAs may act either as *oncogenes* or *tumor suppressors*, Only recently, the analysis of the entire miRNAome has become possible by the development of *microarrays* containing all known human miRNAs. One important finding of such analysis is that specific miRNAs have altered *expression* in tumors compared to normal cells.

Furthermore, miRNA revealed distinct signatures in specific cancer forms, and enabled successful classification of poorly differentiated tumors that were *inaccurately* classified using mRNA profiles. Recognition of differentially expressed miRNAs may help to identify miRNAs involved in specific processes, such as normal differentiation or *carcinogenesis*, and establish their roles. Given the role of miRNAs in regulating gene expression, we have developed several strategies for profiling miRNA expression in human cells.

MICROARRAY TECHNOLOGY

kMicroarray technology is b sed on basic principles of nucleic acid *hybridization*, whereby fragmented NA is attached to a substrate then hybridized with a nown gene or fragment. Typically, *spotted oligos* or cDNA *microarrays* are *hybridized* with labeled cD A probes prepared from polyA+ RNA using standard methods for making cD A libraries. Subsequent to *hybridization*, the captured images are analyzed using software that quantifies the signal of each spot corresponding to individual capture probes.

The intensity of each spot is expected to be proportional to the amount of mRNA present within the analyzed samples. Since the first expression *microarray* report was published in 1995, the microarray *technology* has seen a number of *emerging* approaches. The analysis of miRNAs

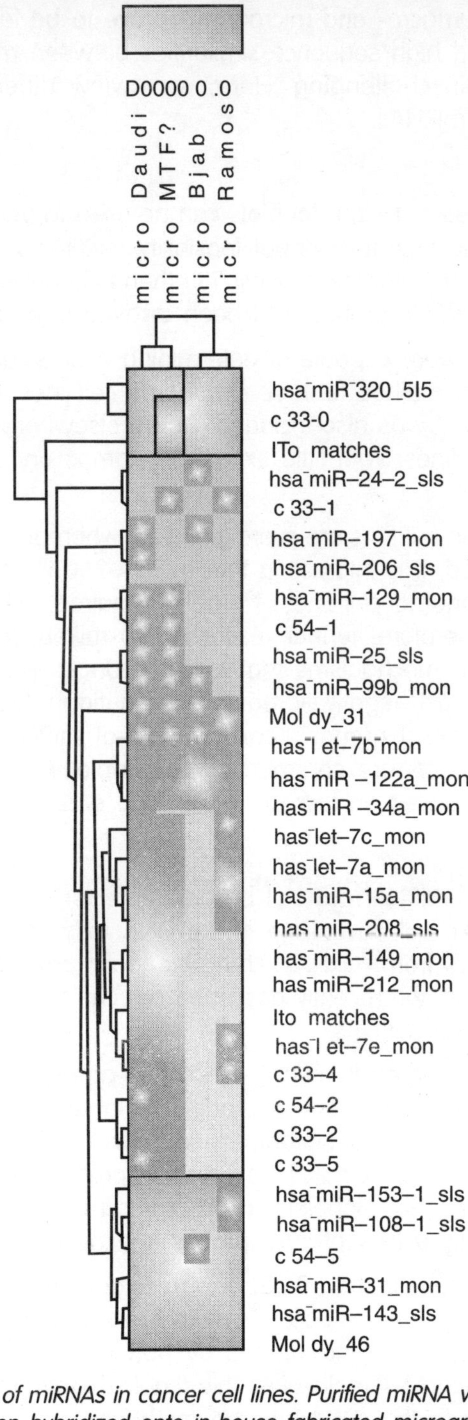

Figure 11.1: Expression profiles of miRNAs in cancer cell lines. Purified miRNA was internally labeled during reverse transcription with Cy5 and then hybridized onto in-house fabricated microarrays containing around 250 human miRNA sequences

expression by oligonucleotides macro- and microarrays prove to be feasible. However, because of the small size of miRNA and high sequence similarities between miRNA families, the miRNA microarray approaches remain challenging. Here, we review different arrays and labeling techniques used for detecting miRNAs.

Analysing Techniques

Macroarrays on nitrocellulose or Nylon dot blots can be used to analyze miRNA expression in mammalian cells. To assess the high-throughput feasibility miRNAs were gel purified from total RNAs, internally Cy5-labeled during reverse transcription then hybridized to a glass slide on which specifically designed multiple miRNA-probes had been arrayed in-house.

The fabricated microarrays were capable of determining expression patterns of at least 250 human miRNAs. We applied the method to three *lymphoma cell* lines (Daudi, BJAB, and Ramos). The breast cancer cell line MCF-7 was also included. Figure elsewhere in ths chapter shows the miRNA cluster of the four cells lines. With little exception, Ramos and BJAB cell lines showed a comparable expression profiles.

The data suggest that certain miRNAs are overexpressed, whereas others are down-regulated. Although further work is needed, it is interesting that miRNA-260 is up-regulated in MCF-7 cells but is down-regulated in lymphoma cell lines. Notably, impaired miRNA expression has been implicated in *tumorigenesis*. Therefore, further research into miRNA functions might lead to an advanced understanding of the mechanisms that lead to cancer. For example, a recent study showed that miR 15a and miR 16-1 negatively regulate the antiapoptotic gene BCL2, which often is *overexpressed* in many types of human cancers. Several miRNA oligo sets or preprinted microarrays and labeling *protocols* are commercially available. For example, mirVana miRNA Bioarrays miRCURY LNA Array, based on locked *nucleic acid oliconucleotides* from *Exiqon*.

Sensitive Probes

Generally, the small size of miRNA may not be adequate for microarray hybridization using direct *labeling* protocols. To take full advantage of the DNA *micro-arrays*, one would like to be able to prepare sensitive probes from miRNAs. Using standard procedures to prepare internally labeled cDNA probes from RNAs, we recently described a technique that allows the preparation of cDNA probes from gel-purified miRNAs.

In this procedure, miRNA recovered from 15% denaturing acrylamide gel slices were reversed transcribed using *hexamer* random primers to produce a labeled probe for analyzing miRNA expression between two cell *populations*. Notably, during cDNA synthesis miRNA can be tagged with *fluorescently* labeled molecules for miRNA expression profiling. In contrast to end labeling using T4 polynucleotide kinase, cDNA labeling incorporates multiple *radioactive* or *fluorescent bases*, and therefore the specific activity of such probes is expected to be high.

Since most miRNAs constitute less than 0.001 % of a total RNA sample, this type of *labeling* is very important for obtaining sensitive and accurate miRNA microarray data. As we have *previously* indicated, purified miRNAs can be 3' end-labeled by polyA polymerase, which adds several radioactive- or fluorescence-labeled ATP molecules. This simple method also provides high sensitivity comparable to that obtained with the cDNA-labeling methods.

In the case of the cDNA labeling, the microarrays should contain the processed miRNA

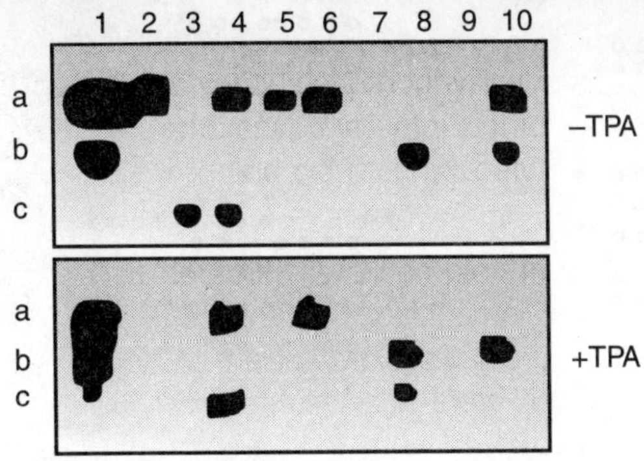

Figure 11.2: Oligonucleotides representing 30 arbitrary miRNA sequences were spotted onto Hybond-N+ nylon membranes and then hybridized with cDNA probes made from small miRNAs prepared from either untreated or TPA-treated HL60 cells. A differential expression profile between samples was found.

sequences, whereas in the case of the 3' end labeling the microarrays should contain the miRNA complementary strands. To reduce the *nonspecific cross-hybridization* of longer mRNA or ribosomal RNA, we purified miRNAs. However, Liu and colleagues *demonstrated* that total RNA can be used directly with no need for miRNA purification.

In this method, total RNA was reverse transcribed with *biotinylated* random primers and hybridized to *oligonucleotide* spotted arrays. Subsequently, miRNA expression was detected using *streptavidin-bound fluorophores*. This method of labeling is crucial for small biological materials. To further asses the feasibility to total RNA labeling, we have *investigated* the expression of miRNAs in biopsies from patients with B cell *lymphomas*.

In these experiments 5 µg of total RNA from each sample were internally Cy5-labeled during reserve transcription and then *hybridized* to the array. Figure elsewhere in this chapte shows an image of a microarray hybridized with Cy-5 cDNA probe prepared from 5 µg of lymph node total RNA. As can be seen, several miRNAs were differentially expressed in this tumor sample. Certain miRNAs are overexpressed (e.g., miRNA let7, miRNA449, miRNA130a, miRNA423), while others are not (e.g., miRNA226, miRNA107).

Therefore, it will be interesting to investigate the biological meaning of the expression data. However, when purified miRNAs from the same sample were used as template for probe preparations, only four miRNA sequences were detected. Similarly, a significant discrepancy was seen with other samples and total RNA prepared from cancer cell lines. In all cases, the number of positive signals obtained with probes prepared from total RNA was significantly higher than those obtained with purified miRNAs.

These observations indicate that the data obtained with total RNA must be interpreted with caution and validated with additional strategies such as Northern and PCR techniques. Some of the technical advances that we described for the first time, such as polyA tailing of miRNAs, are similar to those used by Ambion to label samples for miRNA array hybridization. Also, a polyA tailing protocol was implemented by several other *investigators*. For example, *polyadenylated* miRNAs were tagged with unique capture sequences labeled with either Cy3 or Cy5. After *ligation*, tagged miRNAs were purified and then hybridized to the miRNA microarrays.

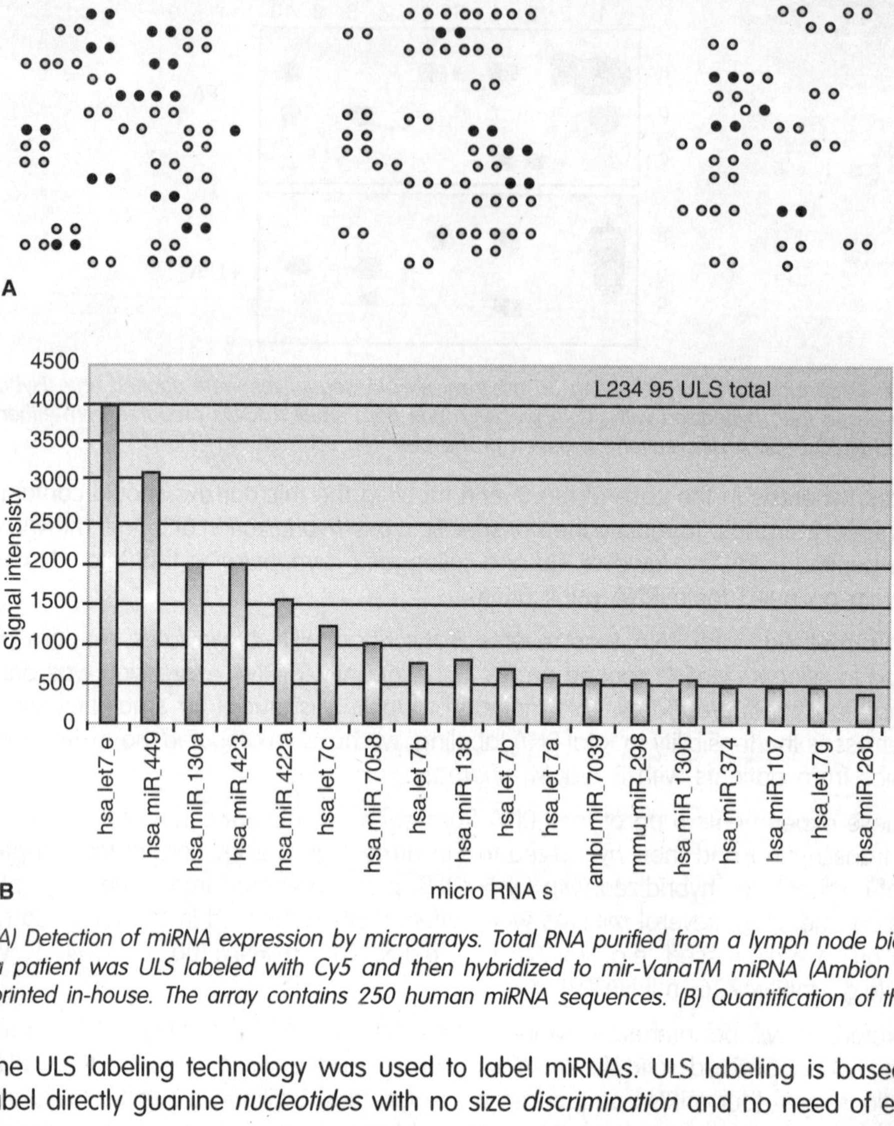

Figure 11.3: (A) Detection of miRNA expression by microarrays. Total RNA purified from a lymph node biopsy from a lymphoma patient was ULS labeled with Cy5 and then hybridized to mir-VanaTM miRNA (Ambion oligo set) microarray printed in-house. The array contains 250 human miRNA sequences. (B) Quantification of the signals shown in A

Also, the ULS labeling technology was used to label miRNAs. ULS labeling is based on an ability to label directly guanine *nucleotides* with no size *discrimination* and no need of enzymes.

Probic Expansion

Generally 50-200 µg of total RNAs is required to prepare miRNA probe by gelpurification for each microarray hybridization. This requirement for large amounts of starting materials severely limits the use of microarrays to profile miRNA gene expression in small specimens of human tissue, such as obtained by core needle or fine needle *aspiration biospsies.*

To overcome this problem, we developed a PCR-based amplification method that allows the analysis of miRNAs gene expression in small biopsies. The amplification method we used is diagrammed in Figure elsewhere in this chapter. In this procedure, purified miRNA were polyA tailed using *E. coli* polyA polymerase and the RNA primed by oligo-dT carrying on its 5' end a T7

RNA *polymerase* promoter sequence. The use of the reverse transcriptase Power Script during cDNA synthesis allows the addition of few *additional* nucleotides, primarily *deoxycytidines*, to the 3' end of the cDNA.

The 3'-tailed first-strand cDNA then is primed with an SP6 RNA polymerase promoter sequence carrying GGG at its 3' end, leading to the generation of double-stranded cDNA. The resulting double-stranded DNA can be either PCR amplified for a desired number of cycles or in-vitro transcribed using T7 (or SP6) RNA polymerase, resulting in linear *amplification* of miRNAs. The protocol is expected to generate large amounts of miRNA starting from few ng of miRNAs.

Figure elsewhere in this chapter shows the hybridization signals using an internally labeled 32p PCR probe, which was prepared from miRNA that had undergone 25 rounds of amplification. In these experiments, the probe was hybridized to *dot-blots* containing 160 human miRNA sequences.

We obtained reproducible results between different samples and no indication of a bias due to the amplification step. Moreover, a hybridization of fluorescently labeled PCR products to a glass slide on which specifically designed miRNA-specific probes have been arrayed in-house revealed satisfying results, thus enabling the use of *microarray* technology for qualitative and *quantitative* studies of miRNAs *expression* in small number of cells.

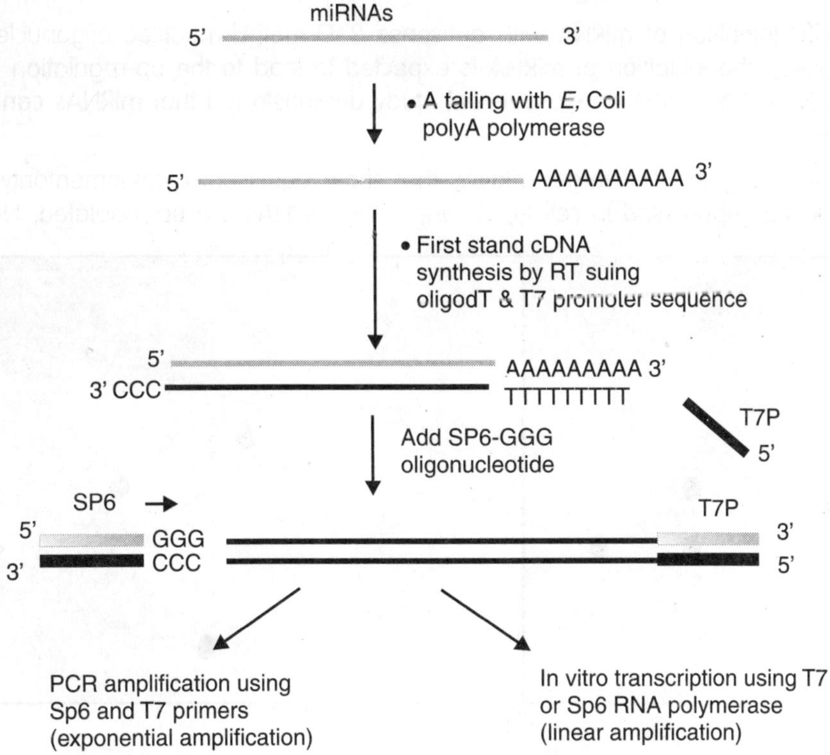

Figure 11.4: Schematic illustration of miRNA amplification.

RECOGNIZING THE SPECIFIC SEGMENTS

Several groups developed computational algorithms designed to predict the target genes of miRNAs. In general, the basis for these *prediction* programs is the degree of sequence *complementarity* between a miRNA and a target mRNA gene. In animal cells, single-stranded miRNAs bind to specific target tnRNAs through partly complementary sequences that are predominantly in the 3' untranslated region with a cDNA probe prepared from PCR-amplified miRNAs prepared from the breast cancer cell line MCF-7.

The result is reproducible for different miRNA preparations (3'UTR). Computational approaches to find target *transcripts* have been hindered by this imperfect *complementarity*. In addition, many known target genes of miRNAs contain several miRNA binding sites and several genes are predicted to be regulated by a single miRNAs.

These observations further increase the *complexity* of target prediction using bioinformatics. Therefore, experimental validation is required for miRNA target genes. To date, different strategies have been used to validate miRNA target genes. Most commonly used are luciferase or GFP *fusion* constructs containing the miRNA target site. In cotransfection experiments, active miRNAs are expected to reduce luciferase activity or GFP expression. Notably, the same strategy was used to identify effective siRNAs in human cells.

Also, specific inhibition of miRNA with antisense 2'-O-methyl-modified oligonucleotides was used. In this assay, the inhibition of miRNA is expected to lead to the up-regulation of its target gene. In addition to translation arrest, a recent study demonstrated that miRNAs can reduce the transcript levels.

These findings suggest that target mRNAs that show significant complementarity to miRNAs are expected to be suppressed in cell types where the miRNAs are up-regulated. Northern blot

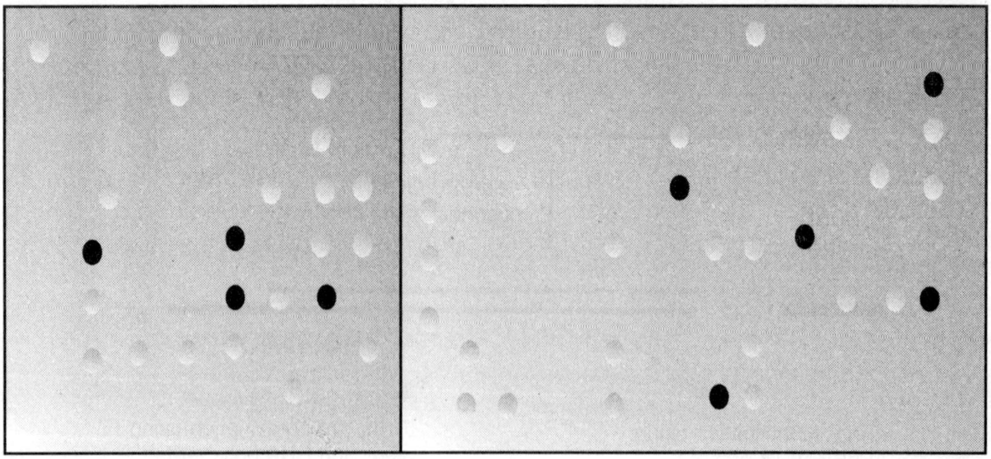

(A) (B)

Figure 11.5: Amplified miRNAs produce significant hybridization signals. Oligonucleotides representing 160 miRNA sequences were spotted onto Hybond-N+ nylon membranes then hybridized

analysis or RT-PCR could be used to measure the expression levels of candidate mRNAs. In addition, the use of specific antibodies to measure the protein levels of candidate genes in cells with high and low expression of the miRNA is recommended.

By transfecting cells with candidate miRNAs, it is possible to see whether the overexpression of specific miRNAs affect the expression of the identified target genes in in vitro and in vivo. However, a more global approach to find target mRNAs is to use proteomics to uncover target proteins that are affected as a consequence of miRNA expression in target cells.

This strategy works only when the cell populations compared are identical except for the artificial increase of miRNA in one. Proteins that are altered as a probable consequence of higher miRNA expression may be purified from gels and *subsequently* sequenced.

Chapter 12

Williams Syndrome

Williams syndrome also called Williams-Beuren *syndrome* WBS; is a *multisystem* developmental disorder that is almost always associated with an approx 1.5-Mb deletion of *chromosome* 7q11.23 (OMIM no. 194050). The deletion was identified in 1993 based on the *observation* of phenotypic overlap with supravalvular aortic stenosis (SVAS), a distinct *autosomal* dominant disorder affecting the *cardiovascular* system.

It has since been shown that SVAS arises because of the disruption of one copy of the *elastin gene*, through either deletion, *translocation* or point mutation, but the genes *contributing* to the remaining aspects of WBS have not yet been definitively determined. The 7q11.23 *chromosomal* region that is commonly deleted in WBS contains more than 25 genes and, as might be expected, is flanked by *chromosome-specific* low-copy repeats (LCRs).

The nature of these repeats are such that in addition to deletions (and presumably the *reciprocal duplications*), inversions of the region are also mediated, one of which exists as a *polymorphism* that seems to *predispose* to the deletion.

The increasing number of genomic rearrangements being discovered to be associated with this region are, therefore, not only expanding the *repertoire* of molecular *diagnostic* tests for WBS and its associated phenotypes, but also impacting on *genetic* counseling for this disorder. To facilitate future molecular genetic studies involving the WBS locus, we are actively *cataloging* chromosome rearrangements involving the WBS-LCRs and associated *phenotypes*.

PHARUMACEUTICAL STUDIES

The WBS phenotype is quite variable, as would be expected for a developmental disorder. but there are a number of core clinical features that are present in the vast majority of affected individuals. These include both *physical* and *neurological* elements that combine to produce a unique and therefore characteristic clinical picture. The frequency of WBS has been estimated at between 1 in 20,000 and 1 in 7500, with the second *statistic* likely a more accurate reflection of the population frequency because it was generated from a larger data set. Physical *manifestations* of WBS usually include involvement of the *cardiovascular* system, most often as a narrowing of the ascending aorta (SVAS) although a generalized *arteriopathy* can lead to vascular stenoses in other vessels, and *hypertension* is common in later life. Stellate irides, flat nasal bridge, short, up-turned nose with anteverted nostrils, long *philtrum*, full lips and lower cheeks, and a small chin are the recognizable *facial* features.

Other *symptoms* include *hernias, visual impairment, hypersensitivity* to sound, chronic otitis media, *malocclusion*, small or missing teeth, *renal anomalies, constipation, vomiting, growth deficiency, infantile hypercalcemia, musculoskeletal abnormalities*, and a *hoarse voice.*

As WBS individuals grow older they may also present with premature graying of the hair, *diabetes* and *impaired* glucose tolerance, decreased bone mineral density, sensorineural hearing loss, and a high frequency of *psychiatric symptoms*. Individuals with WBS usually have mild mental *retardation*, with an average intelligence quotient (IQ) of between 55 and 60, although there is a wide range of recorded values.

The most striking aspect of the WBS phenotype is the distinct behavioural profile, which encompasses a unique combination of both friendliness and anxiety. It is characterized by impaired cognition, hyperreactivity, sensory integration dysfunction, delayed expressive and receptive language skills, and multiple developmental motor disabilities affecting balance, strength, coordination, and motor planning.

In addition, approx 70% of WBS individuals suffer from attention deficit and hyperactivity disorder and there is a high incidence of anxiety and simple phobias. WBS was linked to chromosome 7q11.23 through the occurrence of a part of the phenotypic spectrum, SVAS, as an independent autosomal dominant disorder. SVAS was mapped to 7q11.23 through linkage analysis and the elastin gene *(ELN)* was then found to be disrupted in an individual with a t(6;7) chromosome translocation.

The vast majority of individuals presenting with the classic symptoms of WBS harbor a deletion of 7q11.23. Individuals without rearrangements of 7q11.23, or with rearrangements of other regions of the genome have been reported, but these are likely misdiagnoses or perhaps phenocopies of WBS.

SEQUENCE OF HUMAN CHROMOSOME 7Q11.23

The WBS deletion region is comprised of around 1.2 Mb of unique DNA sequence, including at least 24 protein-coding genes, flanked by large LCRs; one on the centromeric side and two on the telomeric side. These LCRs, which predispose the region to non-allelic homologous recombination (NAHR), are between 300 and 400 kb in size and are comprised of both transcribed and non-transcribed genes and pseudogenes.

The structure of each LCR is complex, but has been subdivided into segments of homologous sequence termed blocks A, B, and C. The centromeric, medial, and telomeric copies of each block can be distin guished from each other by single *nucleotide* differences, or *paralagous* sequence variants, although the sequence identity can be as high as 99.6% over large stretches of DNA. The individual blocks are made up of genes and sequences that are present in duplicated copies,

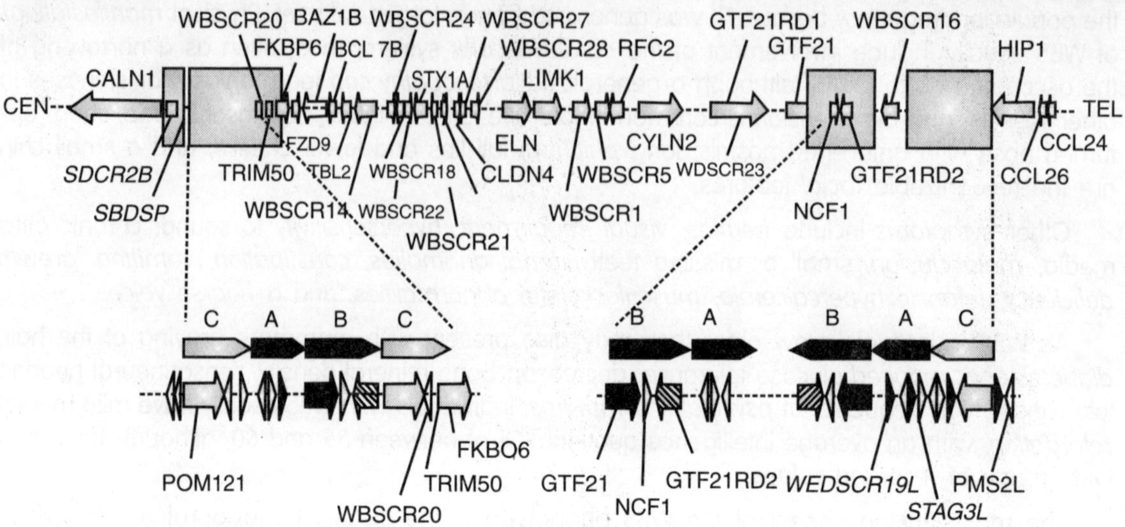

Figure 12.1: The Williams-Beuren syndrome (WBS) deletion region at 7q11.23.

both within the WBS region and at other sites on chromosome 7. These genes include general transcription factor 21 *(GTF21),* GTF21 repeat *domain-containing* protein 2 *(GTF2IRD2), neutrophil cytosolic factor* I (NCFI), stromal antigen 3 *(STAG3),* POM121 membrane glycoprotein *(POM121),* FK506-binding protein *6 (FKBP6),* Williams-Beuren syndrome chromosome region 20 *(WBSCR20),* tripartite motif-containing 50 *(TRIM50),* and a number of highly related *postmeiotic segregation* increased 2-like *(PMS2L)* genes.

Of these, functional copies of *WBSCR20, TRIM50,* and *FKBP6* are disrupted or deleted near the *centromeric breakpoint. GTF2I* is always disrupted or deleted at the telomeric breakpoint, with *NCFI* and *GTF2IRD2* variably deleted depending on the location of the distal deletion breakpoint. Three *STAG3* pseudogenes are present within the LCRs, but the functional copy lies distal to the WBS region at 7q22, along with two other pseudogenes.

Several POM121 transcription units are also present within the LCRs, but it is unknown which of these corresponds to the functional gene. Further complexity is added by the presence of fusion transcripts between POM121 and Zona *pellucida* sperm-binding protein 3 (ZP3), which maps close, but distal, to the WBS region and numerous other POM121-like sequences *scattered* throughout the genome. Intriguingly, POM121-like sequences are also found within the LCRs that flank the 22q11.2 *deletion* region.

Gene clusters of highly similar transcription units, the PMS2L genes are also contained within the LCRs. Because of the large number and high homology of these genes, it is still unknown exactly how many exist and which are actively transcribed, but many different *transcripts* have been *demonstrated. PMS2L* gene clusters are also present in two other locations on human chromosome 7; 7q11.22 and 7q22.

Syntenic Region in Mouse

The gene content and order of the WBS deletion region is conserved within the syntenic region of mouse chromosome 5, band G1. As with other chromosome regions involved in genomic disorders, the LCRs seen on human chromosome 7q11.23 are absent in the mouse, but *interestingly* their positions correspond to *evolutionary* inversion breakpoints.

This coincidence of LCRs and evolutionary chromosome breakpoints has also been reported in the Smith-Magenis syndrome region on human chromosome 17p11.2 and mouse chromosome

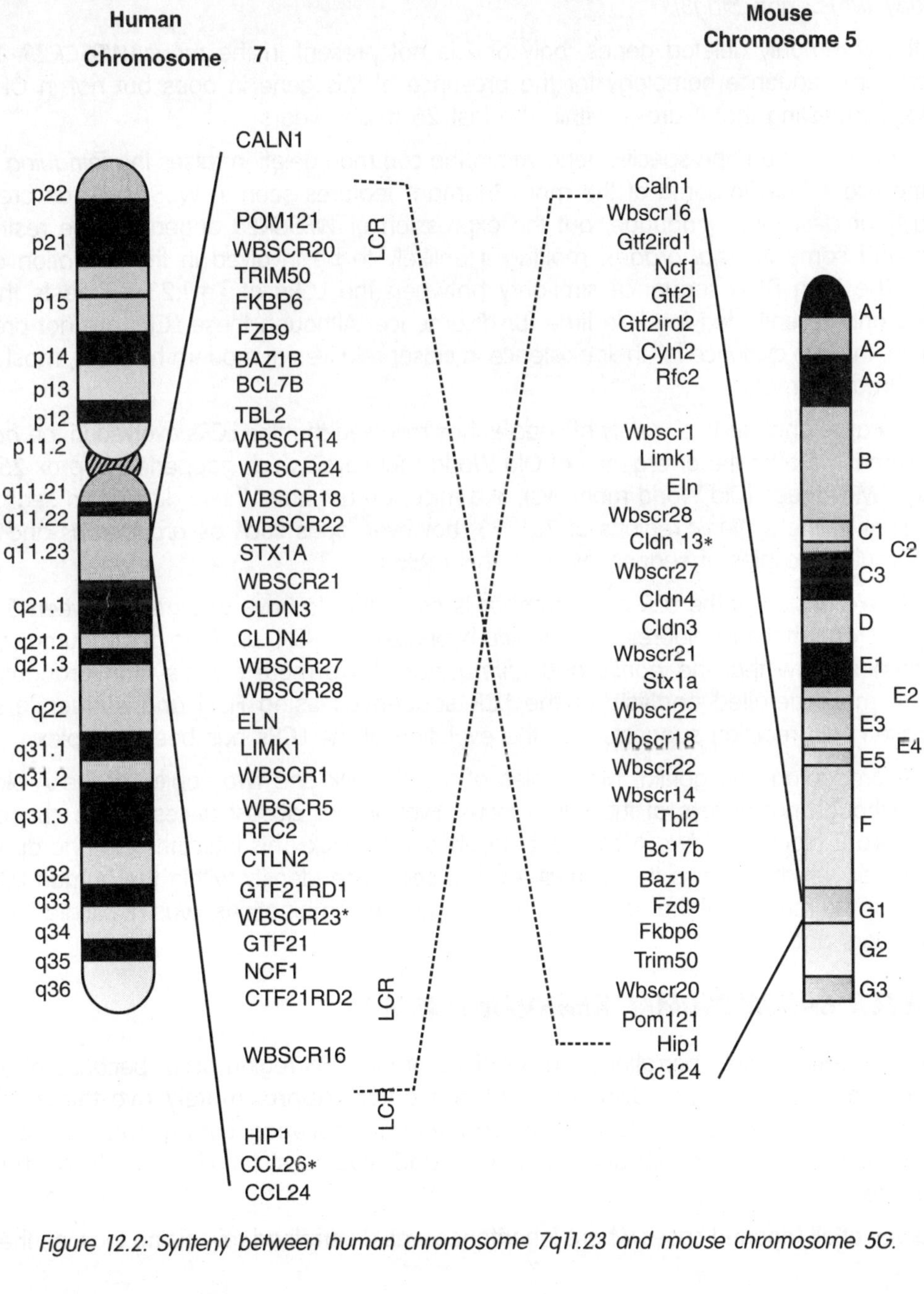

Figure 12.2: Synteny between human chromosome 7q11.23 and mouse chromosome 5G.

11, and in the Prader-Willi/*Angelman* syndrome region on human chromosome 15q1l-q13 and mouse chromosome 7. Because the LCRs are absent in the mouse, so are the multiple *psuedogenes*. Any genes or pseudogenes contained within the human LCRs are single copy in the mouse and located either within the WBS syntenic region (in the case of *Gtf2i, Gtf2ird2, Ncf1, Wbscr20, Trim50, Pom121,* and *Fkbp6)* or within the region of synteny with their *functional* human gene copy *(Pms2* and *Stag3).*

Of the commonly deleted genes, only one is not present in the mouse*WBSCR23*. There is evidence from sequence homology for the presence of this gene in apes but not in Old World monkeys, suggesting that it arose within the last 25 million years.

The presence of an ape-specific gene within the common deletion raises the *intriguing* thought that it may contribute to some of the more "human" features seen in WBS, such as gregarious personality or descriptive language, but the expression of *WBSCR23* appears to be restricted to the skin and some *internal organs*, making it unlikely to be involved in the alteration of brain function. The high DNA sequence similarity between the LCRs at 7q11.23 suggests that they appeared only recently, leaving little time for divergence. Although these LCRs are not present in the mouse, there is evidence for their existence in closer relatives of modern humans, most *notably* the nonhuman primates.

Comparative analysis in a variety of species has revealed that the LCRs evolved in the *hominoid* lineage sometime after the divergence of Old World monkeys, which happened approx 25 million years ago. Macaques (Old World monkeys), like mice, do not have these duplicated segments of DNA present at the *syntenic* regions of 7q11.23, however, apes such as *orangutans* and *gorillas* do show multiple copies of sequences from this region.

Further expansion of the LCR copy number is estimated to have occurred between 5 and 10 million years ago, because fluorescence *in situ* hybridization (FISH) analysis of nonhuman primate chromosomes show that the *gorilla* and *chimpanzee* have more signals than orangutan and gibbon. A more detailed analysis of the LCR sequences using FISH and sequence-specific *polymerase* chain reaction suggests that the evolution of the LCRs has been complex.

In the orangutan and gorilla, two copies of the C-block and two copies of the A-block are present, although not always at the same locus. Even in our closest ancestor, the chimpanzee, there are two C-blocks and two A-blocks, but only one B-block. This indicates that the duplication of the B-block, which contains the highest level of sequence identity within the human LCRs and is the site for the majority of the non-allelic homologous recombinations, was duplicated less than 5 million years ago.

INTER CHROMOSOMAL REARRANGEMENTS

Specific chromosomal aberrations the deletions of the WBS region occur because of unequal meiotic recombination in grandparental chromosomes. Approximately two-thirds of these *recombinations* are *interchromosomal* rearrangements between chromosome 7 homologs, whereas the remainder are intrachromosomal and involve sister chromatids of the same chromosome 7.

A survey of all reported cases where inheritance was determined reveals no bias in inheritance

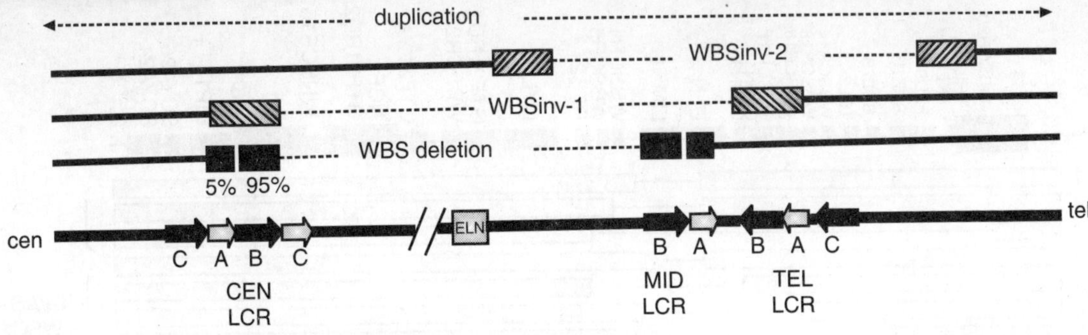

Figure 12.3: Genomic rearrangements of the Williams-Beuren syndrome (WBS) region. All documented, recurrent rearrangements of the WBS region are shown. The block components of the flanking low-copy repeats (LCRs) are shown, along with the relative position of ELN, *which lies at the center of the commonly deleted region. The regions to which the breakpoints have been localized are indicated by boxes.*

of the deletion, as it apparently occurs with equal frequency on *maternally* or *paternally* inherited chromosomes.

Usual Deletions

The polymorphic marker *D7S1870,* which lies within the proximal part of *GTF2I, was* initially shown to be deleted in every patient with a diagnosis of classic WBS. *D7S489B,* however, which lies near Frizzled 9 *(FZD9),* was shown to be present in all such patients. Subsequently, it has been shown that in approx 95% of WBS patients, the deletion occurs between highly homologous sequences within the *centromeric* and medial B-blocks, whereas in the remainder it occurs between sequences in the adjacent A-blocks.

These two block pairs are the only ones that are directly oriented with respect to each other. The predominance off recombination between the B-blocks, rather than the A-blocks is postulated to occur because of the higher sequence identity (99.6 vs 98.2%), more contiguous homology (there are two interstitial deletions of the medial A-block), and the shorter distance between the blocks (1.55 vs 1.84 Mb).

In addition, the active transcription of all three of the GTF2I genes within the testes and in *blastocysts* suggests that they are transcribed in germ cells, which could induce an open *chromatic* formation that is conducive to recombination. The exceptionally high identity between repeat-block sequences copies has made it extremely difficult to identify the exact deletion breakpoints in patients.

Site-specific nucleotide differences (SSNs) between the repeats, also referred to as paralagous sequence variants or cis-morphisms, have been utilized to study the relative copy number of different segments of the B-block, allowing the sites of recombination to be mapped. These elegant studies have revealed a 1.4-kb recombination hotspot within *GTF21,* which accounts for one-third of all deletions arising on a noninverted chromosome. Other breakpoints exist throughout the remainder of *GTF2I, NCF1,* and *GTF21RD2,* with additional hot spots associated with both paternally inherited and maternally inherited deletions.

This translates into a variable number of genes within the deletion; 26 if the recombination occurs in *GTF21, up* to 28 if it occurs within *GTF2IRD2* or the A-block. The inclusion of *NCF1* within

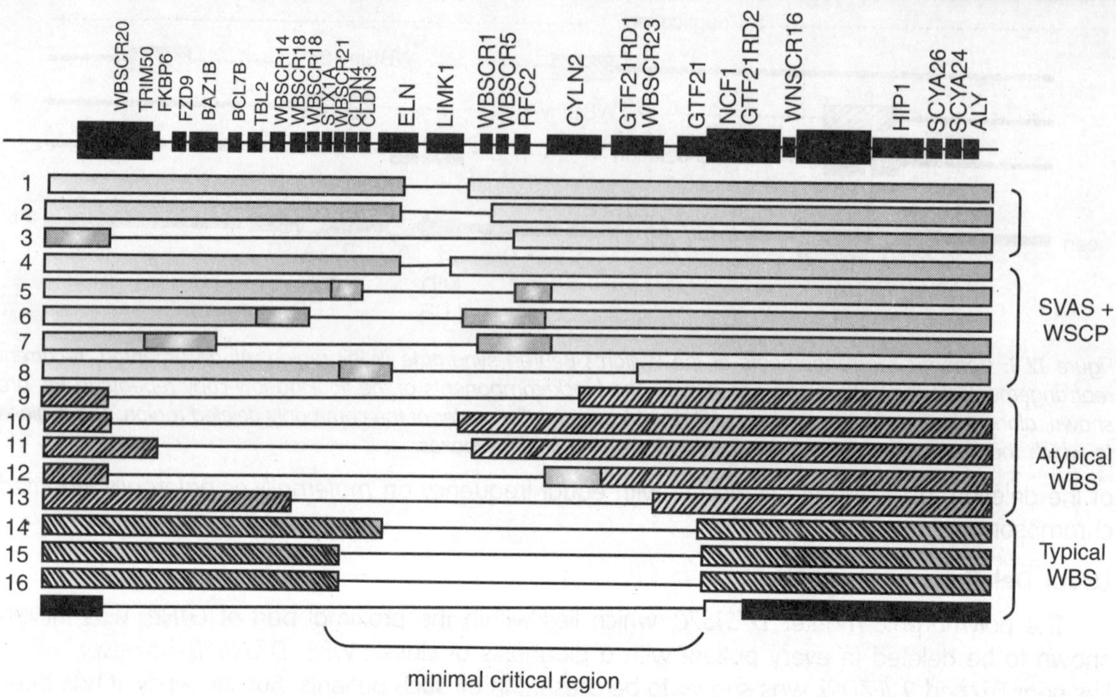

Figure 12.4: Genotype-phenotype correlation in Williams-Beuren syndrome (WBS). The WBS region is shown including all the known genes and the flanking repeats depicted as shaded boxes.

the deletion is supported by two reports of patients with WBS and chronic granulomatous disease (OMIM no. 233700), presumably owing to a mutation of the intact *NCFI* gene.

A recent correlation has been made between the deletion of *NCF1* and reduced incidence of hypertension in WBS. It has also been suggested that hypertension is linked to parent-of-origin, however, this may reflect the predominance of paternally inherited deletions that have recombination breakpoints within the *GTF2I* hotspot, thereby leaving *NCF1* intact.

Smaller Deletions

Although the vast majority of WBS individuals have the common deletion, there are some who carry partial deletions of the region, most being associated with a subset of the WBS *phenotype*. With relatively few patients, none of whom are deleted for exactly the same interval, it is hard to make precise correlations between the genotype and the phenotype.

Even very distinctive characteristics such as the WBS cognitive profile, with its verbal strengths and weakness in *spatial* abilities, do not appear to correlate with deletion size. Initial reports of two families with small deletions and the WBS cognitive profile suggested that hemizygosity for the LIM Kinase 1 gene *(LIMK1)* was responsible for deficits in visuospatial processing. A subsequent study, however, identified three families with deletions spanning *LIMK1* that showed no evidence of *visuospatial* deficits.

Yet a third study identified another three families with deletions of *LIMK1* who did exhibit the WBS cognitive profile. This typifies the difficulties in making accurate genotype-phenotype correlations where patients are seen by different *physicians* in different countries and where often there is only one affected family member being analyzed.

Moreover, for any such genomic examination it will be imperative to scan the entire region for potential chromosomal micro-rearrangements (including noncontiguous ones), *mutations*, or *complex alterations*, before any firm conclusions are drawn. Some general conclusions can be drawn from the patients with smaller deletions. It appears that a minimal deletion region of approx 10-12 genes is necessary for the full phenotypic spectrum of WBS to be expressed.

In addition, it is likely that the more telomeric end of the deletion makes a significant contribution to the developmental delay seen in these patients, because individuals with this region intact do not have significant *cognitive* impairment.

Larger Deletions

A small number of larger deletions of 7q11.23, which include the WBS commonly deleted region have been reported, but the boundaries of most are not yet well delineated. From the published reports, however, it would seem that patients with deletions that extend either *proximally* or distally from the WBS region have a more severe phenotype, often with significant developmental delay and seizures; personal *observation*). It is likely that the characteristic WBS phenotypes are still present, although many of the behavioural and cognitive features are masked by severe mental retardation.

ANOTHER TYPE OF REARRANGEMENTS

Inversion of the WBS Region

The highly related blocks flanking the WBS region are arranged not only in direct *orientation* (*centromeric* and *medial copies*) but also in inverted orientation (centromeric and telomeric copies) with respect to each other. The presence of these blocks in an inverted orientation can mediate unequal *meiotic* recombination events that result in the balanced inversion of the *intervening* region (without the loss of any part of the chromosome).

This inversion, named WBSinv-1, is found in approx 25-30% of parents who transmit a deleted chromosome to their WBS child, and can be readily detected by three-colour interphase FISH analysis or by pulsed-field gel electrophoresis (PFGE) followed by Southern blot analysis. In carriers of WBSinv-l, the inversion can be visualized directly using three differentially labeled probes using FISH analysis of interphase spreads of cells.

Two of the probes lie within the commonly deleted region, while the third lies outside, either proximally or distally, and acts as an anchor with which to orient the other two signals. Using PFGE, a junction fragment can be detected on digestion of genomic DNA with *Noël* restriction enzyme and *hybridization* with a probe for the *GTF21* gene. WBSinv-1 can also be inferred in a parental chromosome, based on analysis of the deleted chromosome in their WBS offspring. For this, analysis of the same SSNs used for *mapping* the deletion breakpoints is used.

If the deletion occurred through interchromosomal recombination between a normal and WBSinv-1 chromosome, the pattern of SSNs is different than if it occurred through either inter-chromosomal recombination between two normal chromosomes 7, or through intrachromosomal recombination. A model has been developed that predicts the possible regions undergoing NAHR based on SSN analysis in WBS families where the inversion status of the transmitting parent has been established by FISH.

The WBSinv-1 inversion is predicted to occur as an intrachromosomal rearrangement during meiosis, or perhaps during mitosis, while the subsequent deletion is thought to be restricted to interchromosomal rearrangements. Analysis of SSNs has narrowed the region of inversion recombination to the B-block or A-block in some individuals, but it is unknown whether NAHR is restricted to these blocks or if it also occurs in the C-block.

This implies that size of the inverted region could be anywhere between 1.8 and 2.5 Mb. Based on this analysis of deletions, it is predicted that the most likely region to undergo unequal exchange during the generation of the WBSinv-1 chromosome would be the B-block, because it shows the highest nucleotide identity.

Alternatively, there may be specific hot spots of recombination associated with this rearrangement, as there are for the deletions. All WBSinv-1 chromosomes that subsequently undergo deletion show recombination within the terminal 38 kb of B-block, which is not present in the centromeric copy.

This requires the recombination event leading to WBS to take place between homologous sequences in the medial and telomeric B-block. WBSinv-1 has also been identified in three individuals with clinical features that show some similarity to WBS. In one such patient, the junction fragment detected by PFGE was smaller (approx 500 kb) than that seen in three WBSinv-1 carriers (approx 600 kb), suggesting that the inversion breakpoints may differ in this patient or that the rearrangement may be more complex (with the simplest definition being that a deletion is also present). Alternatively, the WBSinv-1 breakpoints could be quite variable and the clinical features of these patients may, therefore, be unrelated to the inversion.

Risk Factor

The finding that one-fourth to one-third of transmitting WBS parents are carriers for the WBSinv-1 polymorphism suggests that it predisposes to unequal meiotic recombination. A thorough evaluation of the population frequency is needed in order to accurately assign a numerical risk factor to carrier status for WBSinv-l, but preliminary data do support a significant increase in risk associated with the inverted chromosome.

The population frequency has been estimated at 5% based on the analysis of 200 families with a classic WBS deletion member, where both *transmitting* and *nontransmitting* parents had their WBSinv-1 carrier status determined. The same study took the frequency of WBS to be 1 in 7500 and generated a risk factor of 4.5-fold for an inversion carrier.

This gives an overall chance of having an affected child of 1 in 2000 for WBSinv-1 carriers and 1 in 9200 for noncarriers. A second study looked at the WBSinv- 1 status of the transmitting parent in two families with multiple affected children.

In one family, there were no inversion carriers, but in the second family, both affected siblings had inherited a deleted WBSinv-1 chromosome from their father, who was a carrier for WBSinv-1. Together, these data indicate that the WBSinv-1 polymorphism seems to be a significant risk factor for WBS. Although the increase in risk associated with the inversion is substantial, the absolute risk of having a child with WBS is still well below the risk of fetal loss associated with prenatal diagnosis procedures, making widespread carrier testing unfeasible at this time. However, as prenatal diagnostic techniques improve and their associated risks decrease, carrier testing for WBSinv-1 may become part of a standard screening panel for those planning families.

Clinical Studies

A second micro-inversion of the WBS region, named WBSinv-2, has been identified in several patients with clinical features of WBS, but not in any unaffected individual. One of the patients with WBSinv-2 also carried a deletion of 7q31.2-q32 making the *comparison* of her phenotype to that seen in WBS difficult.

A second patient, however, showed no other genomic rearrangement and presented with a history of developmental delay, short stature, coarse facial features, stellate irides, and a very friendly affect. She also exhibited hypersensitivity to sound, mild joint laxity, a love of music, and had an IQ of approx 50 (phenotype and breakpoints are found at http://chr7.org). WBSinv-2 has breakpoints that lie within apparently unique sequences, one between *LIMK1* and *GTF2IRD1* within the commonly deleted interval, and the other between *HIP1* and *POR*, distal to the telomeric LCR.

The location of the breakpoints does not immediately suggest an LCR-mediated mechanism for inversion. Indeed, because the breakpoints are still only roughly mapped in these patients, it may be that their coincidence is owing to phenotypic selection. A WBSinv-2 breakpoint could directly interrupt at least one gene within the commonly deleted interval. There may also be wider reaching effects on nearby genes, and this could explain the similarity in clinical features.

DUPLICATION TYPE OF ABERRATION

Based on the repeat-mediated mechanism of deletion at the WBS locus, the reciprocal duplication event of 7q11.23 is expected to occur at an equal frequency (which is reflected in the 1 in 7500 to 1 in 20,000 occurrence of WBS). However, analysis of numerous patients with WBS-like features without deletion has not yielded evidence of duplication of the region.

The fact that duplications have not been observed suggests either that there is no associated phenotype, that they are incompatible with life, or that they result in a phenotype that is quite different from WBS.

Reciprocal duplications of both del and (17)(p11.2) and del(22)(q11.2) have been identified, and in both cases the phenotypic features are less severe, particularly in dup(17)(p11.2), which can manifests with any one off a range of neurological or psychological disorders alongside variable physical features. Recently a large duplication of approx 16 Mb spanning 7q11.22 to 7q21.11 was identified in a boy with clinical features quite unlike WBS. This patient presented at the age of 7 months with *hypotonia* and *dysmorphic* features.

At 18 months old he has developmental delay, with greatest deficits in gross and oral motor skills, including expressive language. His facial features are not typical of WBS, with low set ears and a very small mouth, and he has no vocalization pattern and has not been weaned from the bottle. Because this duplication extends much further than the WBS region, it remains to be seen whether a smaller duplication of just the WBS interval would lead to similar or less complex features. The characterization of this patient does tell us that such a duplication is unlikely to result in any

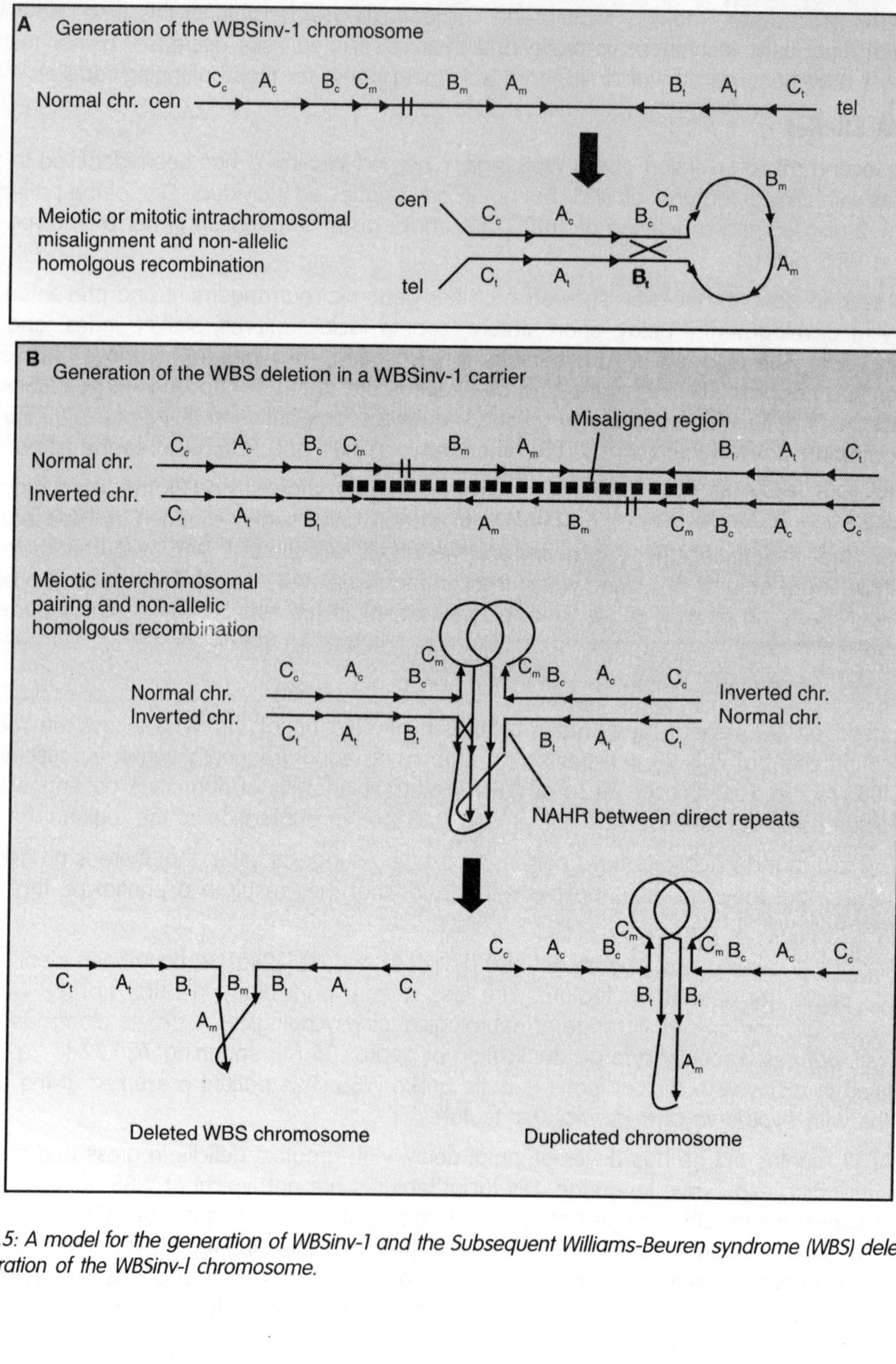

Figure 12.5: A model for the generation of WBSinv-1 and the Subsequent Williams-Beuren syndrome (WBS) deletion. (A) Generation of the WBSinv-I chromosome.

of the *hallmark* features of WBS, and that as such the search for dup(7)(q11.23) should be expanded to other *phenotypic* groups. Somerville et al. describes a reciprocal WBS duplication.

APPLICATIONS

The repeat laden genomic structure of the WBS region predisposes to susceptibility for deletions, inversions, and, quite possibly, other more complex rearrangements. In order to develop the most accurate *diagnostic* tests and fully understand statistical *relevance*, it will be important to resolve all of the possible normal and disease-associated *chromosomal* variants associated with this region.

Unpublished data from our group and collaborators (LA PerezJurado, personal communication) indicate that in addition to the anomalies described in this chapter, there can also be variation in the number of LCRs, smaller rearrangements between the two adjacent LCRs, and duplications at this locus. In addition, there appears to be a background of gene conversion within these blocks, which makes identification of LCRspecific sequences impossible in some individuals. Subregions of the WBS segmental duplications also exist at other locations on chromosome 7, suggesting that additional rearrangements of this chromosome may occur through NAHR. These rearrangements could occur either as *de novo* or inherited germline events, or as somatic changes that may be involved in cancer initiation or progression.

Indeed, more than 30 rearrangement breakpoints associated with malignancies have already been mapped to the 7q22.1 segmental duplication that includes WBS-like LCRs. Linking additional genes to *WBS-subphenotypes* will be an *arduous task*, and will likely necessitate the study of both patient samples and animal models. Several single gene mouse mutants have been generated already, but although some show aspects of WBS, no combination fully recapitulates the full phenotypic spectrum of this disorder. It is likely that a multiple gene deletion knockout will provide a better model for WBS. Ultimately, however, the screening of patients with atypical WBS for smaller deletions or point *mutations* at specific sites, will provide the most definitive evidence of the genes contributing to the underlying molecular basis of this *fascinating* genomic and clinical condition.

CONCLUDING REMARKS

The WBS region on chromosome 7q11.23 is subject to *numerous* genomic rearrangements, including deletions, inversions, and duplications. A commonaly deleted interval is present in the vast majority of individuals, with the calssic features of the disorder and encompasses between 26 and 28 genes, although the *minimal critical* region is thought to span only around 10-12 genes. Smaller deletions and inversions can give rise to subsets of the phenotype, but there is no clear-cut *genotype-phenotype* correlation, and the only gene definitely implicated in any aspect of the clinical features is *ELN*.

There is a polymorphic inversion (WBSine-1) present in between one-fourth to one-third of WBS parents, which predisposes the region to subsequent deletion and may have implications for genetic counseling. Deletion breakpoint mapping has shown that the nonallelic *homologous* recombination that leads to the common deletion often occurs within defined hotspots, some of which are *gender-specific* and one of which occurs only on the WBSinv-1 chromosome.

Chapter 13

Methods of Direct Identification

The direct identification of biological end points is a major advantage of *proteomic* research. The primary sequence of proteins does not depend on the DNA or the mRNA sequence alone, various *posttranslational* modifications, like *acetylation, glycosylation, phosphorylation,* or other types of covalent alterations, can *diversify* proteins. *Protein biomarkers* are among the important tools for detection, diagnosis, treatment, and monitoring of diseases.

The estimated number of different human proteins is about 500,000, and it will be challenging to identify, *catalogue,* and *functionally* assign these molecules. *Serum* is one of the preferred mediums when searching for markers, because it is obtained with low levels of *invasiveness,* it is available in *sufficient* quantity, and it is a source with great information content. Serum is probably the most complex *proteome,* containing other tissue proteomes as subsets. Profiling human serum is *ambitious,* because the serum contains large proportions of albumin (55%) and *glycoproteins* with an *enormous heterogeneity.*

Twenty-two protein groups, such as *albumin, immunoglobulins, haptoglobin, transferins,* and *lipoproteins,* account for 99% of the protein mass in serum. As many as 10,000 unique proteins are estimated to be present within the human serum proteome that span a *dynamic* range of concentrations approximately of $>10^9$.

Therefore, the background matrix represents a very complex milieu, and proteins observed in general analyses, in general, are of high abundance. The ProteinChip System was introduced with the intention of an easy to use, time- and resource-saving method by Ciphergen *Biosystems* (Fremont, CA, www. ciphergen.com). In the meantime the proteomics instrument business was sold to Bio-Rad *Laboratories* (www.bio-rad.com).

Based on the surface-enhanced laser *desorption-ionization* process (SELDI), it combines two methods: solid phase chromatography and time-of-flight mass spectrometry (TOF-MS). This *technology* integrates the complex protein expression on one integrated platform and enables the study of complex serum protein profiles. Chip arrays capture individual proteins out of a probe, which are subsequently resolved by mass *spectrometry* (MS). Varying coatings on the chips allow the binding of different protein classes. ProteinChip arrays are available with different *chromatographic* properties, including *hydrophobic, hydrophilic, anion exchange,* cation exchange, and immobilized-metal affinity surfaces; or they are preactivated for the coupling of *capture* molecules.

Thus, each analysis is a sort of an on-spot fractionation step, which reduces the *complexity* of the serum sample. This kind of fractionating is not be effective enough in selected cases, and the dynamic range of proteins and protein concentrations in human serum sometimes *necessitates* effective *fractionating* methods capable of separating high abundance from low *abundance* proteins.

The sample analysis takes place in SELDI-TOF MS, and data are processed by *complex bioinformatic* software. Further detailed information and images about the SELDITOF MS technique can be found on the *homepage* of BIO-RAD (www.bio-rad. com) and in the literature.

BASIC REQUIREMENTS

For beginners to the SELDI-TOF *technique*, there is a starter kit (ProteinChip SELDI System tarter Kit) available. This includes a selection of *ProteinChip arrays, calibration standards,* energy absorbing molecules (EAM), and a *bioprocessor.*

Probes

Serum samples are obtained in routinely used serum *monovettes.*

Fractionating the Serum

Serum may be used unfractionated or *fractionated.* Any method to fractionate the serum probe is fine. *Expression* Difference Mapping Kits (Bio-Rad Laboratories, Hercules, CA, USA) are especially designed for serum *fractionating* in combination with SELDI TOF analyses (procedures are described in detail in the Expression Difference Mapping™ Kit-Serum Fractionating, www.bio-rad.com), but any method to *fractionate* the serum is suitable.

Protein Chip Arrays

1. IMAC30 is an immobilized metal affinity capture array with a nitriloacetic acid (NTA) surface, with an updated hydrophobic barrier coating.
2. CM 10 is a weak cation exchange array with carboxylate functionality, with an updated *hydrophobic* barrier coating.
3. Q10 is a strong anion exchange array with quaternary amine functionality, with an updated *hydrophobic* barrier coating.
4. H50 binds proteins through reversed phase or hydrophobic interaction *chromatography* with an updated *hydrophobic* barrier coating.
5. H4 *mimics* reversed phase chromatography with C 16 *functionality.*
6. NP20 mimics normal phase chromatography with silicate functionality. 7. Au are gold chips to be used directly for MALDI-based experiments.
8. Mainly used ProteinChip surfaces for serum diagnostic: Metal Affinity Capture

(IMAC3/IMAC30), Weak Cation Exchange (CMIO/WCX2) reversed Phase (H50).

Role of Buffers

1. PBS buffer: *2.7 mM* potassium chloride, 120 mM sodium chloride, and 10 mM phosphate buffer maintained at a pH of 7.0.
2. 100 mM ammonium acetate, pH 4.0 (with or without 0.1 % Triton X-100) for CM10.
3. 100 mM Tris-HCl, pH 9.0 (with or without 0.1 % Triton X-100) for Q 10.
4. 100 mM sodium phosphate, 500 mM NaCl pH 7.0 (activated with either 100 mM copper sulphate or 100mM nickel sulphate hexahydrate) for IMAC30.
5. 10% acetonitrile (AcN), 0.1 % trifluoroacetic acid (TFA) or phosphate-buffered saline (PBS) for H50.

Soaking Energy

Matrix solution: 50% saturated sinapinic acid, in 50% acetonnitrile and 0.5% trifluoroacetic acid, or 50% saturated α-Cyano-4-hydoxycinnamic acid, in 50% acetonitrile and 0.5% trifluoroacetic acid (C30-00004 Kit, or ProteinChip SELDI Starter Kit, Bio-Rad).

Mass Spectrometry

1. SELDI-TOF MS: PBS-11c ProteinChip Reader or ProteinChip System Series 4000 (Bio-Rad).
2. SELDI-QTOF: QStar Pulsar or QStarXL instrument (Applied Biosystems) equipped with a PCI 1000 ProteinChip Interface (Bio-Rad).

Software

Analytical software package: Protein Chip software version 3.1 or later versions (Bio-Rad Laboratories, Hercules, CA, USA).

METHODS

SELDI ProteinChip technology binds distinct groups of serum proteins on the chip surface after adding EAM. The bound proteins finally are read in a SELDI-TOFMS. Generally, the serum is processed according to the technique of each institution, and many different procedures have been described. Regardless of the method chosen, *standardization*, with careful attention to detail, is *imperative* in protein research.

Sampling for Storing

1. Standardize blood collection in a serum separator tube and allow it to clot for 30 min at room temperature.
2. Centrifuge the blood for 10min at 1,000g.
3. Aliquot the supernatant serum before flash *freezing* at -80°C. It also is possible to aliquot the samples later (for example, first thaw) but the quality of the probes *deteriorates* with each freeze-thaw cycle.

Requirement of Unfractionated Serum

1. After thawing frozen samples on ice, spin *aliquots* at 20,000g for 10 min at 4°C.

2. Mix 20 µL of the serum sample with 30 µL of U9 buffer and vortex samples for 20 min.
3. Dilute this sample with 50µL of PBS buffer.
4. If necessary dilute further 1:8 to 1:10 into the respective binding buffer. The probes now are ready to be analyzed. Continue with step 3.4.1.

Requirement of Fractionated Serum

The procedure of fractionating the serum results in *aliquots*, which can be used on the chip. Dilute the fractionated serum samples 1:5-1:10 into the appropriate binding buffers.

Significance

For using the bioprocessor, the minimum sample requirement is 1-10µg of the total protein per spot and a sample volume of 50-200µL. The serum samples now can be profiled with different array surfaces. The most commonly used and widely tested surfaces with serum are Metal Affinity Capture (IMAC3/IMAC30), Weak Cation Exchange (CMIO/WCX2), and Reversed Phase (H50). To find the chip that suits best, different surfaces have to be preevaluated by testing or studying the literature.

1. Add 50-150µL to the bioprocessor and incubate for 30min with shaking on moderate speed.
2. Wash three times with 150 µL binding buffer for 5 min with shaking.
3. Wash twice with 150µL distilled water (1 min each wash) to remove buffer salts.
4. Add twice, 1 µL of 50% saturated matrix solution to the spot surface and allow to air dry in between.

The probes now are ready to be analyzed. If necessary, store the chips in a dark room at room temperature until processing.

Mass Analyzer

The Protein Chip arrays are placed in the SELDI-TOF MS (PBS IIc Protein Chip reader or Protein Chip System 4000) and irradiated with a pulsed UV nitrogen laser. Here, protons are transferred onto the peptides and proteins, which subsequently are accelerated by *electromagnetic* fields through a flight tunnel.

The time of flight corresponds inversely to the molecular mass. The resulting signals are converted to mass-to-charge (m/z) ratios based on the time each species takes to pass through the TOF *mass analyzer*.

1. Set the mass analysis range you want to evaluate.
2. Adjust the laser intensity and some other parameters before measurements.
3. Calibrate the instrument using molecular markers.

Analysis of TOF MS Spectra

Complex software packages, based on *genetic algorithms*, decision trees, and a unified maximum separability *algorithm* are used for data analyses. The concept of an independent procedure, a supervised analysis with a training data set to differentiate between sample classes, is used. The results are visualized in a graph with the mass-to-charge ratio of the sample components on the x-axis and the corresponding signal intensities on the y-axis.

Modern software allows a wide spectrum of different views and spectra for the best *visualization* of the data. Widely used is the Protein Chip Biomarker WizardTM, the EDM package, and the Biomarker Patterns Software (Bio-Rad). More detailed information can be found in the literature or on the homepage of Ciphergen (www.bio-rad.com).

Mass Fingerprinting

Using the SELDI-based technique, identification of the target proteins or suspected biomarkers is the last step. The detailed steps of protein purification and identification are beyond the scope of this chapter. Briefly, each protein of interest needs to be purified or enriched for subsequent analysis. Different methods (mini-spin columns, microplate formatted *chromatographic* devices, preparative scale columns, and antibodies in combination with *preactivated* arrays) can be used.

Finally, the enriched protein is proteolytically cleaved and the masses of the resulting peptide fragments are determined using MS. The obtained data can be submitted directly to protein databases, matching the measured masses with peptide fragments of known sequences (for an overview of protein databases, see http://userpage.chemie.fu-berlin.de/biochemie/protprak.html).

POINTS TO REMEMBER

1. Proteins are easily altered or destroyed, when thawed or frozen. Therefore, serum samples should not undergo more than one or two freeze-thaw cycles prior to the assay. The handling, including the freeze-thaw cycles, should be well documented. This helps in explaining the differences of the later results as artifacts due to sample handling.
2. If serum is likely to be stored for certain period, prepare serum samples as soon as possible soon after collection.

Chapter 14

Block Interchanges

With an increasing number of genomic data (DNA, RNA, and protein sequences) being available, the study of genome rearrangement has received a lot of *attention* in computational biology and *bioinformatics* because of its applications in the measurement of *evolutionary* difference between two species. One of the most *promising* ways to do this research currently is to compare the orders of the *identical* genes in two different genomes.
Different from the *traditional* point mutations (e.g., *insertions, deletions,* and *substitutions*) acting on individual base, various large-scale mutations acting on genes within or among chromosomes have been proposed for measuring the *rearrangement* distance between two related genomes, such as the so-called *intrachromosomal* rearrangements (e.g., *reversals/inversions, transpositions,* and block-interchanges) and the so-called interchromosomal rearrangements (e.g., *translocations, fusions,* and *fissions*).

Block-interchanges, introduced by Christie, are a new kind of rearrangements that affect a chromosome by *swapping* two nonintersecting blocks of genes of any length. Examples of *block-interchange* rearrangements are shown in Figure 14.1.

More recently, the study of block-interchange rearrangements is becoming increasingly important because of its applications in molecular evolution. From the biological perspective, block-interchanges can be considered as a generalization of transpositions (one of the most frequent rearrangements) because the blocks exchanged by transpositions must be contiguous (or adjacent) on the chromosome, whereas those by *blockinterchanges* need not be.

The study of block-interchange rearrangements usually requires to solve a combinatorial problem, called the block-interchange distance problem, which is to find a minimum number of block-interchanges between two given gene orders of linear/circular chromosomes such that these block-interchanges can transform one gene order into another.

Christie first proposed an $O(n^2)$ time algorithm to solve this problem on linear chromosomes using breakpoint diagram approach, where n is the number of genes on chromosomes. Lin et al. later made use of permutation groups in algebra to design a simpler *algorithm* of $O(\delta n)$ time for

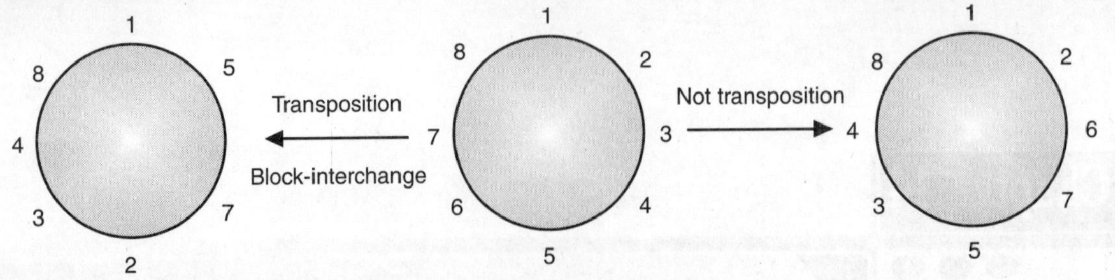

Figure 14.1: Prokaryotic chromosome showing rearrangements. The left rearrangement, swapping two adjacent blocks [2, 3, 4] and [5, 6, 7], is a transposition, but the right one is not because the exchanged [3, 4] and [6, 7] are not adjacent. However, both of them are block-interchanges.

solving the block-interchange distance problem on linear chromosomes as well as on circular *chromosomes,* where 8 is the block-interchange distance that can be calculated in *O(n)* time in advance. They also demonstrated that block-interchange rearrangements seem to play a significant role in the genetic evolution of bacterial species (e.g., *vibrio pathogens*).

Recently, Lu et al. have further *implemented* the algorithms designed by Lin et al. into a web server, called ROBIN (short for rearrangement of blockinterchanges), for allowing the public to conduct the online analyses of blockinterchange rearrangements between two linear/circular chromosomes.

Particularly, not only gene orders but also bacterial-size sequences are allowed to be the input of ROBIN. If the input is sequence data, ROBIN can *automatically* search for the *identical landmarks,* which are the homologous/conserved regions shared by all the input sequences. In this chapter, we shall first introduce the basics of *permutations* and their relationships with genome rearrangements, then describe a simple *algorithm* on the basis of permutations for calculating the block-interchange distance between two given gene orders of linear/circular chromosomes, and finally introduce ROBIN and describe its detailed steps to conduct the analyses of block-interchange rearrangements for chromosomal sequences or gene orders.

DEFINING PERMUTATIONS

In group theory, a *permutation is* defined to be a one-to-one mapping from a set $E = \{1, 2, .., n\}$ into itself, where *n* is a positive integer. For example, we may define a permutation a of the set {1, 2, 3, 4, 5, 6, 7} by specifying $\alpha(1) = 4$, $\alpha(2) = 3$, $\alpha(3) = 1$, $\alpha(4) = 2$, $\alpha(5) = 7$, $\alpha(7) = 6$, and $\alpha(6) = 5$. The above mapping can be expressed using *a cycle notation* as illustrated in Figure 14.2 and simply denoted by $\alpha = (1, 4, 2, 3)(5, 7, 6)$. A cycle of length *k,* say $(a_1, a_2, ..., a_k)$, is simply called *k-cycle* and can be rewritten as $(a_i, a_{i+1}, ..., a_k, a_1, ..., a_{i-1})$, where $2 \leq i < k$, or $(a_k, a_1, a_2, ..., a_{k-1})$.

Any two cycles are said to be *disjoint* if they have no element in common. In fact, any permutation, say α, can be written in a unique way as the product of disjoint cycles, which is called the cycle decomposition of α, if we ignore the order of the cycles in the product. Usually, a cycle of length one in α *is* not explicitly written and its element, say *x,* is said to be *fixed* by α because $\alpha(x) = x$. Especially, the permutation whose elements are all fixed is called an identity permutation and is denoted by 1 (i.e., 1= (1) (2) ... (n)).

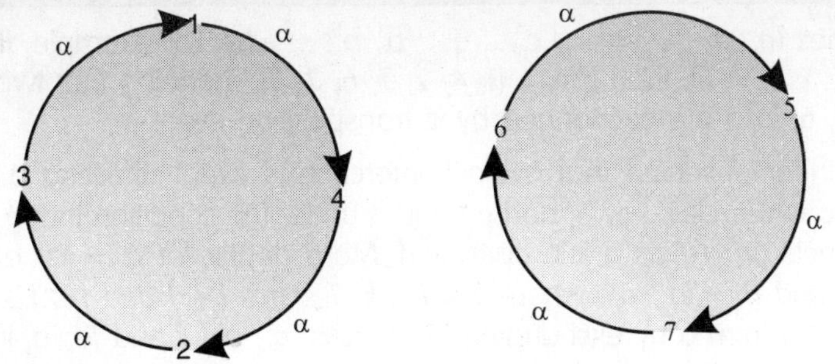

Figure 14.2. The illustration of a permutation α = (1, 4, 2, 3) (5, 7, 6) meaning that α(l) _ 4, α(2) = 3, α(3) = l, α(4) = 2, α(5) = 7, α(7) = 6, and α(6) = 5.

Given two permutations α and β of *E*, the *composition* (or *product)* of α and β, denoted by $\alpha\beta$, is defined to be a permutation of *E* with $\alpha\beta(x) = \alpha(\beta[x])$ for all *x*•*E*. For instance, if we let *E* = {1, 2, 3, 4, 5, 6, 7}, α = (1, 5) and β = (1, 4, 2, 3, 5, 7, 6), then $\alpha\beta$ = (1, 4, 2, 3)(5, 7, 6). If α and β are disjoint cycles, then $\alpha\beta = \beta\alpha$. The *inverse of a* is defined to be a permutation, denoted by α^{-1}, such that *ad* = $\alpha^{-1}\alpha = \alpha^{-1}$= 1.

If a permutation is expressed by the product of disjoint cycles, then its inverse can be obtained by just reversing the order of the elements in each cycle. For example, if α = (1, 4, 2, 3) (5, 7, 6), then a^{-1} – (3, 2, 4, 1) (6, 7, 5). Clearly, $\alpha^{1} = \alpha$ *If* α is a two-cycle. Meidanis and Dias first noted that each cycle of a permutation may represent a circular chromosome of a genome with each element of the cycle corresponding to a gene, and the element order of the cycle corresponding to the gene order of the chromosome.

Figure 14.2, for example, denotes a genome with two circular chromosomes, one represented by (1, 4, 2, 3) and the other by (5, 7, 6). Moreover, they observed that rearrangement events, such as fusions and fissions (respectively, transpositions), correspond to the composition of a two-cycle (respectively, three-cycle) and the permutation representing a circular genome. For instance, let α be any permutation with cycle decomposition of $\alpha_1\ \alpha_2\ \ldots\ \alpha_r$..

1. If ρ – (*x*, *y*) is a two-cycle and *x* and *y* are in the different cycles of α, say $\alpha_p = (a_1 \equiv x, a_2,$ .... $a_i)$ and $\alpha_q = (b_1 \equiv y, b_2, \ldots, b_j)$ where $1 \leq p, q \leq r$, then in the composition pα, α_p and α_q are joined into a cycle ($a_1, a_2, \ldots$, *ai*,$b_1, b_2, \ldots$, *bj*).

   For example, if p = (1, 5) and α = (1, 4, 2, 3) (5, 7, 6), then pα = (1, 4, 2, 3, 5, 7, 6), meaning that two circular chromosomes (1, 4, 2, 3) and (5, 7, 6) in a are merged into one larger circular chromosome (1, 4, 2, 3, 5, 7, 6) by a fusion event p (i.e., p acts on a as a join operation).

2. If p = (x, y) is a two-cycle and x and y are in the same cycle of α, say $\alpha_p = (a_1 \equiv x\ a_2, \ldots,$ $a_i \equiv y, a_{i+1}, \ldots, a_j)$ where $1 \leq p \leq r$, then in the composition pα, this cycle α_p is broken into two disjoint cycles ($a_1, a_2, \ldots, a_{i-1}$) and ($a_i, a_{i+1}, \ldots, a_j$). For example, if ρ = (1, 5) and α = (l, 4, 2, 3, 5, 7, 6), then pα = (1, 4, 2, 3) (5, 7, 6), meaning that a circular chromosome (1, 4, 2, 3, 5, 7, 6) in α is broken into two smaller circular chromosomes (1, 4, 2, 3) and (5, 7, 6) by a fission event p (i.e., ρ acts on a as a split operation).

3. If ρ = (*x*, *y*, *z*) is a three-cycle and *x*, *y* and *z* are in the same cycle of α, say $\alpha_p = (a_1 \equiv x,$ $a_2, \ldots, a_i b_1 \equiv y, b_2, \ldots, b_j\ c_1 \equiv z, c_2, \ldots, c_k)$ where $l \leq \rho \leq r$, theninthe composition $\pi\alpha$, the cycle

α_p becomes $(a_1, a_2, \ldots, a_i, c_1, c_2, \ldots, c_k, b_1, b_2, \ldots, b_j)$. For example, if $\rho = (1, 3, 7)$ and $\alpha = (1, 4, 2, 3, 5, 7, 6)$, then $\rho\,\alpha = (1, 4, 2, 7, 6, 3, 5)$, meaning that two adjacent blocks [3, 5] and [7, 6] in a are exchanged by a transposition event ρ.

In fact, we further observed that a block-interchange event affecting a corresponds to the composition of two two-cycles, say ρ, and ρ_2, and a under the condition that ρ_1 acts on a as a split operation and ρ_2 acts on $\rho_1\alpha$ as a join operation. More clearly, let $\alpha_p = (a_1, a_2, \ldots, a_k)$ be a cycle of α, $\rho 1$ _ (α_1, α_i) and $\rho_2 = (a_h, aj)$, where $1 < i \leq k$, $1 \leq h \leq i - 1$ and $i \leq j \leq k$. Then $\rho_2\rho_1$ a is a permutation obtained from a by exchanging the blocks $[a_h, a_{i-1},]$ and $[a_j, a_k$ in α_p. For instance, if $\rho_1 = (1, 3)$, $\rho_2 = (4, 7)$ and $\alpha = (1, 4, 2, 3, 5, 7, 6)$, then $\rho_2\rho_1\alpha = (1, 7, 6, 3, 5, 4, 2)$, a resulting *permutation* by *exchanging* blocks [4, 2] and [7, 6] in (1, 4, 2, 3, 5, 7, 6).

IDENTIFYING PERMUTATIONS

Given two permutations $\alpha = (a1.\ a2, \ldots, a_n)$ and $\beta = (b_1, {}_{b2}\ b_n)$ of $E = \{1, 2, \ldots, n\}$, the block-interchange distance problem is to find a minimum series of block-interchanges $\sigma_1, \sigma_2, \ldots, \sigma_1$ such that ... $\sigma_t\ \sigma_{t-1} \ldots \sigma_{1\alpha} = \beta$ (i.e., *transforming* α into β), and the number t is then called the block-interchange distance between α and β. Usually, β is replaced with $I = (1, 2, \ldots, n)$. In this case, the block-interchange distance problem can be regarded as a problem of sorting a using the minimum number of block-interchanges and hence this problem is also called .as sorting (a permutation) by block-interchanges.

In fact, the problem of sorting by block-interchanges on circular chromosomes is equivalent to that on linear chromosomes, because the algorithm of solving the former can be used to solve the latter, and vice versa. In the following, we shall introduce an algorithm on the basis of permutation groups to solve the *sorting* circular chromosomes by block-interchanges. In the next section, we then describe how to use it to solve the same problem on linear chromosomes, and vice versa.

Now, we let α and I be permutations representing two given circular genomes. Then it is clear that $I\alpha^{-1}\alpha = I$, intuitively implying that $I\alpha^{-1}$ contains all information needed to transform α into I. Actually, we are indeed able to derive a minimum number of block-interchanges from $I\alpha^{-1}$ required to transform α into I. Moreover, the block-interchange distance between α and I equals to $\delta = (n - f(I\alpha^{-1}))/2$, where $f(I\alpha^{-1})$ denotes the number of the disjoint cycles in the cycle decomposition of $I\alpha^{-1}$. It should be noted that $f(I\alpha^{-1})$ counts also the nonexpressed cycles of length one. For example, if $I\alpha^{-1} = (1, 4, 2, 3)(5, 7)$ and $E = \{1, 2, \ldots, 7\}$, then $f(I\alpha^{-1}) = 3$, instead of $f(I\alpha^{-1}) = 2$, because $I\alpha^{-1} = (1, 4, 2, 3)(5, 7)(6)$, a product of three disjoint cycles.

The algorithm we designed to compute a minimum series of block-interchanges for the transformation between α and I is very simple. It just calculates $I\alpha^{-1}$ in the beginning and then iteratively repeats the following steps for the purpose of finding the ith block-interchange that is represented by a product of two two-cycles, say $(u, v)\ (x, y)$, where $1 \leq i \leq \delta$.

1. Arbitrarily choose two adjacent elements x and y in a cycle of $I\alpha^{-1}$. Then the composition of (x, y) and a leads a to be broken into two disjoint cycles.

2. Find two adjacent elements u and v in $I\alpha^{-1}\ (x, v)$ such that the product of (u, v) and $(x, y)a$ becomes a single cycle.

3. Replace α with $(u, v)(x, y)\alpha$ and recalculate $I\alpha^{-1}$, meaning that the original α affected by a block-interchange $(u, v)(x, y)$ has become a new chromosome $(u, v)(x, y)\alpha$.

Let us take $\alpha = (5, 2, 7, 4, 1, 6, 3)$ and $I = (1, 2, ..., 7)$ for an instance. We first calculate $I\alpha^{-1}$, which equals to (1, 5, 4)(2, 6)(3, 7). Clearly, $f(I\alpha^{-1}) = 3$ and hence we understand immediately that the block-interchange distance between α and I is $\delta = (7 - 3)/2 = 2$, which means that a can be transformed into I using only two block-interchanges. Next, we show how to find these two block-interchanges, denoted by $\sigma_1 = \rho^1{}_2\, \rho^1{}_1$ and $\sigma_2 = \rho^2{}_2\rho^2{}_1$. Initially, we arbitrarily choose two adjacent elements in $I\alpha^{-1}$, say 2 and 6, and let $p^{*1}{}_1 = (2, 6)$. Then $p^{*1}{}_1$ affects a as a split operation because $\rho^1{}_1\alpha = (2, 7, 4, 1)(6, 3, 5)$. Then we continue to choose two adjacent elements 3 and 7 in $I\alpha^{-1}\, p^{*1}{}_1 = (1, 5, 4)\,(3, 7)$ and let $\rho^1{}_2 = (3, 7)$ because $\rho^1{}_2\rho^1{}_1\, \alpha$ becomes a single cycle (5, 6, 7, 4, 1, 2, 3). As a result, we obtain a new chromosome $\rho^1{}_2\, \rho^1{}_1\, \alpha = (5, 6, 7, 4, 1, 2, 3)$ from $\alpha = (5, 2, 7, 4, 1, 6, 3)$ by exchanging blocks (2) and (6) using the first block-interchange $\rho^1{}_2\rho^1{}_1 = (3, 7)(2, 6)$. Next, we proceed with the second iteration by first replacing α with $\rho^1{}_2\rho^1{}_1\alpha$ and recalculating $I\alpha^{-1}$, which is now (1, 5, 4) = (1, 4)(1, 5). Similarly, we arbitrarily choose $\rho^2{}_1 = (1, 5)$ and then let $\rho^2{}_2 = (1, 4)$. Consequently, $\rho^2{}_2\rho^2{}_1\alpha = (1, 2, 3, 4, 5, 6, 7)$ is equal to I, the target chromosome obtained from $\alpha = (5, 6, 7, 4, 1, 2, 3)$ by exchanging blocks (4) and (1, 2, 3) using the second block-interchange $\rho^2{}_2\rho^2{}_1 = (1, 4)\,(1, 5)$.

In the previously described algorithm, there are totally δ iterations to be proceeded with each iteration costing O(n) time. Hence, the time-complexity of this algorithm is $O(\delta n)$, where the calculation of $\delta = (n - f(I\alpha^{-1}))/2$ can be done in 0(n) time in advance.

PROKARYOTIC VS EUKARYOTIC CHROMOSOMES

In the previous section, we have introduced an algorithm to solve the block-interchange distance problem on circular chromosomes by representing permutations as circular chromosomes. Of course, permutations can also be used to represent linear chromosomes. In this situation, a permutation $(a_1, a_2, \ldots, a_n)$ is not equal to $(a_i, \ldots, a_n, a_1, \ldots, a_{i-1})$, where $2 \le i \le n$. In the following, we shall describe how to slightly modify this algorithm to deal with the case of representing $(a_1, a_2, \ldots, a_n)$ and $(1, 2, \ldots, n)$ as linear chromosomes a and I, respectively.

First of all, we add a new element 0 into the *beginning* of α and denote this new permutation by $\alpha' = (0 \equiv a_0, a_1, a_2, \ldots, a_n)$. Then we regard α' as a circular chromosome and apply the modified algorithm (we will introduce later) to α' such that the obtained blockinterchanges, say $\sigma'_1, \sigma'_2, \ldots, \sigma'_\delta$, for optimally transforming α' into $I' \equiv (0, 1, \ldots, n)$ satisfy the property that none of the two blocks interchanged by each σ' contains a_0, where $1 \le i \le \delta$. The purpose of this property is to ensure that for each σ_i affecting circular chromosome $\sigma'_{i-1} .. \sigma'_1\, \sigma'_1$ we can find a corresponding block-interchange σ'_i to affect linear chromosome $\sigma_{i-1}, \ldots \sigma_1\alpha$ such that the blocks interchanged by σ_i are the same as those interchanged by σ_i. Because of this property, $\sigma_1, \sigma_2, \ldots, \sigma_\delta$ can be used to transform linear chromosome α into I and moreover, their number is minimum, because any series of block-interchanges of transforming α into I must also be a series of block-interchanges of *transforming* α' *into* I'.

Now, we describe how to modify the algorithm *in* the previous section such that *none* of the two blocks interchanged by each σ_i contains a_0, where $1 \le i \le \delta$. Notice that when applying the original algorithm to α', the chosen (x, y) and (u, v) in the ith iteration (i.e., $\sigma'_i = (u, v)(x, y)$) may lead one of the exchanged blocks, say B_1 and B_2, to contain α_0. If this situation occurs, then we consider the following two cases to modify the algorithm.

Case 1: B_1 and B_2 are not adjacent. Then we just interchange the roles of *(x, y)* and *(u, v)* by letting *(u, v)* as the split operation and *(x, y)* as the join operation, and then apply *(x, y)(u, v)*, instead of *(u, v)(x, y)*, to $\sigma'_{i-1}, \ldots \sigma'_1\sigma'$. We illustrate its reason by just considering the first iteration of producing σ_1 as follows. Let $(x, y) = (a_i, a_j)$ and $(u, v)\_ (a_k, a_1)$, where $1 \leq i < j \leq n$, $i < k < j$ and $j < 1$. Then $(u, v)(x, y)\alpha = (a_i, \ldots, a_{k-1}, a_1, \ldots, a_{n}, a_{o}, \ldots, a_{i-1}, a_j, a_{j+1}, \ldots, {}_{a}J\text{-}1\ a_k, \ldots, a_{j-1})$, and (x, y) (u, v)α - $(a_k, \ldots, a_{j-1}, a_i, \ldots, a_{k-1}, a_l, \ldots, a_n, a_o, \ldots, a_{i-1}, a_j, \ldots, a_{l-1})$. From the viewpoint of circular chromosome, they are equivalent (i.e., *(u, v)(x, y)* α =*(x, y)(u, v)* α). However, both the blocks interchanged using *(x, y) (u, v)* contains no σ a_o.

Case 2: B_1 and B_2 are adjacent. In this case, we let $B_1 = [a_1, \ldots, a_n, a_o, a_1, \ldots a_{i-1}]$ denote the block of containing a_0 and B_3 be the remaining block. It is not hard to see that either $x = u$ or $y = v$. If $x = u$, then (u, v)(x, y) = (x, v)(x, y), which also equals to (x, y)(v, y). Then we modify the algorithm by applying *(x, y)(v,* y), instead of *(u, v)(x, v), to* $\sigma'_{i-1} \ldots \sigma'_1\sigma'$. Because it leads to the exchange of B_2 and B_3, which is equivalent to the exchange of B_1 and B_2 from the circular viewpoint, and neither B_2 nor B_3 contains a_o. If $y = v$, then $(u, v)(x, y) = (u, y)(x, y)$ that equals to $(u, x)(u, y)$. Then we modify the algorithm by applying (u, x)(u, y) to $\sigma'_{i-1} \ldots \sigma'_1\alpha'$, because it leads B_2 and B_3 to be exchanged.

In fact, we can use the algorithm of sorting linear chromosomes by blockinterchanges to solve the same problem on circular chromosomes as follows. Now, let $\alpha = (a_1, a_2, \ldots, a_n)$ be represented as a circular chromosome, instead of a linear chromosome. Then we can linearize it by removing an *arbitrary* element, say a_1, from a so that the resulting permutation $\alpha' = (a_2, a_3, \ldots, a_n)$ becomes a linear chromosome. The removed element a_1 actually plays the same role as the a_0 we have introduced in the previous discussion. With the same arguments, the optimal solution for linear α' *corresponds* to a minimum series of block-interchanges required to transform circular a into *I*.

ANALYZING GENOME REARRANGEMENTS

Based on the algorithms we described in the previous sections, we have developed a web server, called ROBIN, for analyzing genome rearrangements of block-interchanges between two linear/circular chromosomes. ROBIN takes two or more linear/circular chromosomes as its input and outputs the block-interchange distance between any pair of the input chromosomes, as well as an optimal scenario of the needed block-interchanges.

The input can be either bacterial-size sequence data or gene-/landmark-order data. If the input is a set of sequences, ROBIN will automatically search for the identical landmarks that are the homologous/conserved regions shared by all the input sequences.

Matching by Collinear Blocks

In analyses of genome rearrangements, the commonly used landmarks can be exact matches such as the maximal unique matches (MUMs) as in MUMmer, approximate matches without gaps such as the yielding fragments as in DIALIGN and LAGAN, approximate matches with gaps such as the hit fragments as in BLASTZ, or regions of local collinearity such as the locally collinear blocks (LCBs) as in Mauve. A LCB is a *collinear* set of the multi-MUMs, where *multi-MUMs* are exactly matching subsequences shared by all the considered genomes that occur only once in each of *genomes* and that are bounded on either side by mismatched nucleotides.

In ROBIN, we adopt LCBs as landmarks because each LCB may correspond to a homologous region of sequence shared by all the input *chromosomes* and does not contain any genome rearrangements. Usually, each LCB is associated with a weight that can serve as a measure of confidence that it is a true homologous region rather than a random match.

The weight of an LCB is defined as the sum of the lengths of multi-MUMs in this LCB. By selecting a high minimum weight, users can identify the larger LCBs that are truly involved in genome rearrangement, whereas by selecting a low minimum weight, users can trade some specificity for sensitivity to identify the smaller LCBs that are possibly involved in genome rearrangement.

Simple Web Inter Face

ROBIN can be easily accessed via a simple web interface and it also provides an online help to help users to run it. The input of ROBIN can be two or more linear/circular chromosomes that are either genomic sequences with bacterial size or unsigned integer sequences with each integer representing an identical gene/landmark on all the chromosomes being considered.

If the input is a set of genomic sequences, ROBIN will identify all the LCBs meeting the user-specified minimum weight. The minimum LCB weight is a user-definable parameter and its default is set lo be three times the minimum multi-MUM length. The output of ROBIN is the block-interchange distance between any two input chromosomes and an optimal scenario of block-interchanges for the transformation.

Input Data

Users can choose to input sequence data or gene-/landmark-order data to run ROBIN. If the input is the sequence data, then users can run ROBIN by following the steps described next.

1. Input or paste two or more sequences in FASTA format in the top field of web interface, or simply upload a plain text file of sequences in FASTA format.
2. Specify the minimum multi-MUM length, whose default is set to $\log_e n$, where n is the average sequence length, and the minimum LCB weight, whose default is set to 3* (minimum multi-MUM length).
3. Select the type of chromosomes being considered, which can be either linear or circular.
4. Check email box and enter an email address simultaneously, if users would like to run ROBIN in a batch way. In the batch way, users will be notified of the output via email when the submitted job is finished. This email check is optional but *recommended* if the sequences users enter/upload are large-scale.
5. Click "Submit" button to run ROBIN.

If users choose gene-/landmark-order data as the input of ROBIN, then they just follow the steps s next.

6. Input or paste two or more unsigned gene/landmark orders in FASTA-like format in the bottom field of web interface. The FASTA-like format an example) contains a single-line description followed by lines of unsigned integers, which are separated by space(s), with

each integer representing an identical gene/landmark on the chromosomes being considered.

7. Select the chromosome type.
8. Click "Submit" button to run ROBIN.

ROBIN Output

If the input is sequence data, ROBIN will first show the values of the parameters specified by users, then output the order of the identified common LCBs that shared by all the input sequences, and finally display the calculated distance matrix with each entry indicating the block-interchange distance between any two input sequences.

In addition, users can further view the detailed information of each identified LCB order just by clicking the associated link, such as the position, length, and weight of each LCB and the overall coverage of all LCBs on the chromosome. The position of each identified LCB is denoted by its left and right end *coordinates* on the chromosome.

If both the left and right coordinates are negative values, it implies that the identified LCB is the inverted region on the opposite strand of the given chromosome. It should be noted that users can also view and check the optimal scenario of block-interchanges for any two input sequences by clicking the link associated with the entry of the distance matrix. If the input is gene/landmark data, ROBIN will only show the values of the specified *parameters* and output the calculated block-interchange distance matrix along with the optimal scenarios of block-interchanges.

POINTS TO REMEMBER

1. Block-interchanges can be considered as a generalization of transpositions from the biological perspective. From the computational perspective, the block-interchange distance problem is solvable in polynomial time, whereas the transposition distance problem is still open. As introduced in this chapter, permutation groups in algebra are a useful tool in the design of an efficient algorithm to the block-interchange distance problem. Hence, it is interesting to further study whether or not they can still be applied to the transposition distance problem.
2. Each LCB identified by ROBIN can be considered as a homologous/conserved region shared by all chromosomes without any genome rearrangements to occur. The weight of each identified LCB can be used to measure its confidence of being a true *homologous/ conserved* region. The larger the weight, the greater the confidence. By *selecting* a high minimum weight, users can use ROBIN to identify the larger LCBs that are truly involved in genome rearrangement, whereas by selecting a low minimum weight, they can trade some specificity for sensitivity to identify the smaller LCBs that are possibly involved in genome rearrangement.
3. A phylogenetic reconstruction based on the block-interchange distances can be easily carried out using PHYLIP, a package of programs for *inferring phylogenies*. PHYLIP is distributed by Joe Felsenstein (University of Washington) and available for free from http:/ /evolution. genetics. washington. edu/phylip. html.

Chapter 15

Over Growth Syndrome

Overgrowth syndrome is also known as *sotos syndrome* (sos). It is associated with mental *retardation,* specific *craniofacial* features, and advanced bone age. Since *NSD1 haploinsufficiency* was proven to be the major cause of SoS in 2002, many *intragenic* mutations and chromosomal *microdeletions* (MDs) involving the entire *NSD1* gene have been described.

The sizes of most SoS MDs are identical and a specific genomic *architecture* around these MDs was found. Recently, precise analyses of the low-copy repeats (LCRs) flanking the SoS common deletion showed that the deletion arises through *nonhomologous* recombination (NAHR) utilizing the LCRs, and proved that SoS is a genomic disorder. SoS (OMIM no. 117550), also known as cerebral *gigantism,* was originally reported by Sotos et al. in 1964.

SoS is characterized by overgrowth, characteristic *craniofacial* fea-tures, developmental delay, and advanced bone age. In 2002, *NSD1* disruption was found in a patient with SoS and *haploinsufficiency* of *NSD1* has been shown to be a major cause of SoS. In the Japanese population, about half cases have chromosomal MDs involving the entire *NSD1* gene. The majority of MDs are identical and the NAHR was shown to be a *causative* mechanism. In this chapter, we focus on recent progress in clinical and genetic aspects of SoS. The genomic *architecture* around the common MD will be presented.

DIGNOSTIC CHARACTERS

SoS typically presents with overgrowth, characteristic craniofacial features, developmental delay, and advanced bone age.

Overgrowth

Overgrowth usually starts prenatally and can be observed especially in early infancy. However, overgrowth seems to *normalize* during *childhood* and may not be found in *adulthood.*

Craniofacial Features

Macrocephaly, a high hairline with coarse hair growth, high arched palate, and *prominent* jaw are common findings. The *occipitofrontal circumference* is more than 97th *percentile* in childhood.

Developmental Delay

Developmental delay is a cardinal symptom and can be attributed to central nervous system *abnormalities.* Speech delay and abnormal motor development are commonly observed. Dilatation of cerebral *ventricles* is sometimes noted. In *neuroimaging* studies of 40 SoS patients, *prominence* of the trigone and the *occipital* horns was found in 90 and 75%, respectively.

Advanced Bone Age

Advanced bone age was observed in 31 of 37 (84%) SoS patients and does not appear to be associated with *abnormalities* of *collagen metabolism.*

Others

Cardiac, urogenital, musculoskeletal, and *ophthalmologic anomalies* have been observed also. *Neoplasms* in SoS have been found with a frequency of 2.2-3.9%.

DETECTION OF SYNDROME

SoS has been diagnosed based on clinical *manifestations,* however, the diagnosis is difficult when craniofacial features are less remarkable or are similar to those of other overgrowth *syndromes* with *overlapping* phenotypes. Weaver and Beckwith-Wiedemann syndromes associated with *NSD1 abnormalities* showed similar facial features to those of SoS.

Overgrowth tends to normalize in adulthood, thus, the diagnosis is easier in early life. Objective diagnosis of SoS has been *substantially* improved, since the discovery of *NSD1* mutations in SoS. The major *diagnostic criteria* proposed by Cole and Hughes include pre- and postnatal characteristic overgrowth with advanced bone age, and developmental delay.

However, a recent report by Rio et al. indicated that typical facial *appearance* and *macrocephaly* were consistently recognized but overgrowth or advanced bone age was not always observed in patients with *NSD1* mutations.

Therefore, the *diagnostic* criteria for SoS likely need to be carefully revised after collecting more data on the phenotypic spectrum in SoS patients with known *NSD1* abnormalities. "Sotos-like syndrome" usually has been referred to a large spectrum of patients that do not fulfill the major criteria for SoS.

BIOMOLECULAR APPROACH

Mouse *Nsd1* was originally isolated in 1998 as one of the nuclear proteins interacting with retinoic acid receptors and thyearsoid hormone receptors. The human *NSD1* was identi-fied at the 5q35 breakpoint of a SoS patient with an apparently balanced *translocation* t(5;8)(q35;q4.1).

NSD1 has an 8088-bp open reading frame consisting of 23 exons and is translated into a 2696 amino-acid protein. NSD1 protein has at least six functional domains, a su[var]3-9,enhancer-of-zest,trithorax (SET) domain, two proline-tryptophan-tryptophan-proline (PWWP) domains, and three plant *homeodomain* protein-finger (PHD) domains with nuclear *localization* signals.

The SET domain of NSD1 showed methyl trans-ferase activity for Lys36 of histone H3 (H3-K36) and Lys20 of histone H4 (H4-K20), whose methylation may result in transcriptional silencing of developmentally regulated genes. PHD finger domains are found in a large number of *chromatin* regulatory factors like CBP/p300, and chromatin remodeling protein ACF.

The PWWP domain was initially identified in the protein encoded by the Wolf-Hirschhorn syndrome candidate gene 1 *(WHSC1)*. The essential function is still unknown, however, the PWWP domain is often *associated* with SET domains and is thought to be an essential for development. Thus, these six protein domains are suggested to regulate *chromatin* formations and gene *transcriptions*.

Homozygous knockout of *Nsd1* in mice led to embryonic death at 10.5 days and gastrulation failure, *suggesting* that Nsd1 maybe a protein regulating development especially in an early post-implantation period. The NSD1 is a portion of a fusion protein in a recurrent translocation t(5; 11)(q35;p15.5) found in childhood acute myeloid leukemia.

Six non-SoS cases of acute myeloid leukemia had fusion transcripts of *NSD1* and *NUP98* (nucleoporin 98 gene). SoS has been suggested to be associated with *neoplasms*. It is also important to address whether *NSD1* abnormality in itself can cause *tumorigenesis* in the near future.

NSD1 Mutations

NSD1 mutations in SoS have been reported by several groups. To date, among a total of 91 point mutations (PM) have been identified in 241 patients suspected of SoS. Seventy-one protein truncation mutations and 20 missense mutations have been reported with 62 being de novo. Five protein *truncation* mutations were found in 23 Sotos-like patients.

Protein truncation mutations are spread throughout the entire *NSD1* coding regions; however, *missense* mutations cluster at the 3' part of *NSD1* where most of the known functional domains are located.

In SoS, six missense mutations have been identified in the SET domain, three in the PHD domains and three in the PWWP domains. Thus, among different world populations, *NSD1* mutations have been consistently shown to be the major cause of SoS. Interestingly, the frequency of MDs involving *NSD1 is* quite different.

In the Japanese population, about half of the cases (49/95) had MDs, whereas microdeletions are observed in only 11% of cases (13/118) analyzed in European populations. The reason for this observed difference remains to be elucidated.

CORRELATION BETWEEN GENE AND THEIR PHYSICAL EXPRESSION

Interestingly, some clinical differences between SoS patients with PMs and MDs have been reported. Nagai et al. compared clinical *phenotypes* between 5 PM and 21 MD SoS patients. Both PM and MD cases showed typical craniofacial features. Remarkably, the peak height at younger than 6 years of age and the intelligence quotient/developmental quotient (IQ/ DQ) in patients older than 6 years were significantly different between PM and MD patients.

The values of the standard deviation (SD) scores were 3.3 (PM) and 2.2 (MD), and IQ/DQ

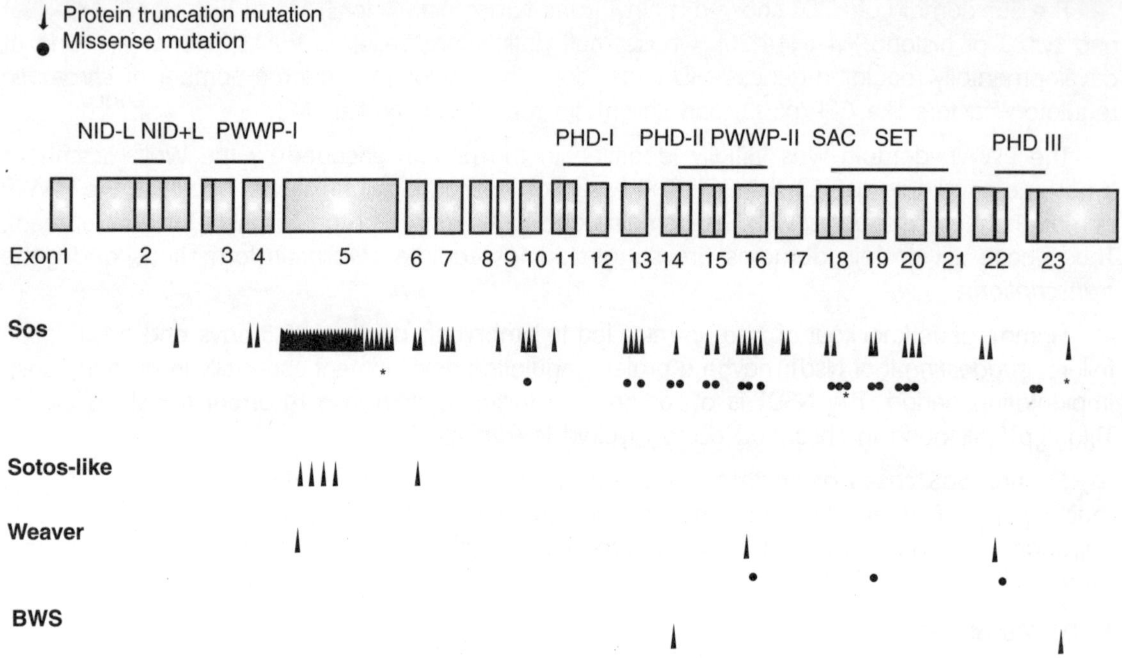

Figure 15.1: Showing the mutations in Sotos syndrome, Sotos-like syndrome, and other syndromes. Each box indicates exon, and nuclear receptor interacting domains, NID^{-L} and NID^{+L}. Start and stop codons are located at exon 2 and 23, respectively. Arrowheads and filled circles indicate protein truncation mutations and missense mutations. Asterisks below arrowhead or circle show familial mutations.

(mean) were 78 ± 12 (PM) and 57 ± 12 (MD), respectively. In addition, MD patients predominantly showed *cardiovascular* and *urogenital abnormalities*, and recurrent convulsions. Rio et al. compared 16 PM and 6 MD patients. Two MD patients showed typical SoS, but the other four patients were diagnosed as Sotos-like syndrome because they did not have overgrowth (height less than +2 SDs), or advanced bone age. Four out of six MD patients had severe mental retardation with no speech at all. *Cardiovascular anomalies* were found in three out of six MD cases.

Among 16 PM cases, 14 were diagnosed as typical SoS and 2 as Sotos-like syndrome owing to the absence of advanced bone age. The degree of mental *retardation* was variable and 2 out of 16 patients had cardiac *septal* defect. Both reports suggested that mental retardation in SoS patients with MD is more severe than in *patients* with PM and *cardio-vascular* complications in SoS patients with MD are more frequent than in those with PM.

Mutations in Other Overgrowth Syndromes

Notably, in six cases of Weaver syndrome, whose phenotype overlaps significantly with SoS and in two cases of another overgrowth syndrome, Beckwith-Wiedemann syndrome (BWS), *NSD1* mutations were identified. In *Weaver syndrome*, 6 cases out of 13 *had NSD1* intragenic PM, suggesting that SoS and Weaver syndrome are allelic.

The majority of BWS is caused by either genetic alterations (11p15 paternal *uniparental* disomy or CDKN1 C mutations) or *epigenetic* defects (*demethylation* of the *KvDMR 1 region of KCNQ 1OT* and hypermethylation of *H19)*. Interestingly, in addition to two BWS cases with *NSD1* mutations, two SoS cases without *NSD1* abnormality showed abnormal status of the 11p15 region (demethylation of *KCNQ1OT* and a paternal isodisomy of 11p15).

These data indicate challenges for proper clinical diagnosis of even well-established overgrowth syndromes. Another possibility is an unknown common pathway among the three syndromes. It is very important to evaluate *NSD1* status in other overgrowth syndromes to elucidate whether *NSD1* mutations are specific not only for SoS.

EVIDENCE IN SUPPORT OF SOS

Fifty MDs have been analyzed using fluorescence *in situ* hybridization and *microarray comparative* genomic *hybridization*. Three different types of microdeletions were delineated, among which, the approx 2-Mb MD I was the most common (found in 46 out of 50 patients). In the other four cases two smaller MDs were recognized.

Highly homologous regions at each deletion breakpoints of the MD I were identified. These LCRs were termed Sotos syndrome distal-repeat (SoS-DREP, approx 429 kb) and proximal-repeat (SoS-PREP, approx 390 kb). Sequence comparisons of SoS-DREP and SoS-PREP revealed that six sequence homology subunits (A-F) of PREP showed more than 96% identity to DREP.

Their sizes of SoS-PREP subunits were 123.6 kb (A), 20.1 kb (B), 62.8 kb (C), 7.8 kb (D), 8.2 kb (E), and 93.9 kb (F) and those of SoS-DREP subunits were 119.1 kb (A), 19.7 kb (B), 68.7 kb (C), 7.8 kb (D), 8.3 kb (E), 82.8 kb (F), and 50.1 kb (C′). Each of the homologous subunits, with the exception of one, is located in an inverted orientation and the order of subunits is different between the two SoS-REPs.

Only the subunit C′ in SoS-DREP is oriented directly with respect to the subunit C in SoS-PREP. These subunits are more than 99% identical. Two recent reports showed that the subunit C′ in SoS-DREP and the subunit C in SoS-PREP, were utilized as a substrate of NAHR of the SoS common deletion.

In addition, the reports indicated that the crossover events occurred in those subunits and that an approx 80% of crossover hotspots were within an approx 3-kb genomic sequence in those subunits. These data established that SoS is a new genomic disorder and an NAHR mechanism is a consistent mechanism for generation of the SoS common deletion as in other genomic disorder reported.

MICRODELETION IN SOS

There are at least 22 genes that map within the common deleted region. Both SoS-REPs contain two genes, *THOC3* and *NY REN-7*. *THOC3* and *NY-REN-7* have open reading frames that are completely conserved in SoS-PREP and SoS-DREP. The *PROP I* gene maps only to SoS-DREP between subunit E and C′.

Among those 22 genes deleted, *NSD1*, the plasma coagulation factor 12 gene *(F12*, OMIM +234000), the prophet of the *PIT-.1* gene *(PROP1*, OMIM +601538), and the xylosylprotein β1,4-galactosyltransferase. polypeptide 7 gene *(B4GALT7*, also known as xylosylprotein 4-β-galactosyltransferase I gene, *XGTP1*, OMIM *604327) may be directly related to human phenotypes.

F12 encodes the coagulation factor XII, also known as Hageman factor. Heterozygous deletion of F12 may result in partial F 12 deficiency, which could present with a slight to moderate bleeding tendency. Low levels of factor XII activity may also be a risk factor for repeated spontaneous abortions or skin ulcers.

A common polymorphism in the 5'-untranslated region of *F12,* the c.46C > T substitution, was found to be associated with low F12 level. In cases of c.46T/T, the value of F12 was remarkably decreased. Soria et al. reported that the 5q33-qter region is a quantitative risk factor for thrombosis using genome wide linkage analysis.

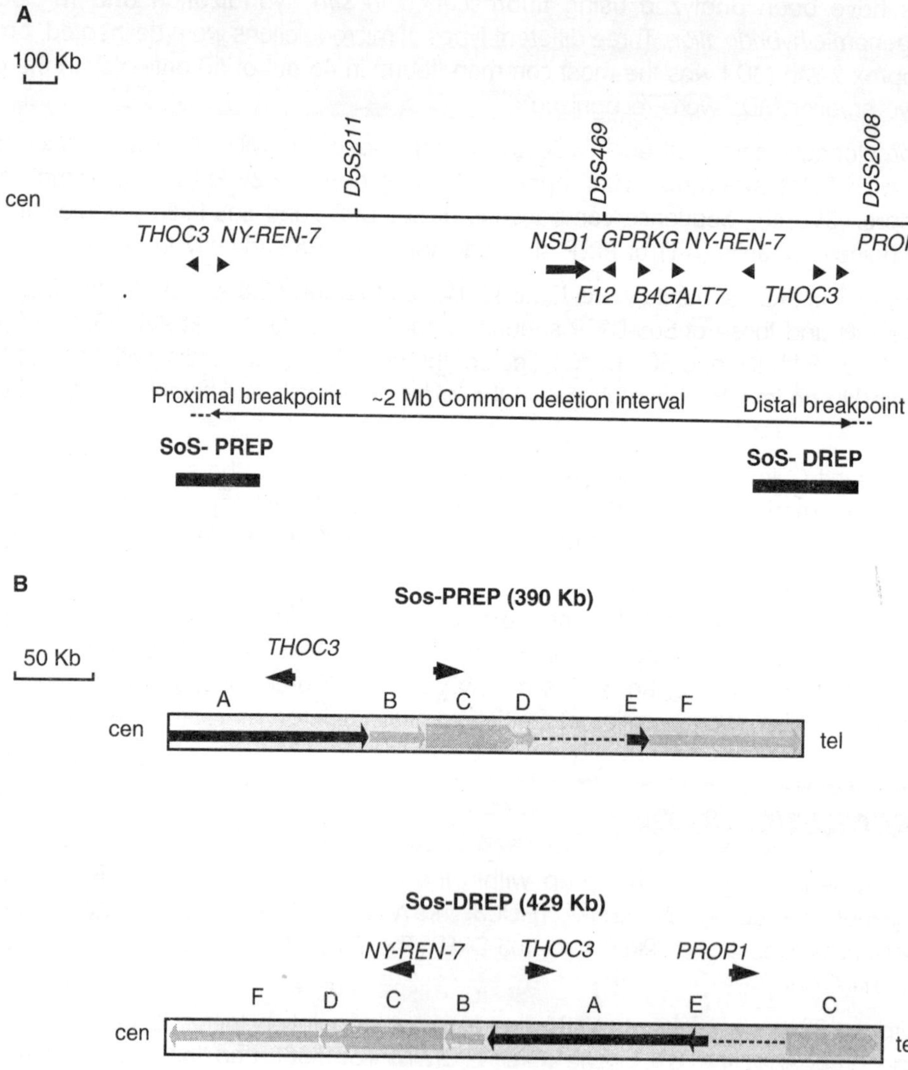

Figure 15.2: (A) Physical map showing microdeletions in Sotos syndrome (SoS) and two low-copy repeat sequences, termed proximal-repeat (SoS-PREP) and distal-repeat (SoS-DREP) at 5q35.

A novel homozygous p.W484C mutation was shown to induce low F12 levels. It is important to evaluate F12 in SoS patients with MDs, although such a risk has not been known in SoS. *Homozygous* or *compound-heterozygous* defects of *PROP1* result in combined pituitary hormone deficiency including GH, PRL, TSH, LH and FSH (OMIM +601538). So far, three SoS cases associated with hypothyearsoidism have been reported.

Unmasking of the recessive *allele* is possible when one allele *harbors* a PM and the other is deleted. It may be worth investigating *PROP1* if hypothyearsoidism is observed. B4GALT7 regulates the synthesis of various *glycosaminoglycans* (GAGs). GAGs are basic components of *heparin/ heparan* sulfate or those of chondroitin sulfate/dermatan *sulfate* and have an important role in the *formation* of various *tissues* and *organs*.

Defects of GAGs may be possibly responsible for the various forms of so-called *mucopolysaccharidoses*. In the progeroid type of Ehlers-Danlos syndrome, compound *heterozygosity* for p.A 186D and p.L206P mutations of *B4GALT7* was confirmed. The father was heterozy-gous for the p.L206P allele and mother *heterozygous* for the p.A 186D allele. Although only one case with such mutations has been reported, carrier status for such PMs in contributions with hemizygous deletion of *B4GALT7* in SoS patients with MDs could contribute to *phenotypic* variability.

GPRK6 encodes G protein-coupled receptor kinase 6 protein (GPRK6) (OMIM *600869), which can regulate G protein-coupled receptors. Using immunohistochemistry, GPRK6 expression was confirmed in striatal *neurons* receiving dopaminergic input and *postsynaptic* D2/D3 *dopamine* receptors were targets of GPRK6.

Investigation of *GPRK6* by gene targeting to create a knockout animal shows higher sensitivity to *psychostimulants* including cocaine and *amphetamine* especially in homozygous mice rather than *heterozygous*, suggesting that such high sensitivity may be related to some potential *psychiatric* diseases in human.

It would be interesting to evaluate for different psychiatric and behavioural aspects between SoS cases with MD versus PM. The influence of the deletion of 21 genes other than *NSD1* needs to be carefully evaluated, as some genes may affect the *severity* of phenotypes in MD patients.

APPLICATION

Rearrangement-prone regions of the human *genome* including LCRs have been challenging to sequence. Validation and mapping of MD breakpoints at the *nucleotide* level should provide further insights into the mechanisms of DNA rearrangement. Functional studies of *NSD1* are required for elucidating pathophysiological aspects of SoS. Intensive molecular analyses of 282 patients with SoS, Sotos-like syndrome, and Weaver *syndrome* revealed the *NSD1 abnormalities* in 168 cases; *NSDI* was intact in the other remaining 114 cases.

Improved methods to detect other types of *NSD1* abnormalities, including partial deletion and nucleotide changes of introns and promoter regions, and more data of clinical phenotypes observed in patients with MDs and PMs should reveal further genotype/phenotype *correlations* and provide insights into SoS *pathophysiological mechanisms*.

Chapter 16

Genetic Disease of Myelin

Gene dosage is a unique mechanism underlying some genetic disorders of *myelin* in the CNS and peripheral nervous systems (PNS). *Alterations* in gene dosage often are a consequence of genomic rearrangements such as duplication and deletion, which respectively result in an increased or decreased production of proteins encoded by the genes within the altered genomic interval.

At least, two myelin genes are known to be dosage sensitive and the changes in their copy number can cause *dramatic* disturbance of myelin development and maintenance. The first and best characterized gene is the *peripheral myelin protein* 22 gene *(PMP22);* duplication of *PMP22* causes CMT1A, whereas the *reciprocal* deletion leads to HNPP. The *PLP1* is the second example of a dosage-sensitive myelin gene.

Altered dosage of *PLPI,* which can arise from gene duplication or gene deletion, causes PMD (MIM 312080). PMD is characterized by arrest of *oligodendrocyte* differentiation and failure to produce myelin in the CNS, resulting in developmental delay usually beginning in the first year of life. PMD patients also develop additional neurological features including *nystagmus, pyramidalm* and *extrapyramidal* signs and *cerebellar* symptoms.

Sequence alterations in the *PLP1* coding regions and splice junctions can also cause PMD *or* spastic paraplegia type 2 (SPG2; MIM 312920), a clinically *distinct* and *milder* disorder that has some similarities to PMD. In this chapter, genomic mechanisms underlying *PLP1* duplication and deletion are compared to those of *PMP22* duplication and deletion.

PLP1 AND PMP22

Both *PLP1* and *PMP22* encode major myelin proteins. Each appears to play an important role in the development and maintenance of myelin, although the exact functions of these proteins remain elusive. PLP1 is primarily expressed in *oligodendrocytes* that form myelin in the CNS, whereas PMP22 is *predominantly* expressed in the Schwann cells that produce myelin in the PNS. *PLP1* spans an approx 17-kb genomic interval on chromosome Xq22.2 and is composed of seven *coding exons.* PLP1 contains 276 *amino acids,* of which 35 residues encoded in the latter half of exon 3 are alternatively spliced out in DM20.

DM20 is predominantly expressed in oligodendrocyte precursors before the initiation of myelin production, whereas major PLP1 expression occurs after birth in conjunction with myelin maturation. *PMP22* is located on chromosome 17p12 and is composed of four coding exons. There are two alternatively spliced non-coding first exons that are transcribed from independent promoters.

Both transcript variants encode the same protein sequence that is 116 residues shorter than PLP1. Expression levels of both *PLP1* and *PMP22* are tightly regulated during development. Modest expression of both genes is observed in *nonmyelin* lineage cells, but the major expression occurs at the stage of terminal differentiation and myelin *formation* in the CNS or PNS. PLPI/DM20 and PMP22 are *tetra-span* membrane proteins with both the amino and carboxyl termini located in the cytoplasm.

During the process of protein synthesis in the endoplasmic reticulum (ER), both proteins are properly folded before being transported through Golgi apparatus to the myelin membrane. PMP22 undergoes *post-translational glycosylation,* whereas PLP1DM20 does not. Disease-causing *amino acid* substitutions in both PLP1/DM20 and PMP22 result in protein *misfolding* and *accumulation* in the ER.

Despite the different glycosylation status of each protein, calnexin, a molecular chaperone in the ER, appears to play a role in quality control of both proteins. Interestingly, abnormal accumulation of mutant PLP/DM20, but not PMP22, evokes the unfolded protein response, subsequently leading to apoptotic cell death.

Nevertheless, accumulation of misfolded proteins in ER leads to toxic gain-of-function effects in both proteins, which is consistent with the observation that patients with point mutations in either *PLP1* or *PMP22* generally have more severe disease.

PMD/SPG2 AND CMT1A/HNPP

PMD/SPG2 and CMT1A/HNPP result from a variety of disease-causing mutations in either *PLP1* or *PMP22* that occur through distinct mechanisms. Genomic duplication is the most frequent mutation in both PMD and CMT1A.

PLP1 duplications account for approx 60-70% of PMD patients. Similarly, more than 70% of CMT1 patients have *a PMP22* duplication. *PLP1* genomic deletion has only been reported in three families, whereas *PMP22* deletion causes the common trait HNPP. Duplication of *PLP1* doubles gene dosage in males owing to X-chromosome hemizygosity.

This contrasts with the autosomal *PMP22* duplication that increases the gene dosage by half. As shown by increased severity in rare patients with either a triplication of *PLP1* or a homozygous duplication of *PMP22,* disease severity is proportionate to gene copy number. Consistent with these observations, mice with homozygous *Plp 1* transgenes have a prominent dysmyelinating phenotype and shortened lifespan, whereas heterozygotes develop a late-onset demyelinating disease. Similarly, *Pmp22* transgenic murine models have shown that the expression level of *Pmp22* correlated with the degree of demyelination. In contrast to the severe phenotype observed

Table 16.1: Similarities and Differences Between PMD/SPG2 and CMT1A/HNPP

		PMD/SPG2	CMT1A/HNPP
Prevalence		< 1 / 100,000	~1/2500
Gene (locus)		*PLP1* (Xq22.2)	*PMP22* (17p 12)
Expression		Oligodendrocyte in	Schwann cells in
CNS myelin		PNS myelin	
Protein		Tetra-span membrane protein	Tetra-span membrane protein
Mutations, frequency and mechanism			
Duplication	Frequency	60-70% of PMD	>70% of CMT 1
	Gene dosage	Twice (in males)	1.5 times
	Severity	Moderate	Moderate
	Cellular pathology	Accumulation in late endosomes	Accumulation in late endosomes
Deletion	Frequency	Rare	>90% of HNPP
	Gene dosage	Null	Half
	Severity	Mild	Mild
	Cellular pathology	Protein deficit	Haploinsufficiency
Point mutation	Frequency	~20%	Rare
	Severity	Mostly severe	Mostly severe
	Cellular pathology	Protein misfolding and ER accumulation	Protein misfolding and ER accumulation
Genomic mechanisms for recombination		NHEJ	NAHR
Clustered breakpoints		Proximal end: no Distal end: yes	Yes
Order of duplication	Tandem/inverted	Tandem	
Origin of recombinant products		Sister chromatid exchange	Crossover between two homologous chromosomes
Involvement of LCRs in the recombination		Probably yes	Yes
Presence of flanking LCRs		No	Yes

with duplications, deletions of either gene results in a milder phenotype. This is especially surprising in *PLP1* deletion, because it results in a *PLP1* null genotype in males. Male patients with *PLP1* deletion show mild PMD or SPG2. Similarly, mice lacking *Plp1* revealed minimal dysmyelinating changes. accompanied by slowly *progressive axonal degradation*.

Heterozygous deletion of *PMP22* results in HNPP, a milder disease than duplication-causing CMT1A. Mice *heterozygous* or *homozygous* for Pmp22 null alleles also recapitulate HNPP with a different degree of severity. In summary, a gene dosage defines the severity for diseases involving both *PLP1* and *PMP22*. It is worth noting that having an extra copy of *PLP1* or *PMP22* conveys a more toxic effect than not having enough of them, which appears to be a unique feature of these two genes.

However, the molecular and cellular bases for the pathophysiology of the gene dosage effect remain largely unknown. Interestingly, recent studies on overexpression of both genes revealed aggregation of proteins within late endosomes coupled with sequestration of myelin membrane components. This common cellular phenotype suggests a similar cellular pathology probably underlies both *PLP1* and *PMP22* duplications.

CHROMOSOMAL ABERRATION

Unlike structural chromosomal *abnormalities* caused by constitutional chromosomal rearrangement, genomic duplications and deletions associated with PMD1SPG2 and CMT1 A/ HNPP are "*submicroscopic*" and not resolved by routine chromosome banding analyses. On the other hand, DNA sequencing cannot detect the gene dosage differences, because the rearrangement does not create a sequence-based abnormality.

Therefore, detection of duplication and deletion require other specific molecular technologies. Interphase fluorescent *in situ* hybridization (FISH) with specific probes that contain the *PLP1* genomic locus with an intrachromosomal control has been applied widely for molecular diagnostics of PMD. Interphase FISH can clearly visualize difference in copy number both in male patients and in female carriers.

Alternatively, quantitative polymerise chain reaction can be utilized as an indirect measurement of gene dosage. Recent advances in real-time polymerase chain reaction quantification technologies have dramatically improved sensitivity, accuracy, and efficiency of such quantitative dosage determinations.

Finally, microarray-based comparative genomic hybridization analysis has emerged as a novel technology that enables detection and determination of the size of the duplicated or deleted genomic intervals in one experiment. Such technologies can provide an enormous amount of information for genomic rearrangements and promote elucidation of the underlying genomic mechanisms.

BIOPROCESS

PMD1SPG2 and CMT1A/HNPP are similar in the types of mutations causing disease, in the relationship between gene dosage and phenotypes, and in the cellular pathology. In contrast, recombination mechanisms for disease-associated genomic rearrangement appear to be significantly different between these diseases. In the majority of genomic disorders, a precise genomic rearrangement commonly results in a uniform recombinant product with clustered breakpoints. The CMT1A duplication and the HNPP deletion represent a prototype for this model of genomic rearrangement.

A 1.4-Mb genomic interval, which encompasses *PMP22* and 20 other genes, is flanked by highly homologous 24-kb low-copy repeat (LCR) units, called proximal and distal CMT1A-REPs. Homologous recombination between nonallelic (or paralogous) copies of CMTIAREPs (i.e., nonallelic homologous recombination [NAHR]) and resolution of its intermediates (i.e., Holliday junctions) results in unequal crossover that causes the CMT1A-associatedduplication and the reciprocal HNPP-associated deletion.

Because the CMT1A-REP repeat-specific genomic feature conveys the susceptibility to NAHR at this locus, the CMT1A duplication and the HNPP deletion recur with high frequency. The same mechanism appears to underlie many other genomic disorders.

The molecular mechanisms for the genomic rearrangements that cause *PLP1* duplication and deletion remain largely elusive, but current knowledge suggests that the involvement of NAHR is unlikely. FISH mapping of the duplicated genomic intervals in individuals with PMD revealed no common recombinant duplicated genomic segments.

It has been thought that the duplications are mainly tandem in nature, but our recent findings also suggest that some are inverted duplications (Lee, move, and Lupski, unpublished data). In contrast to CMT1A, duplicated genomic segments often originate from a single chromosome rather than two homologous chromosomes, as shown by identical haplotypes within the duplicated genomic intervals.

In addition, the origin of the *de novo* duplication events can often be traced to male spermatogenesis. Each of these findings suggests that *PLP1* duplications are commonly generated by sister-chromatid exchange in male meiosis.

Unlike observations in CMT1A and HNPP, extensive genome sequence analysis of this interval identified no flanking pair of tandem LCRs that could serve as substrates for NAHR. Instead, an inverted pair of LCRs was found approx 180-kb distal to *PLP1;* these have been named LCR-PMDA and LCR-PMDB. An analysis of recombination breakpoints associated with the rare *PLP1* deletions has provided a potential model for the genomic rearrangements causing the more common *PLP1* duplications.

The precise mapping and sequencing of recombination junction fragments in three *PLP1* deletion families revealed the involvement of the following complex and distinct processes, including (1) inheritance of an unbalanced insertional translocation, (2) sisterchromatid exchange in male meiosis, and (3) complex rearrangement.

Nevertheless, the products of recombination in each family, when the strand exchange products were analyzed at the nucleotide sequence level, revealed that the recombination process appeared to occur by nonhomologous end-joining (NHEJ). Unlike homologous recombination, NHEJ, a major pathway for repairing double-strand DNA breaks, does not require a stretch of homologous sequences for initiating recombination. Thus, recombination can occur anywhere on a chromosome arm.

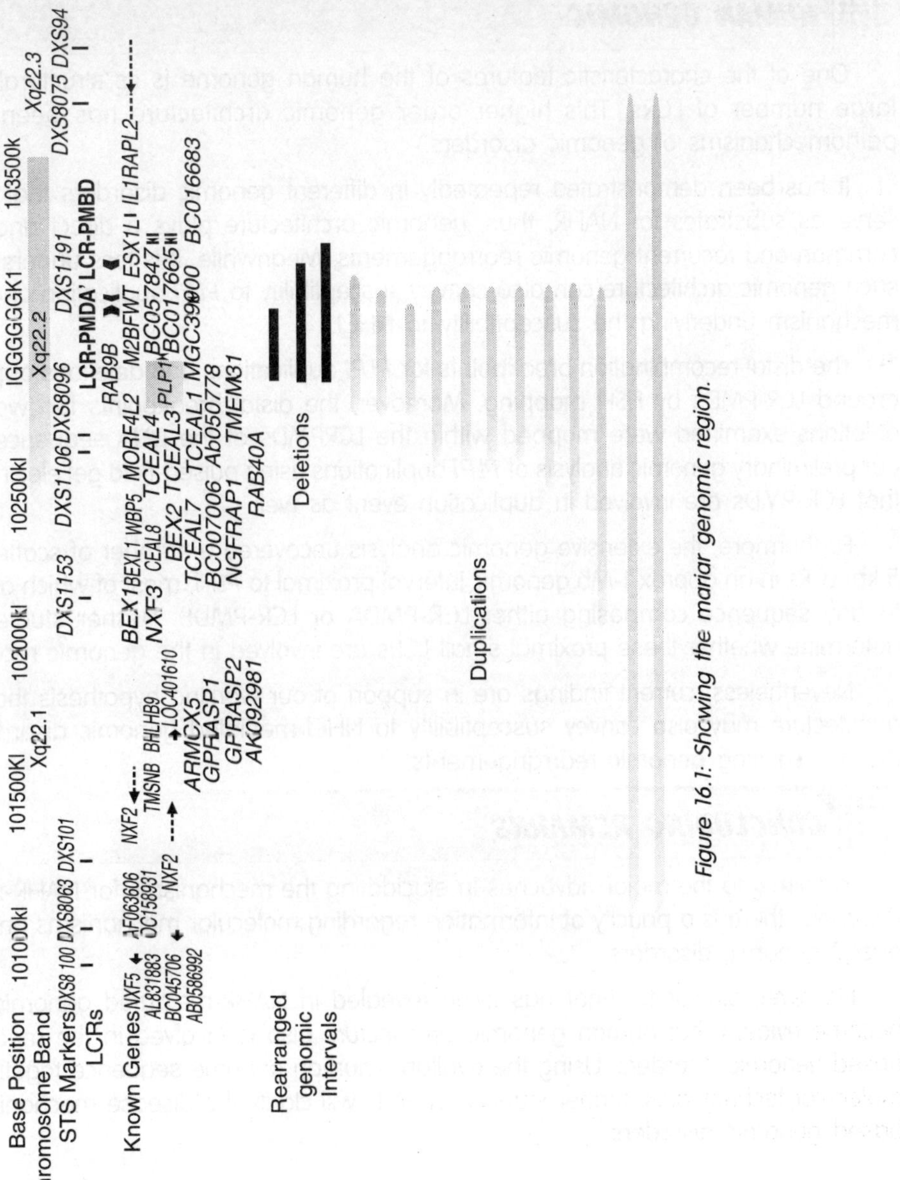

Figure 16.1: Showing the mariar genomic region.

Based on the observation that the recombination breakpoints for *PLP1* duplication are not confined to specific positions but spread across the *PLP1* region and that pairs of tandem LCRs do not flank the duplicated genomic interval, it is likely that NHEJ mediates the genomic rearrangements that lead to *PLP1* duplications. Supporting this hypothesis, DNA sequence analyses of several duplication junction fragments appeared to show absence of homology at the recombination breakpoints, evidence for NHEJ and not NAHR.

HUMAN GENOMIC

One of the characteristic features of the human genome is its structural complexity with a large number of LCRs. This higher order genomic architecture has been implicated in the pathomechanisms of genomic disorders.

It has been demonstrated repeatedly in different genomic disorders that locus-specific LCRs serve as substrates for NAHR, thus, genomic architecture plays a direct and significant role in common and recurrent genomic rearrangements. Meanwhile, it is less understood as to whether such genomic architecture can also convey susceptibility to *PLP1* duplication and, if so, what is the mechanism underlying the susceptibility to NHEJ.

The distal recombination breakpoints for *PLP1* duplications and deletions appear to accumulate around LCR-PMDs by FISH mapping. Moreover, the distal breakpoints for two out of three *PLP1* deletions examined were mapped within the LCRPMDs at the DNA sequence level. In addition, our preliminary genomic analysis of *PLP1* duplications using pulsed-field gel electrophoresis suggests that LCR-PMDs are involved in duplication event as well.

Furthermore, the extensive genomic analysis uncovered a number of scattered, small (approx 5 kb) LCRs in an approx 1-Mb genomic interval proximal to *PLP1,* most of which are not homologous to any sequence composing either LCR-PMDA or LCR-PMDB. Further studies are required to determine whether these proximal small LCRs are involved in the genomic recombination events.

Nevertheless, current findings are in support of our working hypothesis that human genomic architecture may also convey susceptibility to NHEJ-mediated genomic disorders by stimulating disease-causing genomic rearrangements.

CONCLUDING REMARKS

In cotrast to the major advacnes in elucidating the mechanisms for NAHR-mediated genomic disorders, there is a paucity of information regarding molecular mechanisms underlying the NHEJ-based genomic disorders.

However, similar to what has been revealed in NAHR-mediated genomic disorders, it has become evident that human genomic *architecture* also is involved in the mechanism for NHEF-based genomic disorders. Using the available human genome sequence together with advanced molecular technologies, further studies on PMD will clarify the disease mechanisms for the NGEF-based genomic disorders.

Chapter 17

Chemotherapy of Tuberculosis

Approximately 30 years ago, the prevailing opinion was that the light against infectious diseases was all but over. *Optimistic* predictions were made in the belief that a *combination* of improved *public health, vaccination,* and the current variety of *antimicrobial* agents would be sufficient to combat most, if not all, infectious diseases. Synthetic analogs of known *antibiotics* that were effective against a broad range of organisms had been developed, and thus there was little *perceived* need for the development of new drugs.

In fact, until very recently, no new classes of antibiotics had been licensed for clinical use against *bacterial* infections. The search for new drugs was limited to well-known compound classes with insufficient variability to prevent the escalation of clinical resistance With the increasing incidence of infection with multiple-drug-resistant tuberculosis strains. the World Health Organization (WHO) declared a worldwide emergency and *responded* with the directly observed *therapy, short-course* (DOTS) *program.*

Streptomycin, isoniazid, and *pyrazinamide,* which were developed in the 1940s, are still integral parts of the *chemotherapeutic* regimens used today. The most recent introduction of an *antituberculosis agent* into the modern treatment scheme of *tuberculosis* was *rifampin* in the 1960s.

The current treatment recommendations for tuberculosis from the Centers for Disease Control and the WHO for pan-sensitive strains involves a 6-month regimen, beginning with a 2-month (intensive phase) treatment cycle of *isoniazid, rifampin* and *pyrazinamide.* followed by a 4-month cycle of isoniazid and rifampin.

Although protracted, this is considered short-course chemotherapy. Alternative agents such as ethambutol and streptomycin are included in the intensive phase if there is a possibility of drug resistance. The postgenomic era of expression profiling affords insight into the global biological response induced under diseaselike conditions in addition to drug pressure. Through the identification of gene and operon responses, specific pathways can be identified that are unique to a particular disease.

To study bacteria in a disease like state has always been a difficult undertaking to achieve. Recently studies have been performed looking at the global gene response to altered growth states using microarray. These studies have been performed under artificial growth conditions to control the number of variables. In each case, specific changes were made, such as oxygen and nutrient limitation.

Although much information has been obtained at the level of the gene and related metabolic pathway, there is still not a clear picture of the impact this information has on the biology of the bacterium and thus the pathology of the disease. However, when these genes and pathways are analyzed, a picture develops of how the cell alters related *regulons* in order to compensate for the change in growth environment.

The identification of these regulon networks results in information on the metabolome of the *bacillus*. This knowledge can be used to establish gene essentiality, conditional expression requirements, and determine genes *associated* with virulence.

Although this information is extensive for the simple design of *highthroughput* screens, it affords predictive value in terms of the targets affected and the circumventive measures the bacillus uses to avert the pressure of drugs under different growth conditions such as the intracellular environment. To some, high-throughput screening (HTS) has come to mean screening thousands of compounds with a pass—fail criterion. Ironically. none of the current antitubercular drugs could have passed the stringent conditions that are in place today.

Therefore, we now realize that FITS is not intended to screen-out candidates but, rather, define rapidly and systematically the properties of many compounds so that they can be prioritized and further characterized through relevant secondary screens. It is important to clearly define the criteria that will he used to judge the candidates at each step of the drug discovery process.

The overall goal is to make HTS a rational process such that assumptions made in the design do not compromise the quality of the data and therefore *adversely* impact the conclusions drawn from that data.

MYCOBACTERIUM TUBERCULOSIS GENOME

The sequencing of complete genomes has had a profound influence on drug discovery, with its main promise being the potential to understand disease processes at a genetic level and determine optimal targets for drug *intervention*. Thus far over 1 100 complete genomes have been sequenced, including 98 from bacteria and 16 from archaea (National Center for *Biotechnology* Information web site).

This list includes many important human pathogens, although the bias is toward organisms that affect individuals in the developed countries. The complete *M. tuberculosis* genome was added to the list in 1998. It is obvious that this sequence contains information on all possible drug targets encoded therein. Genomic information is now an integral part of the modern drug discovery process and can be used *interpretatively* to understand protein and pathway function, allowing rapid target identification.

This understanding is then used to establish a potential pathway to support the target of interest and virtually *validate* it. In addition to interpretive analysis, the greatest benefit of genome sequence is that it affords predictions toward establishing the metabolism of the *bacterium*. Usually researchers study *M. tuberculosis* on an artificial medium. Although the *transcriptional* regulators and associated networks can be used to predictively develop assays, they may not be reflective of the metabolism occurring in the organism during an infection and disease.

The availability of the entire sequence information provides insight into the global metabolic capabilities and tendencies. which can be used to guide inhibitor design in a manner that may bias development toward relevant aspects of disease. This approach has been deemed "*structural genomics*," and the aim is to structurally characterize proteins and thus pathways from sequence information.

There is considerable effort in the postgenomic era to validate protein predictions with experimental determination via crystallography and NMR spectroscopy and synthetic pathways via experimental enzymology. However, currently the functional *annotation* of proteins and thus *metabolic* roles is based primarily on similarities to homologs in the databases. The primary *assumption* is that proteins with homologous sequences *catalyze* the same reactions.

Although this is generally true, gene sequence cannot always be a reliable way to determine protein activity. Frequently, homologous enzymes do not catalyze the same chemical reaction and the same chemical reaction is not catalyzed by *homologous enzymes.*

Although homologs can maintain function, there is mounting evidence that *divergent* evolution is an active process: a progenitor gene is duplicated and its copy assumes a new function in response to selective pressure. Because the outcome of divergent evolution can be quite different. functional assignment on the basis of homology could be *grossly misassigned.* Although gene and protein homology can provide information about tbe nature of the reaction that is *catalyzed, homology* alone will provide very little information about the specific identity of the reaction.

Therefore, genomic enzymology requires a correlation between structure (sequence) and chemical reaction (*enzymatic* function); neither structure-only-based nor function-only-based analysis will be totally accurate. This could be an underlying factor contributing to the high percentagc (40% in the ease of *M. tuberculosis)* of genes that have no known function. However, regardless of the difficulties associated with predicting all of the diverse chemical reactions present in the life of bacteria, including *M. tuberculoo*rsis, genomic information has already proven invaluable in addressing a host of important research questions.

A primary advantage of the availability of sequence information is that it has effectively made classical "*reverse genetics*" highly abbreviated. With genomic information, genes can be rapidly identified and cloned to evaluate function, and this approach is proving particularly useful when *coupled* with classical biochemical knowledge. Below we provide examples of this principle in the context of target definition and drug discovery.

Pathways that can be exploited as novel and potent antibacterial targets are being mined *in silico* and this information is then applied to actual validation experiments. Identifying genes, predicting protein function and pathway elucidation from *pre-existing biochemical* data, is proving useful in determining the role of *bypothetical* and unknown genes through *linkage.*

This linkage may be established based solely on location in the genome and proximity to known genes or by known function established through *homology* or *experimentation.* Ultimately, the information that is acquired contributes to the elucidation of specific and associated pathways, which may provide complete *metabolome information*. This then can provide clues about essential genes involved with disease, *virulence factors,* and *conditional responses.*

Experiments designed to elucidate specific metabolic responses are rapidly developing into potential options for drug screening. In an attempt to understand the disease process at a genomic level, studies have been performed to explore the altered metabolism of *M. tuberculosis* grown under defined conditions thought to mimic their intracellular niche.

Classical biochemical studies have demonstrated that *M. tuberculosis* can survive extended periods of nonreplicating, stationary phase states, induced by limited growth conditions such as reduced oxygen and nutrient deprivation. However, until recently, these studies have not addressed the bacterial response at the *gene-protein* level. Using microarray and protcome analysis, under nutrient starvation growth conditions, differential gene and protein expression were investigated.

Proiconie analysis derived from bacteria under these conditions revealed alterations in several proteins. Among repressed proteins were the 45-kDa and MPT64 antigens. It is interesting that these very proteins are the ones primarily recognized by the *immune* response of the majority of tuberculosis patients and their contacts.

This phenomenon of the repression of the expression of two important *immunogens* under nutrient starvation may be explained by the fact that these are secreted proteins, and thus the *artificial* nutrient starvation conditions may affect protein secretion. It is not at all clear whether such artificially induced nutrient starvation conditions reflect long-term *persistence* or reactivation of discase.

Several protein spots were shown to increase in intensity under starvation conditions. Among these are two conserved hypothetical proteins that have no identifiable functional motifs or domains when an *in silico* sequence-based analysis was performed. In a more sensitive, perhaps more relevant and global approach to analyze the adaptive response of *M. tuberculosis* to nutrient starvation, the transcriptional profile was monitored at multiple time points using DNA microarrays. *Differential* gene expression was detected for all of the major functional gene classes.

Genes associated with energy metabolism such as *aceAb* were upregulated, whicb may reflect altered carbon sources, although other tvilarboxylic acid cycle (TCA) related genes did not show consistent alteration. In accordance with reduced respiration rates and metabolic shutdown, genes thought to encode proteins involved in a shift from aerobic to *anaerobic* metabolism had altered *expression* profiles.

Genes that are categorized into other functional groups, such as lipid and *polyketide synthesis, translation*, and *transcription*, responded accordingly to the slowdown in bacterial growth rate and metabolism. Of interest was the *altered* expression response of several unknown regulatory genes. and tbis may have tbe greatest potential in understanding the way *M. tuberculosis* responds and postures itself to surviving long-term starvation.

Although a number of these genes have homologs and have been studied outside mycobactetia. the full understanding of their role in metabolic adaptation is as-yet unknown. Sherman and colleagues have also used DNA microarray technology to explore the adaptive response o1 *M tuberculosis.* In these studies gene response to hypoxic growth conditions were investigated. The gene expression changes that were noted were indicative of broad adaptations to reduced metabolic activity. Among the repressed genes, 40-60% (range is due to genes of unknown function) encode proteins with roles in biosynthesis. cell division, and aerobic metabolism.

Genes induced by hypoxia tell a subtler story. Genes classically assigned to play a roles in M. *tuberculosis* latency, such as genes encoding glyoxylate shunt enzymes and alternative sigma-factors, were essentially *unaltered*. However, of *notable* interest was the hypoxic induction of a two-component regulatory operon encoding Rv3l34c1Rv3l33c/Rv3132c.

Providing more evidence that this regulatory operon is involved in hypoxic adaptation is the fact that targeted disruption of the upstream gene Rv3134c, eliminated hypoxic regulation of *hspx*. This is consistant with previous studies which clearly indicate the role of *hspx* in stationary-phase growth and growth in *macrophages*. It is clear that the availability of *genomic* information has fostered the ability to delve into the *intricacies* of the bacterial response to a number of *environmental* changes.

However, the current challenge is how to delineate *expression* profiles to provide an understanding of the disease state. Thus far, *microarray* and protein *profiling* have provided evidence for induction of the *stringent* response and *downregulation* of *aerobic respiration, translation, cell division*, and *lipid biosynthesis*, in response to *hypoxic* and nutrient-limited growth conditions. Induced genes such as *acr*, alanine dehydrogenase, and *fumerate* reductase may facilitate survival under these conditions and therefore may represent relevant drug targets for *persistent* organisms.

In addition, the numerous genes of unknown function induced under these conditions may perform specific mycobacterial functions associated with invasion, pathology, and latent disease and warrant further investigation in terms of their relevance as potential drug targets.

TARGET-BASED DRUG DISCOVERY

Targeted screening allows the researcher to take advantage of the power of *bioinforrmatics*, which can be used to select potential targets from known or previously unexplored *blochemical* pathways. Once the target is selected, molecular biology techniques are used to clone and express recombinant forms of the protein.

The resulting highly purified protein (usually an enzyme or a receptor) is then the foundation for the development of a high-throughput screen. As noted in Table I, this method facilitates rational drug design, since it is often possible to utilize the purified recombinant protein for three-dimensional structural studies, which, in turn, allow structure-activity relationship (SAR) analysis.

With a computational representation of the preferred molecular interaction, drug candidates can be screened, *in silieo,* for their ability to interact within this protein, which streamlines library screening. The power of target-based drug discovery is that very quickly the structure and the putative activity of a protein can be revealed.

Then the specific characteristics can he applied to either understanding the pathway of interest to pursue related steps as targets or to developing an HTS-style drug screen. Since target-based assays are completely synthetic situations in which all of the mechanism kinetics can be predetermined, the assay can inform the researcher very easily about the inhibitory properties of a compound or class of compounds. This information can then be applied to combinatorial chemistry approaches to optimize inhibitory activity as is discussed later. An additional advantage of target-based methods is that they allow detection of inhibitors that are not able to cross permeability barriers.

Although these compounds are obviously not useful as therapeutics because they are not bioavailable nor able to penetrate the bacterium, they do provide useful information that can be applied to discover other novel targets that are to be used to optimize new classes of inhibitors to improve their druglike properties. Target-based screening does have some drawbacks. The method presupposes that a robust assay amenable to FITS can be developed.

Development of assays of this nature can be time consuming and requires a considerable level of expertise. For this reason, target-based screening cannot address potential targets that have no known function. Thus, -40% of the *M. tuberculosis* genome, some of which may he potential targets, is unavailable for target-based drug discovery.

In addition, the essentiality of the potential target must be established *in vivo,* and this is riot

Table 17.1: Characteristics of Target-Based and Genome-Based Whole Cell Screening

Screening method	Advantages	Disadvantages
Target based	Can detect poorly permeable compounds that may be suitable for optimization	Cannot address potential targets of unknown function
	Facilitates rational drug design	Assay development is time consuming (may be difficult to optimize for high throughput assays)
	Highly reproducible	*In vitro* inhibitors need to be converted into drugs
	Compatible with combinatorial chemistry approaches	Targets must be validated in vivo
Genome based whole cell	Selection for compounds that permeate cells	Insensitive to compounds that do not permeate cells
	Can establish antimicrobial properties	No rational basis for compound optimization
	Highly reproducible	Identification of target is difficult
	Historically successful	
	Independent of preconceived targets	May have mixed mechanisms of action
	Can address potential targets of unknown function	Does not predict pharmacokinetic issues
	Compatible with combinatorial chemistry approaches	

a trivial consideration when dealing with *M. tuberculosis 05).* Finally, compounds that inhibit an assay *in vitro* may not have druglike properties or may not he able to cross the permeability harrier imposed by the complex mycobacterial cell wall: turning an enzyme inhibitor into a drug is a very expensive endeavor in terms of both resources and time.

This approach in the context of tuberculosis is exemplified by targeting the L-rhamnose biosynthetic pathway in *M. tuberculosis.* L-Rhamnose plays an essential structural role in the cell wall of *M. tuberculosis* as a key sugar component of the disaccharide "linker unit" (P-GlcNAc(al-3)Rha), which connects the peptidoglycan heteropolysaccharide to the remainder of the cell wall "core." the galactan-arabinan-mycolic acid complex.

In addition to its structural role in the cell wall, rhamnose biosynthesis was also chosen because it does not exist in mammalian cells, and, hence, inhibitors of its biosynthesis should have a lesser chance of being toxic to humans. dTDP-Rhamnose is the precursor of the rhamnose found in the linker unit. and it is known that four enzymatic steps are required for its biosynthesis.

Accordingly, four potential enzymes (RmlA-D) were sought through homology via *in silico* analysis of the *M. tuberculosis* genome. This analysis revealed eight ORFs, which could encode for proteins that catalyze these four enzymatic steps. Only one copy of rmlC (Rv3465) and *rmlD* (Rv3266c) were identified, and, upon cloning, the hypothesized enzymatic activity for each protein was demonstrated.

The situation for *rmlA* was ambiguous because two candidate genes showed possible homology. However, upon cloning and expression, it was shown that Rv0334 encoded an active RmlA, whereas Rv3264c encoded a similar protein, ManB, but not involved in rhamnose synthesis. The situation was not as clear for the identification of the RmlB encoding gene.

Through a series of molecular strategies, Rv3464 was determined to encode *rmlB.* One candidate, Rv3634c, encodes a protein with homology to RmlB, but the fact that its N-terminal sequence is nearly identical to UDP-galactose epimerase (GalE) from *M. smegmutis.* indicated that this gene encodes the GalE protein. Two other possible candidates. Rv3784 and Rv3468c, showed homology to *rmlB;* however, the enzymatic activity of the recombinant protein products could not be established, suggesting that these genes do not encode an active RmlB.

Nevertheless, we cannot rule out the possibility that the soluble recombinant proteins are merely inactive forms of RmlB. Therefore, we assumed that there is only one isoform of RmlB present in *M. tuberculosis,* but this bas not yet been established unequivocally. The genome organization of the MP-rhamnose synthesizing genes in mycobacteria revealed no clues in terms of their catalytic role.

In most other organisms, such as *Escherichia coli, rmlA-D* are on a single operon, often in the order *B, D, A, C.* However, from the results above and the genome sequence it is clear that in *M. tuberculosis* the four genes responsible for dTDPrhamnose synthesis are in three different loci. Thus, *rmlA* (Rv0334) is separate from all of the other genes involved in rhamnose metabolism and appears to he the fourth gene in an operon where the functions of the proteins encoded for by the other three genes are not known. The genes rmlB and *rmlC* (Rv3464 and Rv3465) are the second and third genes in a complex operon consisting of perhaps five genes where the last two genes are part of an insertion sequence and the first gene encodes for a protein of unknown function. Finally, rmlD (Rv3266c) is the first gene of a three-gene operon.

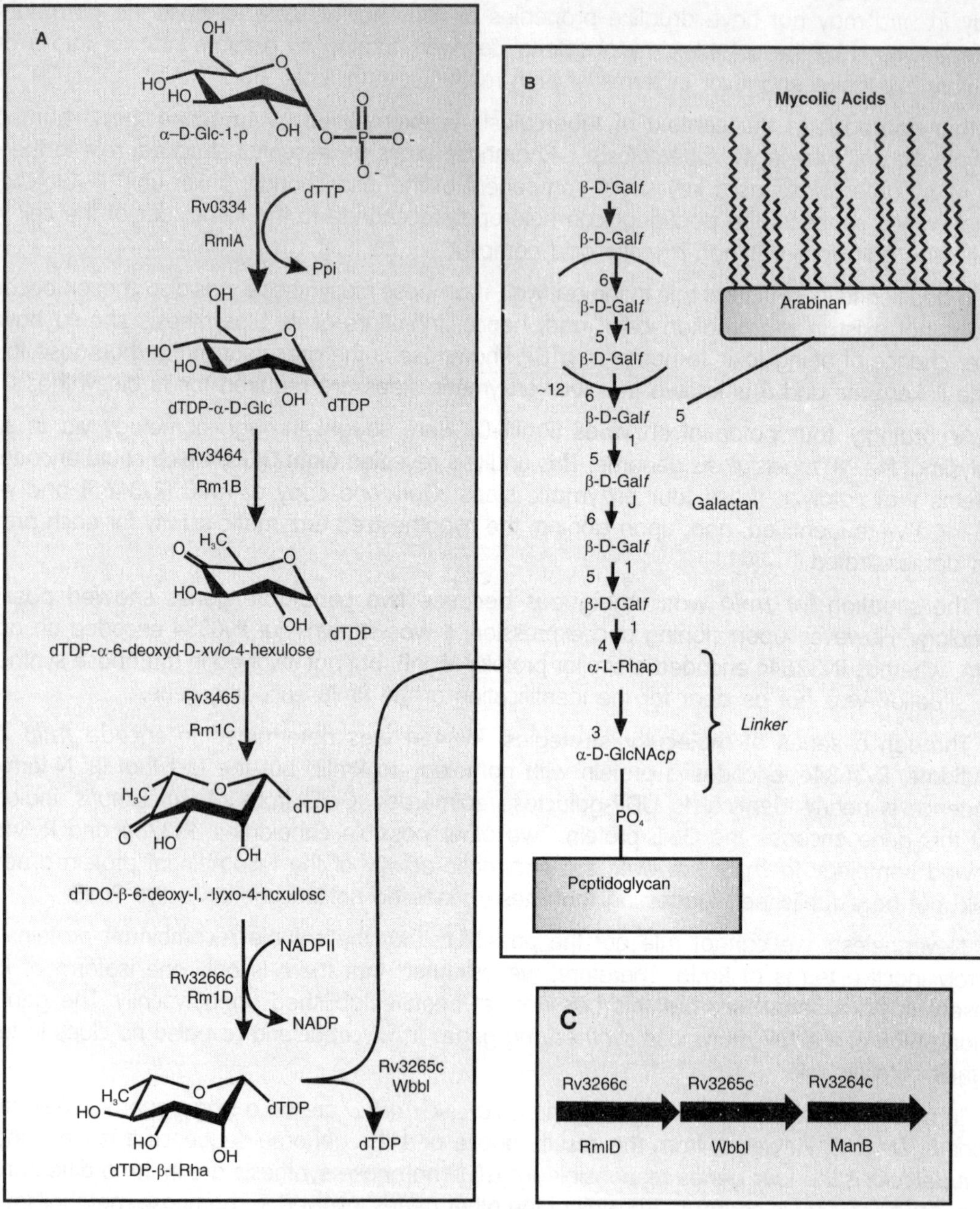

Figure 17.1: Anabolic pathway and proteins associated with rhamnose biosynthesis in M. tuberculosis. *(A) The metabolic conversion of glucose- l-P to the rhamnose nucleotide, which is the direct precursor of cell wall rhamnose. (B) The macromolecular structure of the rhamnose containing "linker region" and arabinoglactan of the cell wall. (C) Operon of genes whose targets are being exploited as novel targets in M.* tuberculosis *drug discovery projects.*

In this case, the second gene is *wbbL,* the rhamnosyl transferase, and the third *is manB* (labeled as *rmlA-2* in the genome sequence). There is some logic in the coordinate expression of *manB* in conjunction with the rhamnose genes, since *manB* is needed for the synthesis of all mannosyl glycolipids and polysaccharides, which, like the rhamnosyl unit, are an important part of the mycobacterial envelope.

The finding, however, that the *rail* genes are so scattered throughout the genome is surprising. With the identification of each enzyme involved in the synthesis of dTDPrhamnose, an assay was then developed. This assay took advantage of the fact that RmlD converts dTDP-6-deoxy-t; lyxo-4-hexulose to dTDP-rhamnose with the concomitant oxidation of NADPH to $NADP^+$, which makes it possible to couple the enzyme activity of each step in a facile microtiter plate assay amenable to screening inhibitors of the rhamnose biosynthetic pathway.

Increased resolution specific to each enzymatic step was achieved by titrating the enzyme of interest within the assay system, thus altering required inhibitor concentrations. Upon analysis, it became apparent which enzymatic step was being inhibited. This target-based assay, developed from genomic information, was used to screen 8000 compounds (supplied by Nanosyn, Tucson, AZ) in an HTS format. Prior to testing, *in silico* analysis of these candidates was performed based on Lipinski's "rule of 5," which categorizes candidates in terms of molecular weight, logP, hydrogen bond donors, and hydrogen bond acceptors.

These characteristics have been experimentally determined and have been shown to correlate to the effective bioavailability of a drug from current orally available medicines. With respect to these four criteria, 89% of the compounds of the library had "druglike" properties and 9% of the compounds were within the acceptable range, satisfying three of the four criteria.

The compounds were also selected for the presence of "druglike" functional groups and the lack of reactive groups such as aldehydes. The candidates that survived were screened at a concentration of 10 μM. Eighteen compounds, which showed activity initially, were rescreened in triplicate.

To confirm the inhibitory activity of these 18 candidates, new samples were reexamined. However, upon rescreening, only 11 of the 18 compounds were inhibitory in the target-based assay. These 11 compounds were then assayed for their ability to specifically inhibit RmlB. RmlC, or RmlD, and all were shown to inhibit RmlC (dTDP-6-deoxy-n-xylo-4-hexalose epimerase). With specific inhibitory activity determined, these drug candidates will now proceed into whole cell-based assays to determine essentiality and inhibitory activity.

POSTGENOMIC SCREENING

Classical antituberculosis drug discovery relied on whole-cell-based assays in which the primary readout was cell death. Although this approach has been responsible for the few anti-TB drugs in use today, the extended culturing times of *M. tuberculosis* (-4 weeks) made high-throughput screening impossible. However, since the genome sequence has become available, whole-cell-based screening has entered the era of HTS.

A number of advances in our understanding of the biochemistry of the tubercle bacillus via genomic information have provided a scientific platform to design quick (-24 hr) bioluminescent-

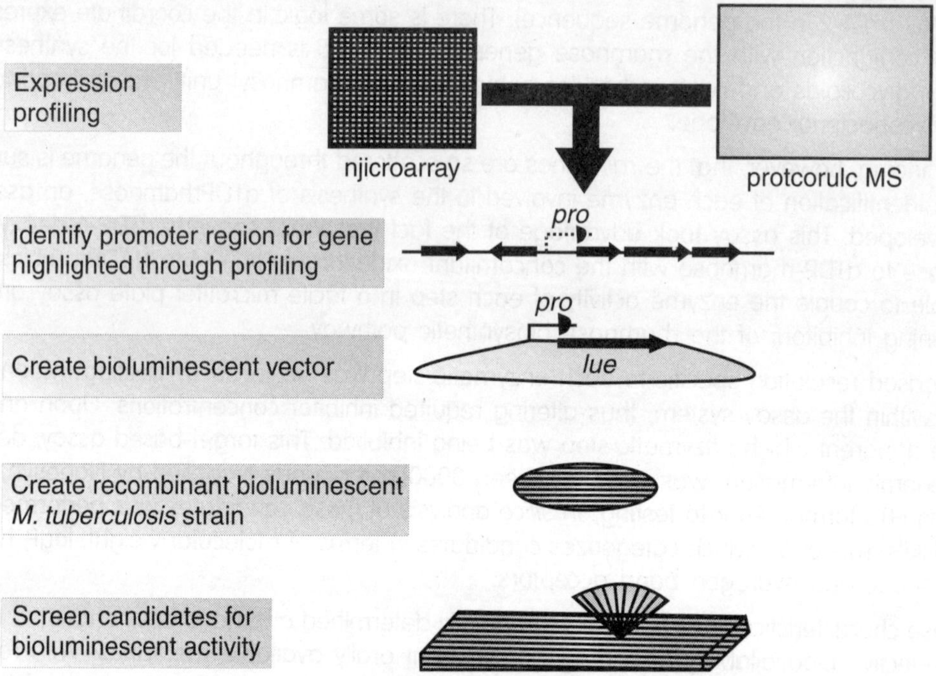

Figure 17.2: Design of high-throughput bioluminescence reporter strains used in M. tuberculosls *drug discovery.*

based HTS assays. A primary breakthrough was the demonstration of the ease of expression profiling tinder stress conditions. The basic rationale in terms of modernized whole-cell-based screen development is that the selective inhibition of a metabolic pathways causes an accumulation of precursors and a concomitant depletion in products.

These gross changes can be expected to selectively induce changes in the transcription of genes, thus altering the expression of the proteins that comprise the affected pathway, particularly before a more generalized stress response ensues. The resulting expression profile then serves as a "biological fingerprint" for that specific metabolic perturbation.

While regulated genes themselves may or may not be targets for drug intervention, they, with the help of software algorithms linking genes with proteins and known functions. can point researchers toward known pathways. In addition, this information can be used to incriminate the affected pathway and classify the target for unknown inhibitors. In terms of drug discovery and screening development. the expression of each gene in the genome or each protein in the proteome needs to be interrogated simultaneously in the presence and absence of the inhibitor.

If an inhibitor of the pathway of interest is not available, then a manipulatable alteration needs to be introduced. One approach that we are currently employing is genetic engineering to construct temperature-sensitive strains of *M. tuberculosis* which can then he studied. This approach allows us to directly perturb the intended target protein, thereby introducing a "druglike" effect. Another strategy that has had limited success is the introduction of an inducible promoter to control the gene of the target protein. The limitation in the case of *M. tuberculosis is* attributable to the lack of an available good promoter that can be turned off when needed. Currently, microarray technology is at tbe forefront of expression profiling in tuberculosis research because it is able to report on differential expression on a global scale that has not been realized by proteomic techniques.

A typical *M. tuberculosis* array is fabricated with either amplicons or oligonucleotides and covers only the predicted ORFs. In terms of expression profiling of *M. tuberculosis* and drug discovery, the utility of transcriptional response profiling has been validated with the use of the *M. tuberculosis-specific* inhibitor, isoniazid.

Although isoniazid has been thoroughly studied with arrays, it illustrates the important aspect of interpreting microarray data for screen development. The mode of action of isoniazid, arguably. has been studied more than any other *antimicrobial* agent. Although the actual proteomic target has been debated it is agreed that the affected macromolecules are the mycolic acids.

When the transcriptional response of *M. tuberculosis* to isoniazid treatment was studied using DNA microarray technology, a subset of genes involved in fatty acid synthesis were differentially regulated. Many of these genes were assigned through biochemical studies and homology to fatty acid metabolism.

The list includes genes of the FAS II operon Rv2243-2247, and associated genes *fbpC* (antigen 85c), *fadE23,* and *fadE24.* in addition to isoniazid associated genes such *as efpA* and *ahpC.* To confirm and validate the microarray results, real-time polymerase chain reaction (PCR) was performed.

These experiments confirmed the expression profiles of all the genes tested. Our understanding of the molecular consequences of isoniazid on mycolate biosynthesis provides confidence that many of the responsive genes encode proteins that are relevant to the mode of action of the drug. Accordingly, microarray hybridization provides evidence useful in predicting the mode of action of a novel compound based on a physiologically derived interpretation of its expression response to that compound.

Specific drug-responsive promoters identified by expression profiling serve as sensors of the intracellular conditions that are characteristic of the drug's activity which can then be exploited for the design of screens to identify novel compounds that exert similar effects on the bacteria. Building on this approach, we have experimentally characterized the bacterial transcriptional response to over 20 inhibitors that affect several aspects of cell wall and macromolecular biosynthesis and cell division.

Although many inhibitors clearly show similar transcriptional profiles, it is not always easy to rationalize the results from other inhibitors. Oftentimes, the highly induced genes are conserved hypothetical genes with no functional assignment. In such cases, it is hard to link a specific gene with the pathway affected by the inhibitor.

Nonetheless, information is still generated that can he used as the basis for a mode of action specific high-throughput screen. We have been able to categorize compounds on the basis of common gene inductions. This not only indicates which drugs may affect similar biosynthetic pathways but also allows for the design of reporter strains of *M. tuberculosis* that can be used in an HTS format to rapidly evaluate the MIC of a molecule and characterize its possible mode of action. One has to remember, as previously discussed, that the premise of whole-cell, genome-based screens is that one is monitoring a unique phenotypic response, a "biological fingerprint" and not necessarily the targeted pathway.

ANTITUBERCULOSIS DRUG

Modern combinatorial chemistry has already contributed to the modern drug discovery revolution. It is now possible to rapidly produce drug candidates with high levels of diversity and complexity in marked contrast to rational drug design of just a decade ago in which chemists selected specific molecules for synthesis.

This new chemical approach leads to compound libraries which have the ability, in theory, to sample all possible three-dimensional space. Barry and colleagues have been instrumental in pioneering the use of combinatorial chemistry and expression profiling to discover new classes of inhibitors for *M. tuberculosis.* The beauty of this approach is that it capitalizes primarily on genomic sequence information which is readily available.

The approach is centered on the idea that from genetic information, screens can he designed to report on the activity of drugs against a specific target, strictly from the reactive response of the bacterium. The advantage of this approach is that reporter strains (based on transcriptional response) can be designed very easily, and these are amenable to HTS automation.

Progress is limited by classical structural methods such as crystallography. To validate the whole-cell-based screening approach, Barry and colleagues used the known antimycobacterial drug ethambutol and interrogated the bacterial response to its inhibitory action. Although the target of ethambutol has been identified as arabinose polymerization in the cell wall, its specific protein target or targets are still unknown.

In addition, ethambutol's potency could he improved; unlike other mycobacterial inhibitors, the toxicity profile is relatively low. These two factors allow for vast improvement of the drug's activity with a lesser chance of toxicity. With information of the transcriptional response to ethambutol treatment, a bioluminescent reporter strain was created and validated using ethambutol as a control.

In addition to developing an ethambutol specific whole-cell-based HTS, a directed compound library was synthesized using combinatorial chemistry strategies. Ethambutol is a symmetrical diamine, and the directed library focused on exploring the activity of asymmetrical molecules around the diamine scaffold. The overall goal was to identify better inhibitors with the same mode of action as ethambutol using only genomic information.

Numerous candidates were identified using HTS formatted MIC and bioluminescent assays that shared the same mode of action of ethambutol, but had better minimal inhibitory concentrations. The inhibitory activities of several of the candidates were confirmed using classical biochemical techniques and tested in the animal model of infection to evaluate efficacy. From analyzing the structure of all of the molecules in the library (compounds with and without inhibitory activity) in terms of their activity it was possible to develop SAR information.

Another round of molecular optimization can then be imitated biased by the SAR information to further improve the potency and the pharmacokinetic characteristics of the compounds. These first-generation molecules demonstrated the power of this approach in the development of new antimycobacterial drugs.

COMBINATORIAL LIBRARIES

The convergence of chemistry and genomic information in combination with automation technologies has fostered a new approach to drug screening whicb utilizes both whole-cell-based and target-based approaches. In the scheme in Figure elsewhere in this chapter, a desired phenotypic change is sought by screening initially through high-throughput low-resolution whole-cell assays.

These assays are designed based on unique patterns associated with specific pathway inhibition as determined by expression profiling. It is important to reiterate that the expression pattern may not implicate the actual proteomic target: it is simply readout of the bacterial response to the metabolic pressure.

These screens have the ability to accommodate large libraries of diverse compounds and drastically reduce the number of candidates progressing through the discovery process. The "hits" identified from the whole-cell-based screens are then interrogated with higher resolution assays designed around specific enzymatic targets.

These high-resolution assays provide high-quality data and establish inhibitory kinetics. Once the library has been narrowed to a small fraction of compounds with the desired effect, they are prioritized by surrogate animal models and finally evaluated for efficacy in an actual animal model.

This drug screening process is designed so that the high-throughput low-resolution assays which are robust enough to handle large compound libraries are used before low-throughput high-resolution assays are used to confirm and bias the procession of candidates into model systems of infection. As the number of potential drug candidates increases, so does the need for quick, efficient, and high-volume screens that generate data of high quality.

The current testing management problem Is that screening is not conducted on the same candidate a million times but on a million different candidates at the same time. The purpose of the drug discovery process is to find and explore significant differences between compounds. Therefore, the screening process has to be designed to categorize quickly the *inhibitory* characteristics of the drug candidates.

The largest challenge in screening is not just the large member libraries, hut the low hit rates associated with them. With the latest advances in high-throughput screening using multiwell plates, the initial testing volume can he much greater than 10,000 candidates a day. Modern primary screens are designed to accomplish two things: first, to establish a minimal inhibitory concentration for each candidate and, second, to provide clues as to the potential target.

This can he accomplished by multiplexing the assays. The advantage of multiplexed whole-cell-based screens is that a couple of requirements are built into the assay. The drug candidate has the strict requirement to be able to subterfuge the mycobacterial cell wall and he metabolically stable within the cell. The primary benefit of whole-cellbased screens is to streamline the process and reduce the screening burden on target-based screens. Potential drug candidates coming through the whole-cell-based assays can he prioritized and biased into the target-based assays based on their specific inhibitory profile. With the information on MIC, tbe ability to kill the bacterium is established and a reasonable idea of target pathway is known.

However, the absolute target is not known and therefore the specific kinetic details have not been established. Although, in simplistic terms, this detailed information may not be necessary, it is required for predictive studies in terms of resistance mechanisms and potential microbial spectrum.

There are a number of advantages in target-based screens, particularly if they are not responsible for the primary categorization of potential drug molecules. They provide better resolution in terms of target identification and they are an additional method of target validation. Whole-cell-based assays and target-based assays are very compatible strategies in the quest to rapidly characterize potential molecules as drug candidates.

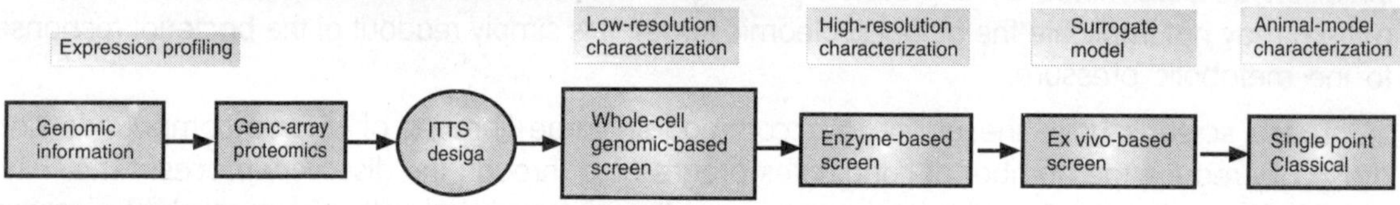

Figure 17.3: The modernized M. tuberculosis genome-based whole-cell screening aḷproach.

They exploit both genomic information and expression profiling. A screening strategy that uses these two approaches in tandem has the ability to establish essentiality through MIC, elucidate targets, and establish kinetic information of the inhibited reaction. Despite all of these advantages, each approach has its disadvantages. Whole-cell-based screens can be insensitive; many active compounds have associated toxicity issues in general and possible mixed mechanisms of action.

Target-based assays have problems associated with turning *in vitro* inhibitors into antibacterial drugs and genetic validation of essentiality. However. in combination, these two approaches compliment each another, and the disadvantages associated with each are essentially eliminated.

CORRELATING IN VITRO ACTIVITY WITH IN VIVO

The greatest need in the search for new antituberculosis drugs is a surrogate screening system for animal testing. Animai models of infection are used in the final stages of the drug development cycle to assess the antimicrobial efficacy in a model of human disease. Recent experience with library testing confirms historical evaluations of the extremely poor correlation between *in vitro* drug potency and *in vivo* drug efficacy.

The reasons for this porn correlation may be due to (i) altered physiology of the bulk of the bacterial cells growing ,in macrophages; (ii) the simultaneous existence of several discrete populations of bacteria each with varying susceptibilities to the drugs being tested; and (iii) the extremely complex pharmacokinetics involved in delivering the drugs to the site of infection, namely the alveolar macrophages.

A large number of compounds are being discovered and synthesized that show promise in *in vitro* assays against *M. tuberculosis.* In fact, many more *in vitro* lead molecules are being generated than can reasonably be assessed through traditional animal models of efficacy.

Table 17.2: Characteristics of Animal Models of Infection.

Screen	Advantages	Disadvantages
Surrogate system	Establishes antimicrobial properties under bacterial metabolism conditions associated with disease Reproducible Prioritizes the procession of candidates Compatible with combinatorial chemistry approaches *In vitro* inhibitors need to be converted into drugs	No rational basis for compound optimization Targets must be validated *in vivo*. Identification of target is difficult Many antimicrobials are also toxic to eukaryotes Assay development is time consuming
Single-point animal model	Can detect poorly bioactive compounds Highly reproducible Prioritizes lhe procession of candidates Compatible with combinatorial chemistry approaches Fast and relatively easy	Cannot address properties of bioavailability Many potential inhibitors arc also toxic to animals
Classical animal testing model	Detects for compounds that permeate cells Can establish antimicrobial efficacy in animal model Addresses properties of bioavailability	Not compatible with combinatorial chemistry approaches Time consuming Many potential inhibitors are also toxic to animals

Therefore, the development of reliable surrogate models for chemotherapy of human disease that has a moderately high-throughput format is an urgent matter. Bioluminescent screening systems to address these questions have been attempted using the same basic approaches as the whole-cell assays.

Currently, murine-derived macrophage cell culture is used as a possible testing surrogate. More recently, screens have been developed with bacteria derived from infectious animal tissue, such as the lung or spleen of a mouse. Since the bacteria have established an actual infection, they have also adopted the proper physiological state to cause disease and persist.

The purpose is to establish this bacterial metabolic state so that drug candidates can be screened rapidly and easily in a system that mimics the bacterial state of disease. Although current approaches have utilized macrophage lines, it k best if this metabolic state could be induced in artificial medium.

This does not seem to be possible at the moment, but, with genomic information and ongoing studies on *in vivo* metabolism of *M. tuberculosis,* a manipulatable, quorum-sensing-based system may he possible in the future. It is important to remember that the primary quality rum a surrogate animal system is that it establishes the bacterium in a metabolic state similar to that of true infection.

This should not be confused with developing a system that tries to mimic the site of infection. After the efficacy of a compound has been addressed in the surrogate model of infection, it is then tested in an animal model of infection. Due to the number of candidates that can proceed through a whole-cell-based discovery strategy to this step, low-cost, low-resolution, single-time-point animal experiments are useful.

These animal tests are based on a single-time-point analysis, such as 40 days of infection. The animals begin therapy at 2 weeks postinfection, and the gross effects of drug efficacy are determined at the time of harvest. It is important to begin therapy before visible pathology is apparent.

The bacterial burden, which reflects the efficacy of the drug candidate. can be qualitatively determined via gross pathology, and the quantities of bacteria can be determined through approaches such as bioluminescence and molecular tagging, if desired. This approach is intended, like the other screens, to bias the procession of candidates, sifting the most promising compound to the top of the pile.

Candidates that show promising activity can be further analyzed *via* classical animal model experiments designed to determine the efficacy profiles of potential drugs over several weeks, consisting of multiple time points that provide high-quality statistical data.

INHIBITION OF AN ENZYME

In terms of drug discovery, it is paramount to have functional knowledge derived from genomic sequence or experimental data to fully understand the inhibitory mechanism. With the genomes of so many organisms completed, it is becoming possible to globally compare organisms with similar disease niches and pathology.

This genomic—metabolic linkage among pathogens is proving to be the aspect that is most promising to understanding the weakness of organisms in general. With this metabolic information, combinatorial libraries can be designed which are biased for unique activity specifically against reactions, not necessarily unique enzymes. The next step to optimizing drug activity is to marry mechanistic and chemical information with the predicted function of the target protein based on genomic information and homology so that chemical libraries can be tailored to specific enzymatic chemistry.

At the moment, whole-cell-based bioluminescent screening strategies have the potential to characterize large sets of candidates against a single enzyme within a pathway. This is slowly producing hits and resulting in the identification of novel inhibitors. However, they usually lead to a single target, and, with the treatment pressure by current drugs, they are rapidly losing their potency due to resistance.

Thus, years of work and resources can be lost because a drug that has been designed against a single enzyme which undergoes a mutation event renders the drug useless. Perhaps the next step is to use genomic sequence to inform drug design against specific enzymatic reactions that are represented numerously within the metabolism of an organism or multiple organisms.

This would effectively make it impossible for an organism to acquire a single polymorphism that would provide *resistance. Microorganisms develop resistance at a frequency as low as* 10^5*; if a* number of targets were inhibited simultaneously, this mutation frequency would be multiplied based on the number of targets. Organisms have the ability to reach high numbers in an infection, which makes this frequency a very real number.

The basic rationale is not to design inhibitors against a reaction catalyzed by a specific enzyme, but rather against a specific reaction catalyzed by protein analog super families. Analysis of the mycobactetial genome sequence reveals that the organism has undergone massive gene duplication. This is particularly evident by the number of fatty acid genes, related polyketide syntheses (PKS), and glycosyl transferases that are responsible for cell wall construction and accessory molecules.

Although all of these related enzymes catalyze similar reactions utilizing a variety of related substrates, they clearly produce very different products. This has essentially led to the characteristic heteropolymeric cell wall of the tubercle bacillus, which is made up of diverse polysaccharides and unique lipids. Once a gene is duplicated, it is thought that it undergoes evolution to fulfill a new metabolic role. It has been proposed that metabolic pathways evolve backward. When the substrate of an enzyme in a pathway is limited, a new enzyme evolves to supply that substrate from an available precursor.

Accordingly, evolution is constrained to retain binding specificity, because the original enzyme and the newly evolved enzyme must rctain the ability to bind the same substrate/ product. The specific mechanism of the reactions need not he conserved. Ultimately, the progenitor and new enzyme may catalyze the same reaction from slightly different substrates or catalyze a slightly different reaction from similar substrates, each case utilizing the same type of intermediate.

In *M. tuberculosis,* an example of the first idea may he the ketoacyl syntheses KasA and KasB and possibly the related domain of fatty acid synthase, Type I (FASI). Each one catalyzes the same reaction under the same condilions, the primary difference being substrate specificity; KasA specifically catalyzes the condensation of short-chain lipids and KasB condenses long-chain lipids.

If molecules can be tailored to mimic the mechanistic state of the condensation in this case, then all three enzymes in addition to PKSs could be potential targets. The family of enzymes responsible for the introduction of heterogeneity into mycolic acids may represent the second scenario.

It has been suggested that these enzymes introduce different functionality into the hydrocarbon chain under a common transition stale. In this case, the substrate and intermediate are the same; the primary difference in the reaction is how the reaction proceeds, therefore resulting in a different functional group.

A synergistic strategy utilizing genomic information and evolutionary principles can he expected to provide a more efficient solution to drug development than the more traditional approach of studying a single enzyme at a time.

CONCLUDING REMARKS

It was once thought that the "Holy Grail" in understanding the molecular basis of infectious disease was to know the total genetic sequence of the infectious organism with the promise of an understanding of the disease and pathology of each infectious agent at a genetic level. However, it is becoming obvious that deciphering and understanding all of the information to the point of assigning function is the largest hurdle.

Having the genomic sequence of a bacterial organism allows one to mine the genome for information beyond predicting ORFs. The big push now is in correlating expression profiles with metabolic outcome. At the moment, there is a tremendous amount of raw sequence data and related information.

Considering all of the preexisting biochemical information and differential expression data from DNA based arrays and proteomic studies, each piece of DNA sequence can now he coupled with this information. Our aim with this chapter is not to rehash the themes of the past, but rather, with history, to spur all of us toward new directions and goals that push the limits of science, to nudge us to step out of the scientific comfort "pale" and explore new dimensions within the drug discovery arena in the postgenomic era.

Therefore, the primary goal of genomics-based research in the M. *tuberculosis* drug discovery process is to identify new drug targets. Drugs emanating from these novel targets will provide great benefits in terms of treatment of MDR strains.

This is obtainable because the efficiency of identifying targets from the genome has increased markedly, replacing the old process of identifying proteins, cloning them, and establishing a level of validation of their potential utility. This barrier has essentially been removed as a consequence of the completion of the *M. tuberculosis* genome and the development of *in silico* analysis tools.

All proteins now become potential targets and can he prioritized on the basis of information indicating essentiality to cell survival, conserved range in pathogens, and presence in humans.

Chapter 18

Analysing AtT20 Pituitary

Whole animal systems can be useful for understanding the phenotypic effect of *exposure* to *toxicants*. The use of whole animal systems for expression *profiling, genomic* or *proteomic*, can be complicated due to the fact that they are composed of many different cell types that may be differentially affected by environmental conditions. As a result, the use of cell culture systems is often required for initial studies. The AtT-20 cell line was derived from a mouse pituitary *adenoma*. It expresses *pro-opiomelanocortin* (POMC), the precursor of several bioactive hormones, including the stress hormone adrenocorticotropin (ACTH), the satiety hormone α-melanocyte-stimulating hormone (α-MSH) and the *analgesic* hormone β-endorphin.

These hormones are generated by multiple internal cleavages of POMC by PC 1. and PC2, two members of the family of calcium-dependent *subtilases* named proprotein convertases (PCs). PC 1 and PC2 are widely expressed in *neuronal* and *endocrine* tissues. Their *intracellular* activities are modulated by the levels of co-resident specific *inhibitors*, proSAAS for PCI and 7B2 for PC2.

Together, these enzymes undoubtedly influence the neuroendocrine peptidergic system that regulates many important biological functions including growth, energy *homeostasis, appetite* and *satiety, nociception* and *analgesia* and, *possibly, cognition* and *behaviour*.

Genetic deficiency for PC 1 and PC2 has been shown to affect some of these functions in mice and humans. PC 1 and PC2 enzymatic activities are sensitive to potential environmental *pollutants*, most particularly to calcium *chelating* agents such as EDTA and to heavy metal ions such as Hg, Cu and Zn.

It is not known how their inhibition by these common pollutants would alter the *proteome* of neuroendocrine cells. Detection of such alterations could be a measure of the potential disruptive effect of a PC-inhibitory compound on the *neuroendocrine peptidergic* system. The AtT20 cell line has been extensively used to study the *biosynthesis* of PC I and PC2 and their roles in POMC processing into multiple *bioactive* peptides.

It represents therefore a valid model to evaluate the effects of pollutants on the peptidergic neuroendocrine cells. As a first step towards determining the usefulness of proteomics as an assessment method for *neuroen-docrine peptidergic* disruption, we examined changes induced in the *proteome* of AtT20 cells by overexpression of proSAAS, the PC 1-specific inhibitor [AtT20(proSAAS) cells].

Two-dimensional (2D) gel profiling of AtT20 and AtT20(proSAAS) cells revealed many differences in the relative protein *abundance*. Mass *spectrometric* analysis identified two proteins found in greater abundance in AtT20(proSAAS) cells as proSAAS and Ephrin type A receptor 2 (EphA2). EphA2 identity was confirmed by 2D *immunoblotting*. This receptor *tyrosine kinase* is known to play a role in neural outgrowth and in cancer.

BASIC REQUIREMENTS

Culture

1. Mouse pituitary AtT20 and AtT20(proSAAS) cell lines. AtT20(proSAAS) were originally obtained by transfection into AtT20 cells of a proSAAS expression vector containing a neomycin resistance gene, followed by selection for cells expressing this gene by their resistance to the cytotoxic drug G418.
2. Dulbecco's modified Eagle's medium (DMEM) (GibcoBLR, Grand Island, NY) supplemented with 10% fetal bovine serum (FBS) (Wisent Inc., Quebec, Canada) and 30µg/ml gentamicin sulfate (Gibco/BLR), with or without 500 µg/mL of G418 (Gibco/BLR).
3. Phosphate-buffered saline (PBS): 1.35 M NaCl, 28 *mM KCl,* 80 *mM* Na_2HPO_4 $.7H_20$,15 *mM* KH_2PO_4, pH 7.2.
4. Versine: PBS containing 0.1 mM EDTA.
5. Buffer B: 10 mM N-2-hydroxyethylpiperazine-N-2-ethanesulfonic acid (HEPES)-KOH at pH 7.4, 10 mM KCl, 1.5 MM MgCl2, 0.5 mM sodium EGTA, l mM dithiothreitol (DTT).
6. Protease Inhibitor Cocktail (PIC) tablets (Roche Diagnostics GmbH, Mannheim, Germany). One tablet is dissolved in 10 mL of buffer B.
7. Corning cell lifter (Sigma-Aldrich, St. Louise, MO).
8. Syringe (3-mL) and needle (26-gauge).
9. Protein Assay Kit (Bio-Rad, Hercules, CA).

Gel Electrophoresis

1. IPGphor strip holder, 18 cm (Amersham Biosciences, Piscataway, NJ).
2. IPGphor isoelectric focusing system (Amersham Biosciences).
3. Hoeffer DALT Electrophoresis System (Amersham Biosciences).
4. Immobilized pH gradient (IPG) strip of pH 4-7L, 18 cm (Amersham Biosciences, Uppsala, Sweden). Store at -20 °C.
5. Rehydration buffer (RB): *7 M* urea, *2 M* thiourea, 4% CHAPS, 1 % DTT.
6. Bio-Lyte 3/10 Ampholyte (Bio-Rad). Store at 4 °C.

7. Mineral oil (Bio-Rad).
8. IPG strip equilibration buffer 1 (EB 1): 50 mM Tris-HCl, pH 8.8, 6 M urea, 30% (v/v) glycerol, 2% sodium dodecyl sulfate (SDS), 1% DTT, 1×10^{-4}% bromophenol blue.
9. IPG strip equilibration buffer 2 (EB2): 50 mM Tris-HCl, pH 8.8, 6 *M* urea, 30% (v/v) glycerol, 2% SDS, 4% iodoacetamide, 1×10^{-4}% bromophenol blue.
10. Acrylamide-bis-acrylamide solution (30.8%T) (neurotoxin, work with gloves!): 30% Acrylamide, 0.8% bis-acrylamide. Store at 4 °C.
11. 10%SDS.
12. 10% Ammonium persulfate.
13. *N, N,* N'N'-Tetramethyl-ethylenediamine (TEMED, Bio-Rad).
14. SDS electrophoresis buffer: 25 mM Tris, 192 mM glycine, 0.1 % SDS.
15. Low melting agarose (LMT) (Gibco/BLR). Dissolve 1 % LMT in SDS electrophoresis buffer by heating in a microwave oven. Keep at 50°C before use.
16. Silver staining solutions: fixative: 30% ethanol, 5% acetic acid; sensitizer: 0.02% sodium thiosulfate pentahydrate; silver reagent: 0.2% silver nitrate; Developer: 4% potassium carbonate, 0.025% formaldehyde (37%); stop reagent: 4% Tris base, 2% acetic acid.
17. Umax Powerlook 1100 scanner (Umax Technologies, Dallas, TX).

MALDI-TOF MS

1. Trypsin (cat. no. V5111, Promega. Madison, WT).
2. ZipTip C_{18} pipet tips (Millipore, Bedford, MA).
3. Matrix solution: 10 mg/mL of α-cyano-4-hydroxycinnamic acid in 50% acetonitrile containing 0.1% trifluoroacetic acid.
4. Voyager DE^{Tm}-PRO matrix-assisted laser desorption ionization time-of-flight (MALDI-TOF) mass spectrometer (PerSeptive Biosystems, Framingham, MA).
5. MALDI plate (PerSeptive Biosystems, Framingham, MA).

Immunoblotting

1. Nitrocellulose membrane (Bio-Rad).
2. Towbin buffer: 25 mM Tris, 192 mM glycine, 0.1 % SDS, 20% (v/v) methanol.
3. 1× Tris-buffered saline (l×TBS): 10 mM Tris-HCl, pH 8.0, 150 mM NaCl.
4. TBS-T: 1× TBS, 0.1 % Tween-20.
5. 5% Skim milk solution: TBS-T, 5% skim milk.
6. ECL^{TM}anti-rabbit IgG, horseradish peroxidase linked $F(ab')_2$ fragment (from donkey) (Amersham Biosciences).
7. Western Lightning Chemiluminescence reagent Plus (PerkinElmer, Boston, MA).

PROCEDURES

Sampling

1. Plate AtT20 and AtT20(proSAAS) cells in 150-mm dishes (three dishes per cell type), in DMEM medium. Let them grow to confluence at 37°C in a 5% CO_2-95% air humidified atmosphere. The culture medium for AtT20(proSAAS) was supplemented with G418 at 500 pg/mL to maintain selection.
2. Wash cell monolayers 3 times with 10 mL of PBS each time.
3. Cover cells with Versine; scrape them off the plate with a cell lifter and transfer the suspensions into centrifuge tubes.
4. Sediment cells by centrifugation at 1300g at 4°C for 5 min. Remove and discard the supernatants.
5. Centrifuge again as described above to remove the remaining Versine.
6. Suspend cells into 0.5 mL of PIC-containing buffer B and disrupt them by 30 pull/push through a 26-gauge needle attached to a syringe.
7. Centrifuge the resulting lysates at 1000g at 4°C for 7 min. Transfer supernatants into fresh tubes and discard pellets.
8. Centrifuge supernatants at 10^5g at 4°C for 30 min; transfer the supernatants (cytosolic fraction) into fresh tubes and keep the pellet (microsomal fraction).
9. Add 3 volumes of acetone to each cytosolic fraction and place at -20°C for 1 h to allow protein precipitation. Centrifuge at 15,800g at 4°C for 15 min. Discard the supernatants and keep the pellet.
10. Dissolve each pellet of cytosolic or microsomal proteins in 500 µL of RB by sonication.
11. Determine the protein concentration by the Bradford method using the Bio-Rad Protein Assay kit.

Electrophoresis

Isoelectric Focusing (IEF)

1. Dissolve 0.8 mg of cytosolic proteins or 1 mg of microsomal proteins in 340 mL of RH buffer containing 0.5% Bio-Lyte 3/10 Ampholyte and 0.5% (v/v) of 0.1% bromophenol blue.
2. Pipet each mixture into an IPG strip holder.
3. Remove the protective cover from an 18-cm long IPG strip of pH 4-7L. Place it with gel side down in the IPG strip holder. Make sure that the acidic end of the IPG strip faces toward the pointed end of strip holder.
4. Dropwise add 800 µL of mineral oil to both ends of the strip holder. Tilt the strip holder to let the mineral oil cover the entire IPG strip. Close the lid.
5. Rehydrate the IPG strip on the IPGphor System at 25°C for 12 h.

6. Perform IEF at 25°C at 50 µA per strip with the following settings: 500 V for 1 h, 1000 V for 1 h, 8000 V for 8 h and, finally, hold at 500 V until electrophoresis in the second dimension.

SDS-PAGE

1. Prepare 1-mm thick, 19 ×23 cm, 12%T polyacrylamide gels in the Hoeffer DALT Gel Caster. Vertical slab SDS gels are cast in a Multiple Gel Caster. For assembly of the gel cassette, refer to manufacturer user manual (Hoefer DALT System User Manual, Amersham Biosciences). To cast 25 slab gels, mix the following solutions: 600 mL of Acrylamide/bis-acrylamide solution (30.8%T), 375 mL of 1.5 M Tris-Cl, pH 8.8, 15 mL of 10% SDS, and 15 mL of 10% ammonium persulfate. Add water to 1500 mL. Before gel casting, add 215 µL of TEMED.
2. Equilibrate each IPG strip in 10 mL of EB-1 solution with shaking for 10 min, and then in 10 mL of EB-2 for another 10 min.
3. Place the strip on top of the vertical SDS gel and seal it with 1 % LMT agarose.
4. Perform electrophoresis overnight, under 90 V constant voltage in the Hoeffer DALT Electrophoresis Tank (Amersham Biosciences) cooled with circulating water bath at 12°C.
5. Unload the gel cassette. Make sure that the gel sticks to one of the plates when opening it. Place the gel into a glass tray containing the fixative for silver staining.

Methods of Staining

Silver Staining

The following steps are carried out with shaking.

1. Soak the gel in the fixative for 30 min.
2. Change the fixative and shake for another 30 *min.*
3. Wash the gel in water 5 times for 5 min each.
4. Incubate the gel in the sensitizer for l min.
5. Wash the gel in water 2 times for 1 min each.
6. Incubate the gel in the silver reagent for 30 min.
7. Wash the gel in water 2 times for 1 min each.
8. Incubate the gel in the developer until protein spots appear.
9. Stop the development with the stop reagent for 30 min.
10. Wash the gel in water 3 times for 5 min each.

Capturing of 2D Image

Silver-stained gels are scanned on Umax Powerlook 1100 scanner and the digitized images are saved in a computer.

Silver Gel Destaining

1. Excise the gel spots of interest.

2. Mix 30 mM potassium ferricyanide and 100 mM sodium thiosulfate in a 1:1 ratio.
3. Cover the gel pieces with 100 µL of the above solution.
4. Shake at room temperature until the silver stain is removed.
5. Wash the gel pieces with several changes of 500 µL of water each to remove the yellow reagent.
6. Incubate the gel pieces in 200µL of 200 mM ammonium bicarbonate with shaking for 20 min.
7. Wash the gel pieces in 500 L of water twice for 1 min.
8. Centrifuge and remove the water.

Hydrolysing Trypsin

1. Incubate the gel pieces in 200 µL of CH_3CN for 10 min. 2. Centrifuge and remove CH_3CN.
3. SpeedVac the gel pieces for 15 min.
4. Incubate the gel pieces in 100 µL of 100 mM NH_4HCO_3, 10 mM DTT at 56°C for 45 min.
5. Centrifuge and remove the DTT solution.
6. Add 100 µL of 100 mM NH_4HCO_3, 55 mM iodoacetamide and incubate at room temperature in the dark for 30 min.
7. Centrifuge and remove iodoacetamide solution.
8. Wash the gel pieces in 100 µL of 100 mM NH_4HCO_3 with shaking for 10 min.
9. Centrifuge and remove NH_4HCO_3 solution.
10. Dehydrate the gel pieces in 100 µL of CH_3CN at room temperature for 5 min.
11. Centrifuge and remove CH_3CN.
12. Wash the gel pieces in 100 µL of 100 mM NH_4HCO_3 with shaking for 5 min.
13. Centrifuge and remove NH_4HCO_3 solution.
14. Dehydrate the gel pieces in 100 µL of CH_3CN at room temperature for 5 min.
15. Centrifuge and remove CH_3CN.
16. SpeedVac the gel pieces for 15 min.
17. Swell the gel pieces in 20 µL of 50 mM NH_4HCO_3 containing 500 ng of trypsin on ice for 30 min.
18. Remove remaining trypsin solution from the tube and add 50 mM NH_4HCO_3 without trypsin to cover gel pieces.
19. Digest proteins in gel at 37°C overnight.
20. Centrifuge and transfer the supernatant into a new tube.
21. Add 50 µL of 50% CH_3CN, 0.1% trifluoroacetic acid to the gel pieces and sonicate in a cold water bath for 3 min.

22. Centrifuge and transfer the supernatant to the primary supernatant.
23. Repeat the steps 21 and 22 twice more.
24. SpeedVac the supernatant until the volume is smaller than 10 μL.
25. Add 0.1 % trifluoroacetic acid to 10 μL.

MALDI-TOF MS of Tryptic Peptides

1. Before MALDI-TOF analysis, purify tryptic digest peptides with ZipTip C_{18H} pipet tips following the manufacturer's instructions.
2. Elute the purified peptides onto the MALDI plate with 2 μL of matrix solution.
3. Perform mass spectrometry in a delayed extraction and reflectron mode with the following instrument setting: 20 kV acceleration voltage, 150 ns delay time, 70% grid voltage, 0.05% guide wire voltage, 128 laser shots, laser intensity of 2000, and a mass gate from 800 to 2800. Trypsin autodigestion products of 842.51 m/z and 1045.56 m/z are used for internal mass calibration.
4. The masses of monoisotopic peaks are chosen to identify the proteins by searching Swiss-Prot and TrEMBL databases using the search engine PeptIdent (http://www.expasy.ch/tools/peptident.html) with a mass match within 50 ppm to the corresponding theoretical mass.

2D Immunoblotting

1. Repeat the 2D gels as described under the heading electrophoresis.
2. Transfer proteins spots from the unstained gels onto nitrocellulose membranes in a Hoefer SE 600 tank containing the Towbin buffer, according to the manufacturer's user manual.
3. Block the membranes with 5% skim milk solution.
4. Incubate the membranes in 5% skim milk solution containing anti-EphA2 primary antibody at room temperature for 1 h.
5. Wash the membranes in TBS-T 4 times for 5 min each.
6. Incubate the membranes in 5% skim milk solution containing HRP-conjugated donkey IgG against rabbit IgG for 1 h.
7. Wash the membrane in TBS-T 4 times for 5 min.
8. Reveal immunoreactive spots by Western Lightning Chemiluminescence reagent Plus as described in the manufacturer's protocol.

POINTS OF REMEMBER

1. Complete solubilization, desegregation, denaturation and reduction of proteins in the sample are critical for the success of protein separation by 2D gel. Therefore, protein pellet is dissolved into RB by sonication for at least 3 min. Since urea can hydrolyze to isocyanate at elevated temperatures, resulting in protein modification by carbamylation, the sonication is carried out in a cuphorn sonicator cooled with an ice water bath.

2. IPG strip holder is made up of ceramic. Handle with care.
3. To avoid contamination, work with gloves. Do not trap bubbles under the IPG strip.
4. Water used here and for preparing solutions must have a resistivity of more than 18 MO*-cm.
5. Glass tubes, 20-cm long, with a cap are ideal for the equilibration.
6. Avoid trapping bubbles between IPG strip and SDS gel.
7. Silver staining is one of the most sensitive visulaization methods. For high-quality results, use high-purity reagents and avoid *contamination* by wearing gloves.

Chapter 19

Immunological Diagnosis of Diseases

The improvement of diagnostic procedures is an important goal for biotechnology and should lead to high-order parallel assay systems based *exclusively* on *multiple* molecular *criteria* from *microarrays*. Increasingly *rapid*, *sensitive*, and specific tests coupled with improved *prognostic* stratification will reduce health care costs and increase *therapeutic* success for a broad range of diseases.

Although single-protein biomarkers, such as prostate-specific antigen (PSA), *carcino-embryonic* antigen (CEA), and *cancer antigen* 125 (CA-125), have wide clinical *applicability*, the use of *unitary* markers typically is limited by their low specificity and sensitivity. Such markers are more useful for monitoring the response to *therapeutic interventions* than in helping determine the initial diagnosis.

One method for improving the sensitivity and specificity of biomarker-based diagnosis is to use assays where multiple protein biomarkers are analyzed *simultaneously*. Antibody microarrays constitute one of several technologies that enable the multiple parallel analyses of proteins derived from clinical samples.

Monoclonal antibody microarrays have the advantage of requiring tiny amounts of antibodies, high sensitivity, *amenability* to *automation* and high-throughput analysis, and the potential for hundreds to thousands of antigens to be analyzed in parallel. Antibody microarrays were pioneered by MacBeath, Haab, and their associates and subsequently were applied to the development of novel diagnostic methods. More recently, monoclonal antibody microarrays have been utilized to obtain larger data sets from tissue and blood samples.

In addition to diagnostic applications, antibody microarray technology is emerging as a useful tool in target discovery and proteomics. One such application has been the comparison of proteome expression signatures in *malignant* and normal samples. It should be possible to construct antibody microarrays that target proteins selected from the entire human *proteome* (estimated at approximately 100,000 primary proteins), however, for reasons of economy and practicality, the content of the array may be restricted to a defined set of targets. Ideally, the target set should have differential expression in a range of different disease states, to make the test applicable to more than one *pathological* process. The human cluster of differentiation antigens constitute one such candidate set, being united by their common expression at the surface of leucocytes where the majority perform functions critical to the human *immune* response.

As there is no a priori method for defining candidate sets, selection and validation is empirical and determined from testing for unique patterns against a broad panel of *pathologies*. Whereas CD antigens have proven to be useful in a broad range of disease situations, alternative approaches may be equally valuable. The human CD antigens are a well-characterized set of molecules expressed at the cell surface of *leucocytes* and other cell types.

As the function of leucocytes is restricted principally to immunity, changes in the expression of CD antigens provide the basis for a general-purpose diagnostic platform. The parallel measurement of multiple CD antigens contrasts with the more limited surface *immunophenotyping* provided by flow cytometry-based techniques.

Limited CD antigen profiling combined with morphological, cytochemical and cytogenetic data forms the basis of the classification system for human *leukemias*. CD antibody microarray-based diagnosis (DotScan microarrays) has been shown to concur with established diagnostic methods used in acute *myeloid leukemia*, HIV, and *colourectal* cancer.

The extensive immunophenotype obtained with DotScan microarrays may enable *sub-classification* of these diseases, for example, into drug responders and *non-responders*, and promises to facilitate the identification of new disease entities.

BASIC REQUIREMENTS

Sampling

1. Histopaque-1077 (Sigma-Aldrilch Pty. ltd. St. Louis, MO).
2. Phosphate-buffered saline (PBS; Sigma-Aldrich).
3. K3EDTA vacuette tubes (Greiner Bio-one, Catalogue number 455036), or lithium heparin (BD vacutainer, catalogue number 367885).

Suspensions Making

1. Hanks buffer (Sigma-Aldrich).
2. Collagenase (Sigma-Aldrich) 2 % (w/v) in PBS, stored for weeks at -20'C.
3. Roswell Park Memorial Institute 1640 (RPMI) medium (Sigma-Aldrich).
4. Stainless steel sieve, a standard tea sieve may be used.
5. 50µm Filcon filter membrane (BD, Franklin Lakes, NJ).
6. 0.1 % (w/v) DNAse in PBS (Sigma-Aldrich), stored for weeks at -20'C.

DotScan Microarrays

1. DotScan microarrays,.
2. 1 mM EDTA in PBS (Sigma-Aldrich).
3. Roswell Park Memorial Institute 1640 (RPMI) medium (Sigma-Aldrich).

4. Heat-inactivated human AB serum (Sigma-Aldrich).
5. Fetal bovine calf serum (FCS; Gibco, Grand Island, NY).
6. 4 % (v/v) formaldehyde in PBS (Sigma-Aldrich). Make up fresh as required. Formaldehyde is very toxic and should be handled with care.
7. Humidity chamber suitable for one to six slides. It is important that the slides are incubated horizontally.
8. Bovine serum albumin (BSA; Sigma-Aldrich, catalogue number A3902).
9. Fluorophore-labeled cell-type-specific antibody or antibodies for detection of captured cells.
10. Blocking reagent (e.g., serum) from the same species as the antibodies used in item 8.
11. Medsaic DotScan array scanner for optical scanning of DotScan arrays or Molecular Imager FX Pro (Bio-Rad, Hercules, CA) or Typhoon 8600 Variable Mode Imager (Amersham-Pharmacia, Castle Hill, NSW, Australia) for scanning DotScan arrays stained with fluorescently labeled antibodies.
12. Medsaic DotScan software.

PROCEDURES

The DotScan microarray platform is based on the capture of whole live cells on nitrocellulose microarrays containing 82 monoclonal antibodies with specificities against different CD antigens. Using whole cells for the detection of CD antigens is advantageous, since cell surface markers often are *hydrophobic* and aggregate when analyzed in solution.

Sampling

Obtaining a homogeneous single-cell suspension is essential. Mononuclear leucocytes are purified from samples of peripheral blood, as red cells interfere with the capture of leukocytes. For samples obtained from solid tumor tissue, it is essential to remove cell aggregates that interfere with cell capture and fluorescent scanning, resulting in false positives. Furthermore, the proteolytic and mechanical tissue *disaggregation* results in DNA release, making the solution viscous and hampering the cell capture. Therefore, cell *suspensions* are treated with DNAse.

Analyzing Blood

1. Collect the blood sample in a K3EDTA vacuette tube, or a lithium heparin vacutainer.
2. Dispense 3 mL of histopaque-Ficoll into a centrifuge tube.
3. Carefully overlay 5 mL of blood onto the histopaque-Ficoll.
4. Centrifuge at 800 g (without brakes) for 15 min at room temperature in a swing out bucket rotor.
5. Remove the top layer (the plasma fraction) from the tube.
6. Carefully transfer the mononuclear leucocytes from the Ficoll/plasma interface to a 10-mL centrifuge tube.
7. Wash the cells with 10 mL of PBS and centrifuge at 400g for 5 min at room temperature.

If necessary, use lysis buffer and/or DNAse to remove red blood cells and/or DNA.

8. Resuspend the cell pellet in 10mL of PBS.
9. Centrifuge at 400 g for 5 min at room temperature.
10. Resuspend the cell pellet in 10mL of PBS.
11. Stain 10 μL of the cells with trepan blue and count the number of living cells using a light microscope in a manual counting chamber (Burker chamber).
12. Centrifuge at 400 g for 5 min at room temperature to pellet the cells. Proceed immediately to Sect. elsewhere in this chapter.

Preparation Smear from Solid Tumors

1. Immediately following the biopsy, place the tissue samples in a cold Hanks buffer to maintain cell viability, which is essential for cell capture.
2. Cut the samples into 2 mm strips, or 2 × 2 mm cubes.
3. Incubate the tissue with 2 % (w/v) collagenase in RPMI 1640 at 37°C for I h.
4. Force the semidigested tissue through a fine stainless steel sieve to disaggregate it into a cell suspension.
5. Remove cell aggregates with a 50-μm Filcon filter membrane to obtain a suspension of single viable cells.
6. Add 0.1 % (w/v) DNAse to cells for 20 min at room temperature to digest DNA from lysed cells.
7. Centrifuge at 400g for 5 min at room temperature.
8. Resuspend the cell pellet in 10mL of PBS.
9. Centrifuge at 400g for 5 min at room temperature.
10. Resuspend the cell pellet in 4-10 mL of PBS.
11. Stain 10μL of the cells with trypan blue and count the number of living cells using a light microscope in a manual counting chamber.
12. Centrifuge at 400g for 5 min at room temperature to pellet the cells. Proceed immediately to Sect. elsewhere in this chapter.

Cell Capturing Cells

Cell capture on antibody dots of the microarray is an active process involving live cells, where CD antigens "cap" to the interface between the cells and dots. Evidence for such a time-dependent capping process has been obtained from experiments in which cells were captured by CD antibodies *immobilized* on *Biacore chips* (L. Bransgrove and R. I. Christopherson, unpublished). It therefore is essential that cells are viable for capture on the *DotScan microarray*.

Leukaemia Cells

1. Resuspend the leukocytes in 300μL of PBS containing 1 mM EDTA.

2. Prewet the microarray with PBS and carefully wipe away excess PBS around the nitrocellulose surface.
3. Carefully apply the cell suspension to the microarray.
4. Incubate the slide in a humid chamber at room temperature for 30 min.
5. Gently wash off unbound cells with PBS.
6. Fix the DotScan array for > 20 min in PBS containing 4% (v/v) formaldehyde.

Preparation of Single-Cell Suspension

1. Resuspend the single cell suspension (4 × 10^6 cells;) in 300 μL of RPMI containing 10% (v/v) FCS.
2. Prewet the array in PBS and carefully wipe away excess PBS around the nitrocellulose surface.
3. Add the cell suspension to the array.
4. Incubate the array in a humid chamber at 37 °C for 1 h.
5. Gently wash off unbound cells with PBS.
6. Fix the DotScan array for 15 min in PBS containing 4% (v/v) formaldehyde.

Identification

Captured cells can be detected using either an optical scanner, where the average optical density of the spot is measured, or by staining the cells with a *fluorescently* labeled antibody and using a laser scanner with the appropriate excitation and emission filters. The latter method allows for *simultaneous* detection of several cell populations captured on a single DotScan micro array.

Visual Scanning

1. Wash the fixed DotScan microarray twice in PBS for 20 s each.
2. Carefully remove excess fluid from the microarray surface and thoroughly wipe the back of the slide until clean and dry. The *nitrocellulose* surface must not be allowed to dry out during scanning.
3. Scan the slides using a Medsaic DotScan array scanner.

Fluorescence Multiplexing

1. Wash the fixed array twice in PBS for 20s each.
2. Block the slides by the addition of 300 μL of 1 % (w/v) BSA in PBS containing 5 μg/mL mouse IgG serum.
3. Incubate at room temperature for 30 min in a humid chamber.
4. Apply 1-5 μg/mL of fluorophore-labeled antibody or antibodies in 150μL of PBS containing 1 % (w/v) BSA and 5μg/mL mouse IgG serum. If there are several different antibodies directed at different antigens, the spectral properties of the combination of fluorophores must be considered .
5. Incubate in the dark in a humid chamber at room temperature for 30 min.

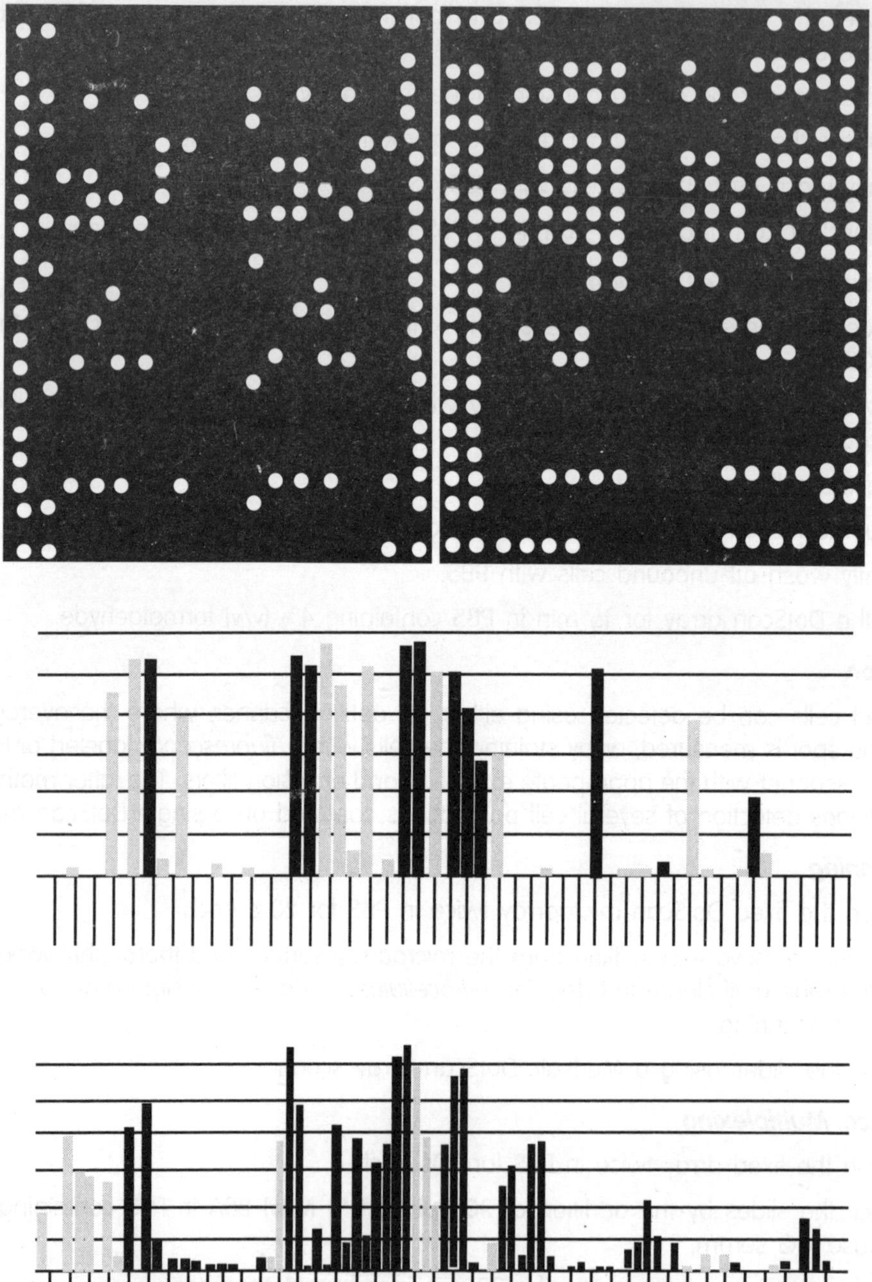

Figure 19.1: Mononuclear leukocytes from a patient with acute myeloid leukemia (A) and a healthy control sample (B) captured on CD antibody microarrays. The DotScan microarray consists of two duplicate arrays, and the signal intensities are averaged by the DotScan software to produce the corresponding bar charts for acute myeloid leukemia (C) and a healthy control (D).

6. Wash the microarray by immersion in PBS at least five times (keep it immersed for 20 s for two of the immersions). Keep away from light.

7. Scan the slides using a laser scanner. Select appropriate laser wavelengths and filter settings for the particular set of fluorophore-labeled antibodies used.

Detecting Disease Signatures

Dot patterns of cells captured on the DotScan microarray are quantified using Medsaic DotScan software. The software quantifies the intensities of each dot from the digital image file and compiles *expression* profiles as bar charts showing the average intensities above background on an 8-bit grayness scale from 1 to 256.

For images generated with scanners other than the Medsaic scanner, for example, files from fluorescent *multiplexing* assays, the image file needs to be converted to a bitmap-file format prior to analysis with the DotScan software. To identify expression signatures associated with disease, a set of matched samples from *healthy* and diseased tissues (or healthy and sick individuals) should be analyzed, preferably with at least 20 samples per group.

Component Analysing

The data obtained from DotScan analysis is highly (up to 82) multidimensional, and for visualization purposes, it may be projected onto two or three dimensions using principal component analysis (PCA). PCA is a method for reducing the *dimensionality* of a data set by performing a

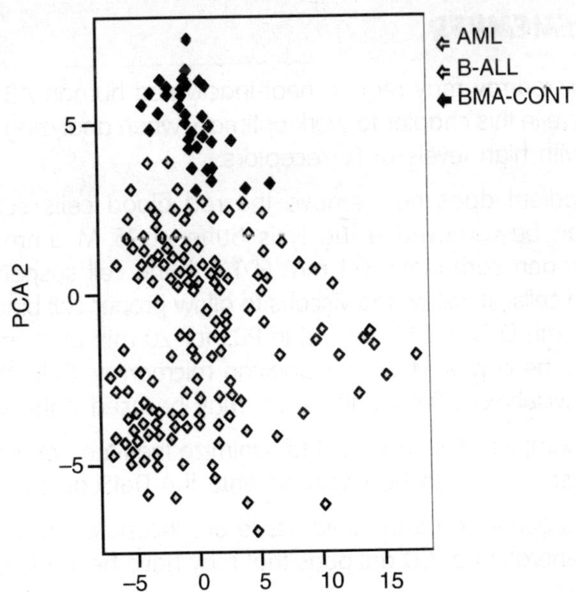

Figure 19.2: Two-dimensional principal component analysis of CD antibody microarray data from bone marrow aspirates of acute myeloid leukemia (AML, n=75), B-cell acute lymphoblastic leukemia (B-ALL, n=73), and healthy controls (BMA-CONT, n=36). The symbols denote individual samples, the axes represent scores in the principal component space for the first and second principal components.

covariance analysis among factors and is suitable for data sets in multiple dimensions. For high-dimensional data, it is preferable to perform principal component analysis first utilizing a large number of components, followed by a *discriminant* analysis that improves separation.

The immunophenotypic data then can be visualised in a two- or three-dimensional plot, using the first two or three discriminant functions and used to identify sample populations and outliers in the sample groups.

Identifying Therapeutic Antibodies

Monoclonal antibodies constitute the most rapidly growing class of human therapeutics and, currently, the second largest class of drugs after vaccines. The majority of the *monoclonal* antibodies currently approved by the FDA are directed toward CD antigens or other cell surface antigens, such as *Rituxan* (Rituximab, Mabthera, anti-CD20), Mylotarg (anti-CD33), and Campath-1H (anti-CD52). It has been estimated that more than 150 other monoclonal antibodies against different targets are in various stages of clinical trial.

For example, the following therapeutic antibodies target surface antigens overexpressed on colourectal cancer, *Edrecolomab* targets EpCAM (epidermal cell adhesion molecule), huAb A33 binds to the antigen A33, and Cetuximab binds to the epidermal growth factor receptor (EGFR). Inclusion of more therapeutic antibodies on the DotScan microarray would provide a more extensive immunophenotype and enable selection of appropriate therapeutic antibodies for patient-specific treatment.

POINTS OF REMEMBER

1. The DotScan microarray may require heat-inactivated human AB serum (2%) at the steps in Sect. elsewhere in this chapter to work optimally when analyzing acute myeloid leukemias or other cells with high levels of Fc receptors.
2. If the Ficoll gradient does not remove the red blood cells sufficiently, the remaining erythrocytes can be removed using lysis buffer (0.15 M ammonium chloride, 10 mM potassium hydrogen carbonate, 0.1 mM EDTA). If the cell suspension contains too much DNA (from lysed cells), it will be too viscous to allow proper cell binding. The cell suspension can be treated with DNAse (0.1 % [w/v] in PBS for 20 min at room temperature). The cells must be alive to be captured on the DotScan microarray. Cells that have been frozen at -80°C in dimethylsulfoxide/fetal calf serum may be used if the viability is retained.
3. For solid tissue samples, it is important to minimize the time between surgery and analysis, as the cells must be alive to be captured onto the DotScan microarray.
4. Cell suspensions generated from solid tissue are incubated in 37°C for 1 h on the arrays to allow for regeneration of CD antigens that may have been affected by the collagenase/ DNAse treatment.
5. If the DotScan microarray is to be stained with a fluorescent antibody, the fixation time should be minimized, as several CD antigens are fixation sensitive. If possible, choose an antibody or CD marker that can be used for staining fixed tissue sections (this information generally is provided by the antibody supplier). As a note of caution, data from spots that

correspond to the same antibody as the staining antibody (i.e., the CD3 dot from a CD3-staining experiment) may be affected by the capping effect described in Sect.elsewhere in this chapter.

6. It is advisable to block the DotScan slides before staining to avoid the nonspecific binding of the fluorescently labeled antibodies to the fixed cells. Speciesspecific antibodies, or serum, should be used for this purpose, such as mouse IgG or mouse serum for blocking the slides prior to staining with a fluorescently labeled mouse antihuman anti-CD antigen antibody.
7. Fluorescence multiplexing using several antibodies with different *fluorophores* must be done with matching combinations of fluorophores. A combination of Alexa488, Alexa532, and Alexa647 works well with most scanners. However, both the wavelengths of the lasers and the *bandwidth* of the available filters have to be considered.

Chapter 20

Virulence Determinants

Microbial pathogens possess a repertoire of virulence determinants, each of which makes a unique *contribution* to bacterial fitness during an *infection*. Understanding expression of the genes involved in infection remains the Holy Grail of research in bacterial *virulence*. The development of molecular biological technologies, including recent progress in microbial genomics and bioinformatics, coupled with molecular genetic approaches, has facilitated the dissection of a *plethora* of bacterial *regulatory* systems implicated in *virulence*.

However, the complexities of bacterial gene expression in vivo during infection cannot be addressed solely by in vitro experiments and must be coupled to in vivo analysis. The infected host represents a complex and dynamic environment, which may vary during the infection process. One may assume then, that bacterial pathogens modulate gene expression in response to the conditions *encountered* in vivo, via bacterial *sensors* responding to a variety of host factors and stimuli.

In addition, discrete cellular steps *presumably* permit the controlled expression of genes essential for *initiation* and maintenance of infection. To gain overall understanding of an infection, we must determine the proportion of the bacterial genome expressed in vivo and the subset of genes essential for *infection*.

This is a formidable task requiring several complementary techniques, such as in vivo expression technology CIVET), signature-tagged *mutagenesis* (STM), elegant *bioinformatics* analysis, functional genomic analysis via proteomics using two-dimensional gel *electrophoresis* (2DE) differential display, DNA and protein arrays, and chip technology.

There is also an urgent need for new technologies for analysis of genes expressed in vivo, as the current tools have significant limitations. In this chapter, we focus on the recent progress and adaptations of STM, a technology that has become accessible to an increasing number of laboratories, especially as screening can now be done rapidly by *automated* high-throughput robotics using PCR instead of hybridization. The design of tags has been simplified, and several different miniTn5s, each with a unique *phenotypic* selection, can now be utilized.

These modifications of STM, developed in our laboratory, have been applied to the study of *Pseudomonas aeruginosa PAO 1,* the 6.3-Mb genome of which has been completely sequenced. The strategies of PCR-based STM are described using the *P. aeruginosa STM* project ongoing in our laboratory as an example. We also discuss the significance of in vivo screening. STM is a mutation-based screening method that uses a population of bacterial mutants for the identification of multiple virulence genes of microbial pathogens by negative selection.

The technique depends upon in vivo selection of virulent organisms, while those mutants whose virulence genes are inactivated will not persist at infection *foci* in models of infection. This elegant technique, developed by Holden and *colleagues,* allows a relatively rapid, unbiased search for virulence genes, using animal hosts to select against strains carrying mutations in genes affecting virulence, among a mixed population of mutants.

To avoid the typical labor-intensive screening of individual mutants in the first steps, each mutant is tagged with a different and unique DNA signature; the power of STM is that it allows large numbers of different mutants to be screened simultaneously, in the same animal host. In the classical STM protocol, a comparative hybridization technique is used, and employs as mutagens a collection of *transposons,* each modified by the *incorporation* of a different DNA tag sequence.

The concept is that when the *tagged* mutagens integrate into the bacterial chromosome, each individual mutant can be distinguished from every other, on the basis of their unique *molecular* tags. The first tag collections were designed as short DNA segments containing 40-bp variable regions flanked by invariant arms that facilitated the coamplification and radioactive labeling of the central portions by PCR, and which were subsequently used as hybridization probes. Since STM is an en masse screening technique, mutants are usually kept in arrays, using *microtiter* plates.

Colony or DNA dot blots are prepared from these mutants, and compared by hybridization to DNA prepared when the same pool of strains has been passed through the animal host. Technically, PCR is used to prepare labeled probes representing the tags present in the inoculum (input) and recovered from the host (output). Hybridization of the tags from the input and output pools to the colony or dot blot permits the identification of mutants that failed to grow in vivo, because these tags will not be present in the output pools.

These mutants can then be identified and recovered from the original arrays, and the nucleotide sequence of DNA flanking the mutation site can be determined. Hence, STM has and will benefit immensely from the sequence data of all bacterial *genomics* projects. In the original method, tags were incorporated into a mini-Tn5 transposon, and their *suitability* was checked prior to use, by hybridization of amplified labeled tags to DNA colony blots of mutants.

Mutants from the in vitro pool whose tags failed to yield clear signals on autoradiograrns were discarded, while those that gave good signals were assembled into new pools for screening in animals. This careful analysis was done prior to STM, so as to *diminish* the inherent problems of hybridization caused by *problematic* tags.

The STM method was subsequently modified to avoid the prescreening process, where a series of *Tn917* transposons were selected prior to mutagenesis, on the basis of efficient tag amplification and labeling, and lack of cross-hybridization to other tags. Since in this case the

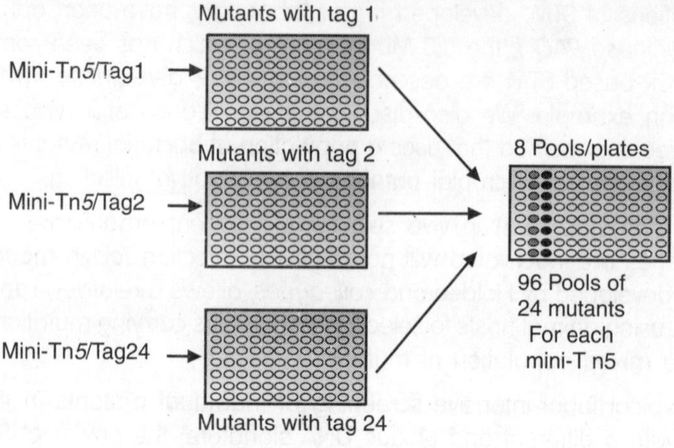

Figure 20.1: Graphical preseutation of arrayed libraries using 24 signature tags and three mini-Tn5 derivatives. In each of the 72 defined libraries, the 96 selected mutants have the same tag, but it is inserted at different locations in the bacterial chromosome. After pooling, the end result is one library containing 96 pools of 72 mutants (3 × 24 mutants with unique tag for each mini-Tn5).

identity of each tag is known, each modified transposon can then be used to generate an infinite number of mutants. It can also be screened using plasmid or tag DNA rather than colony blots, which increases the sensitivity and specificity of the assay.

An additional modification of the technique introduced 96 tags at a disrupted *URA3* locus of *Candida glabrata*. Different mutants were created by insertional mutagenesis using a plasmid that simultaneously complemented the *URA3* mutation, and the DNA flanking the mutations was cloned by *plasmid rescue* in *Escherichia coli*. Variations of the original STM method using various transposons have subsequently been applied to many different *bacterial pathogens*.

Certain transposons are inappropriate tools for genetic analysis of some species of bacteria. For example, Tn5 and its derivatives do not transpose in Gram-positive bacteria, or very poorly in certain Gram-negative bacteria, such as the Pasteurellaceae. However, the recent development of new transposons for genetic analysis, and development of an in vitro transposition assay, should permit the application of STM to a large variety of organisms.

Suitable properties for transposons to be used in STM include ease of manipulation in vitro, high frequency of transposition, random insertion in the genome of the host, and controlled frequency of insertions as with mini-Tn5s. Obviously, STM cannot provide information about genes that are essential. for survival per se, but are inactivated by insertional *mutagenesis*. Conversely, an insertion within a *nonessential* gene that is crucial for virulence would not be expected to affect the growth rate of cells *in vitro*.

SIGNATURE TAGGED MUTAGENES BASED UPON PCR

As in the original STM method, PCR-based STM is divided into two major steps. The first *in vitro* step involves the construction of specific and defined DNA tags, which are used for the preparation of a library of tagged mutants. This first phase comprises: the design of *oligonucleotides* for use as DNA tags; the cloning of these tags into a transposable element that will produce

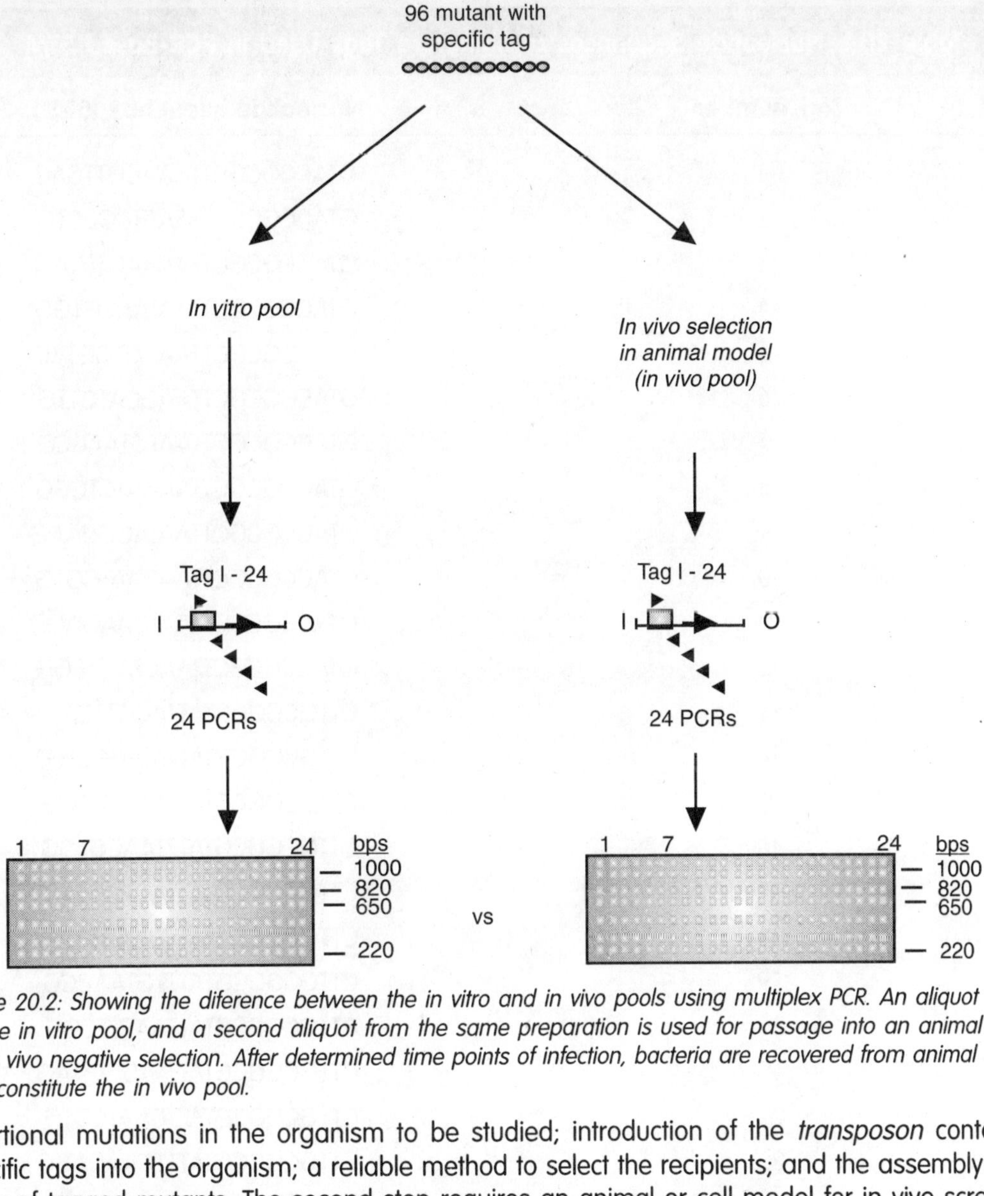

Figure 20.2: Showing the diference between the in vitro and in vivo pools using multiplex PCR. An aliquot is kept as the in vitro pool, and a second aliquot from the same preparation is used for passage into an animal model for in vivo negative selection. After determined time points of infection, bacteria are recovered from animal organs and constitute the in vivo pool.

insertional mutations in the organism to be studied; introduction of the *transposon* containing specific tags into the organism; a reliable method to select the recipients; and the assembly of an array of tagged mutants. The second step requires an animal or cell model for in vivo screening of the library.In this second phase, experimental design needs to take account of the power and limitations of the models used, and the number of rounds of STM screening and the number of animals or cell cultures ro be utilized.

STM also involves a systematic characterization of mutants selected in vivo. Crucial data include the DNA sequence around the site of the insertion mutation (which may be obtained by cloning the transposon marker or by RT-PCR) and confirmation that the tagged mutation is the cause of virulence *attenuation*. The final phase, and the ultimate goal of STM, is to characterize genes

Table 20.1: Nucleotide Sequences of Signature Tags Used in PCR-Based STM

Tag number	Nucleotide sequence (5′-3′)
1	GTACCGCGCTTAAACGTTCAG
2	GTACCGCGCTTAAATAGCCTG
3	GTACCGCGCTTAAAAGTCTCG
4	GTACCGCGCTTAATAACGTGG
5	GTACCGCGCTTAAACTGGTAG
6	GTACCGCGCTTAAGCATGTTG
7	GTACCGCGCTTAATGTAACCG
8	GTACCGCGCTTAAAATCTCGG
9	GTACCGCGCTTAATAGGCAAG
10	GTACCGCGCTTAACAATCGTG
11	GTACCGCGCTTAATCAAGACG
12	GTACCGCGCTTAACTAGTAGG
13	CTTGCGGCGTATTACGTTCAG
14	CTTGCGGCGTATTATAGCCTG
15	CTTGCGGCGTATTAAGTCTCG
16	CTTGCGGCGTATTTAACGTGG
17	CTTGCGGCGTATTACTGGTAG
18	CTTGCGGCGTATTGCATGTTG
19	CTTGCGGCGTATTTGTAACCG
20	CTTGCGGCGTATTAATCTCGG
21	CTTGCGGCGTATTTAGGCAAG
22	CTTGCGGCGTATTCAATCGTG
23	CTTGCGGCGTATTTCAAGACG
24	CTTGCGGCGTATTCTAGTAGG

Each 21-mer has a T_m of 64°C and permits PCR amplification in one step when the three primer combinations are used for multiplex screening. Two sets of consensus 5¢-ends comprising the first 13 nucleotides have higher **Gs for optimizing PCR. Twelve variable 3¢-ends define tag specificity and allow amplification of specific DNA fragments. The set of 24 21-mers representing the complementary DNA strand in each tag are not represented and can be deduced from the sequences present.

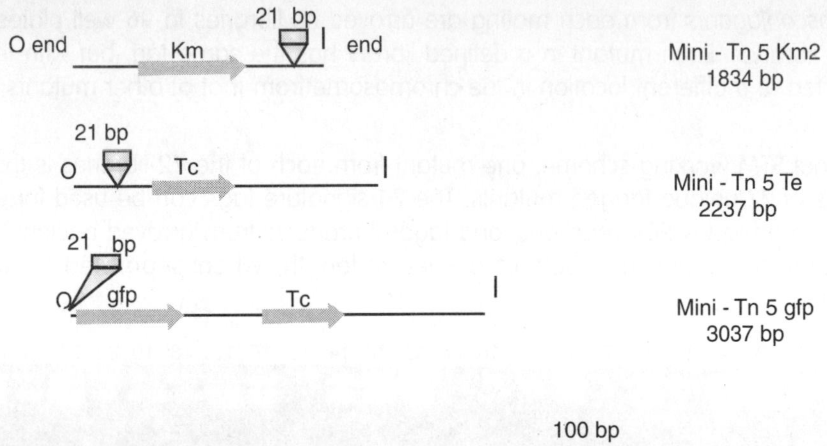

Figure 20.3: Maps of the mini-Tn5Km2, mini-Tn5Tet, and mini-Tn5Tetgfp transposons used. The elements are represented by thick black lines; inverted repeats are indicated as open vertical boxes; and genes are indicated by arrows. Abbreviations: I and 0, inverted repeat ends; Km, kanamycin; Tc, tetracycline; gfp, green fluorescent protein.

shown to play a significant role in virulence, and to define their function and role in pathogenicity. Here, we describe the rationale for PCR-based STM and demonstrate typical results using *P. aeruginosa* as a model.

Mechanism of STM

The PCR-based STM method that we have developed uses 24 pairs of' complementary 21-bp oligomers, the first twelve pairs corresponding to ones described by Lehoux et al.. These pairs are annealed to generate 24 double-stranded DNA tags, which are then cloned into each of the three plasmid-located mini-Tn5 derivatives: mini-Tn5Km2, mini-Tn5Tc, and miniTn5TcGFP.

These Tn5-derived mini-transposons contain genes encoding kanamycin resistance (Km), tetracycline resistance (Tc), and green fluorescent protein (GFP), respectively, have unique cloning sites for tag insertion, and are flanked by 19-bp terminal repeats (the I and the 0 ends).

The transposons are located on the pUT plasmid, a R6K-based suicide delivery plasmid. This procedure yields 72 (24 tags × 3 mini-Tn5 vectors) tagged transposons. There is a precedent for limiting the number of tags used. When *Salmonella Typhimurium* was inoculated into the *peritoneal* cavities of mice, pools of 96 different mutants gave reproducible *hybridization* signals after 3 d of infection, whereas pools comprising 192 mutants did not.

Problems with the complexity of the pool tags have also been described for *Vibrio cholerae,* where it was necessary to reduce the complexity of an orally inoculated pool to 48 mutants, in order to obtain reproducible results.

Creating Libraries of Tagged Mutants

Each of the 72 signature-tagged transposons is replicated in *E. coli,* and is then inserted separately into the chromosome of the host bacterium that is to be investigated, to generate insertional mutants. This is usually achieved by conjugation at high frequency, with transconjugants (i.e., mutants) selected using the appropriate mini-Tn5 markers, Km, Tc, or Tc combined with GFP.

Ninetysix *transconjugants* from each mating are arrayed as libraries in 96-well plates. There are, therefore, 72 libraries. Each mutant in a defined library has the same tag, but with the likelihood that it is inserted at a different location in the chromosome from that of other mutants in the same pool.

For the final STM working scheme, one mutant from each of the 72 libraries is then picked to form 96 pools of 72 unique tagged mutants. The 24 signature tags can be used for specific DNA amplifications in multiplex PCR reactions, and tagged products from arrayed bacterial clone pools can be compared as DNA products of a specific length, when separated by agarose gel electrophoresis.

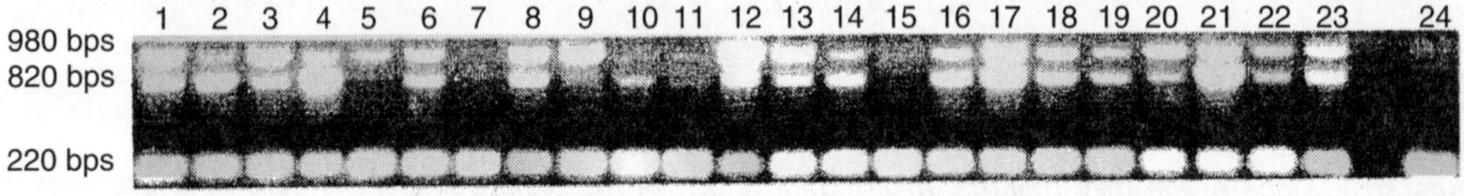

Figure 20.4: Agarose gel electrophoresis separation of multiplex PCR amplified products obtained from the in vivo pool of 72 tagged-transposon mutants. Lane 1 to 24, 10 μL of a multiplex PCR reaction using pUTKana2, tetR1, and pUTgfpR2 primers with Tag 1 to 24 respectively.

Signature Tagged Mutants in Animal

STM usually involves the use of an animal model or cell line to mimic aspects of a particular infection, although these models cannot necessarily reproduce faithfully all the conditions associated with the natural disease.

Several parameters need to be considered, including the number of different mutant strains to be used in a pool, the route of administration, the size of the inoculum, and the incubation period. An additional consideration is the use of different animal hosts for screening the same pool of STM mutants. The inoculum size required to initiate infection will determine the complexity of the mutant pool. On the one hand, the pool must contain sufficient cell numbers to initiate infection reliably.

On the other, the inoculum size must not be too high, or the host's immune system may be overwhelmed, resulting in the growth of mutants which would not otherwise be detected, and the possibility that animals might die of shock. Other important parameters in STM include the route of inoculation and the time-course of a particular infection.

Also, certain gene products that are important, directly or indirectly, for initiation or maintenance of infection, may be niche-dependent or expressed specifically in only certain host tissues. If the duration of the infection in STM in vivo selection is short, genes important for establishment of the infection will be found, while if the duration is long, genes important for maintenance of infection may be identified.

Importance of Mutant Strains

When a potential STM mutant has been obtained, it is essential to confirm rapidly that the tagged mutant is the cause of attenuation, even after two rounds of in vivo screening. Our recent approach is to do this first, even before the actual DNA sequence flanking the insertion point is known.

The degree of virulence attenuation is a prerequisite, and we use the more sensitive and increasingly popular competitive index (CI) and growth index (GI) tests. The analysis is done with an STM mutant strain combined as single or mixed infections with the wild-type parent strain to describe the kinetics of growth, and in certain cases to identify the time and body site where the virulence defect is apparent.

In a mixed infection, the ability of the strains to initiate infection or colonize the host provides a measure of their relative virulence. The CI and GI can then be defined as the output ratio of mutant to wild-type bacteria divided by the input ratio of mutant to wild-type bacteria. The CI is thus a quantitative value for the degree of attenuation of a mutant strain, with the CI of a wild-type strain versus a fully virulent derivative being approx 1.0.

APPLICATIONS OF STM

As STM is increasingly used and parameters are better defined in different models of infection, several routes of inoculation and different animal models may be used for studying the same organism. We present here the protocol that we have used with *P. aeruginosa.*

Studying Chronic Lung Infection

For studying *P. aeruginosa,* we have used a rat model of chronic lung infection, which was adapted for this work. Female Sprague-Dawley rats (140160 g) are inoculated into the left lobe of the lungs with a suspension of agar beads containing 10^5 bacterial cells (the in vitro pool); after 7 d, rats are sacrificed, and homogenized lung samples are plated on medium supplemented with chloramphenicol (5 μg/mL).

To identify mutants not recovered after the in vivo passage, screening is done by multiplex PCR using bacterial colonies recovered from the lung homogenates. The 21-mers numbered 1 to 24 in Table elsewhere in this chapter were used as a first primer in combination with a primer selected in the Km, Tc, or GFP gene from each transposon.

Colony PCR amplification products obtained from the in vivo pool are compared with the in vitro pool, and mutants that give positive results from the in vitro pool, but are absent from the in vivo pool, are kept for further analysis, involving identification and a second round of in vivo screening.

Technically, this is confirmed as absence of a PCR product. In our studies, 96 pools of 72 mutants, forming a collection of 6912 mutants, were maintained in the rat, and caused chronic lung infection. From these 6912 mutants, we identified 214 attenuated mutants whose tag did not give a PCR amplification product from the in vivo pool. Mutants attenuated in vivo were further tested by assessing their growth on minimal media (M9 plates).

From the collection of 214 mutants, 5 auxotrophic mutants were identified and not studied further. We arrayed 14 new groups with the 214 mutants from the first screening, and *29 P. aeruginosa* STM mutants previously screened once to confirm the STM attenuated phenotype. When necessary, we completed a group with the wild-type strain PAO1 in order to maintain 72 clones in each group.

The in vivo selection and detection of mutants were done as described above. From the 214 mutants initially identified, we retained 42 highly attenuated mutants, whose tags did not give any PCR amplification product from the in vivo pool.

These results showed that to identify and obtain the most significant and highly attenuated mutants, a second round of in vivo screening is a prerequisite. The genomic DNA from these *P. aeruginosa* STM mutants was isolated, digested with *PstI,* and cloned into pTZ18R (Amersham Pharmacia Biotech). The recombinant plasmids were introduced into *E. coli* DH5α by electroporation. Selection of recombinant plasmids was based on the cloning of the Km, Tc, or Tc-Gfp marker; plasmid DNA was prepared and the sequences flanking the transposon markers were analyzed to determine the insertion endpoints. To determine the GIs of the *P. aeruginosa* STM mutants, the in vitro growth rates of the parent PAOI strain and mutants derived from it were measured in the rat model of lung infection.

Bacteria were recovered 1- and 7-d postinfection from two animals, by direct plating of lung tissue homogenates on medium supplemented with chloramphenicol. Animals used as controls were injected with strains PAO1 and PAO909 (a purine auxotroph of PAO1). For example, nine mutants found to be attenuated are listed in Table 20.2.

Among the nine in vivo attenuated mutants, growth rates varied from low to high as reflected by the index values; the least attenuated mutant had an index of 0.294. Mutants that were considered to be clearly attenuated had a growth rate index of 0.1. or less. An additional confirmation that we use in combination with the G1 to show that the tagged mutation is the cause of attenuation of virulence, is complementation with a functional allele. In this case, the wild-type gene of the STM mutant identified is cloned as a PCR product and introduced in the STM mutant.

In a first step, we use RT-PCR to demonstrate in vitro expression and proceed with a G1 analysis of the wild-type, the STM mutant, and the mutant strain expressing the complementing allele. The obvious result expected will show that the mutation and the virulence phenotype are linked. Gene analysis can indicate whether the inactivated gene is part of an operon, and that

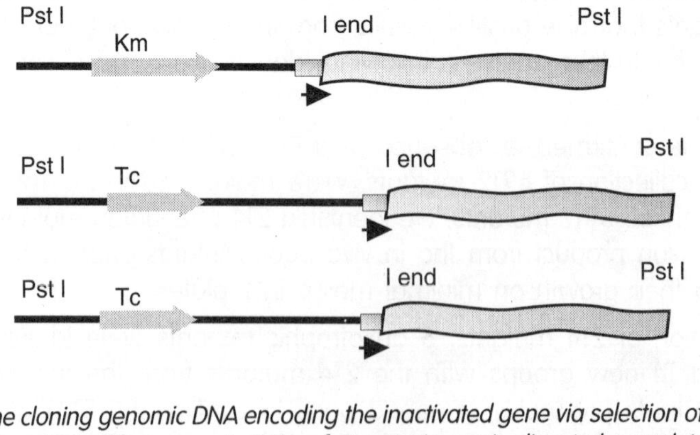

Figure 20.5: Showing the cloning genomic DNA encoding the inactivated gene via selection of the antibiotic resistance genes inserted in STM mutants. The genomic DNA of P. aeruginosa is digested completely with PstI *and cloned DNA fragments are selected on the basis of encoding either Tc or Km resistance. The cloned genomic DNA is sequenced using primers annealing in the I end of each transposon DNA fragment (I-end sequencing primer: 5¢-AGA TCT GAT CAA GAG ACA G-3').*

Table 20.2: In Vitro and In Vivo Growth Index of *P. aeruginosa* STM Mutants

Strains/(genes inactivated)	*In vitro GI*	*Relative GI*
PAO1	1.0	1.0
PA0909	1.1	0.0036
G20T2 (Clp protease, ClpB, *Y. enterocolitica*)	1.5	0.0006
G13T 12 (Molybdate, binding precursor, ModA, *P. aeruginosa*)	1.2	0.025
G19T 12 (Unknown)	2.0	0.01
G10T7 (Unknown)	1.7	0.176
G30T 12 (Hypothetical 64 kDa protein, YfaA, *E. coli*)	1.3	0.154
G56T2 (Hypothetical 51 kDa protein, YgdH, *E. coli*)	2.4	0.25
G 18T2 (Transporter protein, AcrB, D, F, Vc 1663, *V. cholerae*)	1.7	0.294
G94T2 (ABC transporter ATP,YbjZ, *E. coli*)	1.5	0.0267
G38T4 (Heat shock protein, HtpG, *P. aeruginosa*)	5.8	0.034

The disrupted open-reading frames (ORF) in each STM mutant were: PAO090 (G20T2); PA1863 (G13T12); Pa5441 (G19T12); PAO082 (G1OT7); PA4491 (G30T12); PA41 15 (G56T2); PAO0158(G18T2); PA0073 (G94T2); PA 1596(G38T4). Identity values were obtained by comparing the amino acid sequences deduced from the complete ORF.

STM attenuation of virulence may be due to a polar effect. In this case, inactivation of the downstream gene(s) is essential to confirm their implication in the decrease of virulence by further testing with the CI tests. To demonstrate allelic complementation of STM mutants, we will use as an example the G 18T2 mutant (PA00158) described in Table elsewhere in this chapter.

Mutant G1 8T2 corresponds to an insertion in an ORF (PAO158) encoding for a protein of 1,016 residues and having 32% identity with a putative ORF VC 1663 of V. *cholerae,* 31% identity with AcrD of *Rickettsia prowazekii* and CzrA of *P. aeruginosa,* and 45% identity with a membrane protein which is a putative cation efflux protein (HP0969) of *Helicobacter pylori,* designated Hef. The protein PAO158 has been classified as a probable multidrug efflux system component encoding

Figure 20.6: Confirming expression of modA *from the complementing allele. Lanes: 1, wild-type; 2,* modA *STM mutant complemented; 3, STM mutant.*

for a probable RND efflux transporter. In *H. pylori, hefF* encodes an RND pump protein homolog. The role and function of RND multipledrug efflux systems in intrinsic antibiotic resistance has been demonstrated in several human pathogens, including *E. coli, Haemophilus influenzae, Neisseria gonorrhoeae,* and *P. aeruginosa.*

The corresponding mutant has been shown to be mildly attenuated, with a GI value of 0.294 in Table 20.2. Several pseudomonads possess an intrinsic resistance to many front-line antibiotics, due mainly to their low outer membrane *permeability* and active efflux of *antibiotics*.

We have found by STM that this efflux system may be implicated in virulence in vivo; this system could presumably be active against the host defense system by pumping out toxic molecules.

As proof of concept, we have cloned and expressed the *modA* gene as determined by RT-PCR, and shown complementation of ModA in vitro and in vivo using the mutant G 13T12 (PA 1863). The complemented strain G 13T 12/pMON-PA 1863 was used and compared with the G 13T 12 strain.

We observed that the complemented strain was able to grow 3.4 times faster in vivo than the strain containing the m*odA* insertional mutation (GI of 0.24 compared with 0.07, respectively) (data not shown). These results showed that the complementation with *modA* gave a significant recovery to the wildtype phenotype.

This indicated that the disruption of *modA* in the mutant G 13T 12 was responsible for the attenuated phenotype obtained by STM, and confirmed by the in vivo GI experiments. Indeed, insertional inactivation of *modA* gave a significant attenuation of virulence in vivo, but attempts to restore the mutant G13T12 to the wild-type phenotype were only partly successful (GI of 0.6 instead of 1.0).

CONCLUDING REMARKS

The methods described in this chapter have been the first important step in using STM for the elucidation of which genes are essential in *P. aeruginosa* for the maintenance of an infection in the rat lung model. At this stage of the project, we have shown that PCR-based STM can identify typical virulence factors. Indeed, we have isolated *P. aeruginosa* STM mutants with defects in alginate biosynthesis, pilin formation, biofilm production, elastase and protease production, and twitching motility; these have all been shown to have some role in maintenance of infection in vivo.

In the last few years, hybridization-based STM has been widely used, and has given remarkable insights in the virulence of bacterial pathogens, such as *Salmonella, Vibrio, Streptococcus,* and *Staphylococcus.* The PCR-based STM approach that we present here should be a useful addition to the repertoire of technologies for in vivo functional *genomics* and the identification of function of new virulence factors.

Chapter 21

Mutation and Environmental Toxicants

Numerous synthetic chemicals and industrial by-products are introduced into the environment that possess *genotoxic* activity. These chemicals, or their metabolites produced by biotransformation reactions occurring in most eukaryotes, can react with DNA, forming covalently bound adducts that *distort* the double helix and eventually activate repair mechanisms that cause single-or double-strand breaks in the *helix* backbone.

Ionizing radiations transferring their energy into a living organism are also well known to induce gene mutations and chromosome aberrations by a free radical mediated mechanism capable of directly causing DNA strand breaks. Other chemicals do not react with DNA, but may bind to proteins forming *microtubules*, including those of the metaphase spindle that organize chromosome *segregation* at *mitosis* and *meiosis*.

These chemicals, which include antineoplastic drugs but also widely used pesticides, may induce errors of chromosome *segregation* and *produce* cells with missing or extra chromosomes. The characterization of the genotoxicity of chemical and physical agents is one of the main areas of environmental *mutagenesis* research, together with the determination of dose-effect relationships and of the numerous variables that affect the response of a given cell/organism to a certain mutagen exposure.

The ultimate goal of this discipline is to prevent the occurrence of genetic changes in somatic and germ cells, additional to those produced by endogenous *mechanisms*, that can lead to severe degenerative or *heritable* diseases primarily in humans, but also in wild and domestic animals and plants.

An important contribution of environmental mutagenesis is also to generate knowledge needed to set scientifically sound chemical and radiation exposure limits to human populations and to regulate labeling and classification of commercial products. To achieve its goals, environmental mutagenesis makes use of in vitro and in vivo laboratory models and field, or so-called *biomonitoring* studies, carried out mostly in human populations and more rarely in other natural living organisms.

In all of these studies, the frequencies of appropriate biomarkers of genotoxic effects are measured in exposed cells or individuals and compared with those of unexposed control groups. Gene mutations, chromosome *aberrations*, and *aneuploidy* are all genetic changes causally related to cancer development, and high frequencies of chromosome aberrations in human blood lymphocytes were retrospectively shown to correlate with an increased risk of cancer. Structural aberrations essentially comprise fragments of broken chromosomes and various types of rearrangements between *nonhomologous* chromosomes.

Aneuploidy includes loss and gain of whole intact chromosomes. In most cases, these aberrations do not persist for long times either because, at cell division, they are *intrinsically* unstable, such as are acentric fragments or dicentric chromosomes, or produce gene dosage effects leading to cell death. Nevertheless they represent an early biomarker of *genotoxic* exposure or genetic instability.

Conversely, some structural and numerical chromosome changes may be stable through cell division and compatible with cell survival, such as reciprocal translocations and particular aneuploidies; such genetic alterations may directly contribute to the multistep process of cell *transformation* and tumor development. Structural and numerical chromosome aberrations can be detected by optical microscopy in *metaphase* and, more recently, also interphase cells.

Traditionally, metaphase chromosomes stained by DNA specific dyes were scored for numerical and structural changes. Chromosome banding techniques could be applied to help detection of aberrations, such as centromeric- (C-) banding, in the case of mouse acrocentric chromosomes. Reciprocal translocations could not be identified by this approach unless they produced gross karyotypic changes. Numerous techniques, all based on the principles of fluorescence *in situ* hybridization (FISH), were developed in the past few years to improve the cost-effectiveness, speed, and sensitivity of chromosome aberration scoring.

Chromosome painting denotes the technique based on metaphase chromosome FISH with whole chromosome specific probes. Some chromosomes are hybridized with fluorescent probes, and all of them are counterstained with DNA specific fluorescent dyes such as 4,6-diamidino2-phenylindole (DAPI). Chromosome rearrangements are detected as colour switches between painted and unpainted sequences.

Chromosome painting was initially developed to detect stable aberrations as long-term biomarkers of past *radiation* exposure in humans. Later on it was applied to experimental studies on mechanisms of aberration induction and, with the development of mouse chromosome probes, to the quantification and characterization of *persistent* aberrations experimentally induced by radiation and chemical mutagens.

Further developments of chromosome painting were based on a cocktail of probes specific for each chromosome conjugated with different fluorescent dyes and computerized image analysis capable of recognizing each one of the 24 human or the 21 mouse different chromosomes. By this technique, complex rearrangements involving more than two chromosomes could be detected. Detection of target sequences is achieved not only on metaphase chromosomes, but also in interphase nuclei.

The possibility of detecting specific sequences in interphase is of particular interest for cases in which chromosomes are difficult if at all possible to prepare, such as sperm micronuclei, or low proliferation tissues. A useful application of interphase FISH is that with pan-centromeric probes that hybridize close to the functional centromeric sequences of all human or mouse chromosomes.

This FISH technique has been extensively applied to characterize the chromosome content of micronuclei, small chromatincontaining bodies surrounded by a nuclear membrane that are produced after chromosome breaks or chromosome loss at anaphase.

Centromeric FISH-positive micronuclei are, in fact, interpreted as containing whole chromosomes lost at anaphase because of *spindle* damage or other types of *mitotic* defects, while negative micronuclei represent acentric chromosome fragments produced by chromosome breakage.

Alternatively, whole-chromosome containing micronuclei were identified via immunofluorescent detection of kinetochore proteins with CREST antibodies. The *simultaneous* application of CREST and FISH labeling to micronuclei allowed evaluation of possible detachment of *kinetochore* proteins from the centromeric DNA sequences. Human chromosomes are characterized by chromosome-specific *pericentromeric* sequences.

Probes for these sequences mark specific chromosomes in interphase nuclei. These probes can be used to trace chemically induce chromosome nondisjunction in pairs of nuclei of binucleate cells: when damage to any component of the mitotic machinery affects chromosome segregation, in fact, two sister *chromatids* may segregate to the same daughter nucleus.

This event will give rise, at the same time, to one cell with a missing and one cell with an extra chromosome that can be easily detected by counting the number of specific probe signals in each nucleus. Finally, telomeres are specialized functional elements of eukaryotic chromosomes required for chromosome stability maintenance. Telomeres consist of repetitive DNA sequences and specific proteins. All vertebrate *species* have the same basic telomeric DNA sequence. Telomere function is impaired when telomere length changes either below or above the species-specific range. Nonfunctional *telomeres* cause end-to-end chromosome fusion producing various types of stable and unstable structural aberrations.

Quantitative FISH (Q-FISH) is a FISH-based approach developed to measure telomere length with high resolution on a single chromosome Fig. 1. Representative illustration of 2D profiles of silver-stained proteins from the cytosolic fractions of AtT20 and AtT20(proSAAS) cells. Numerous proteins spots were more stained in one profile that in the other.

The most prominent are indicated with arrows. Note the circled clusters of high molecular weight molecular proteins found in more abundance in AtT20(proSAAS) cells. Their abundance may be due, directly or indirectly, to reduced activity of PC1 caused by proSAAS overexpression.basis.

ELEMENTARY APPROACH FOR FISH

FISH is a method for localizing and detecting specific nucleotide sequences in morphologically preserved tissue sections or cell preparations by hybridizing the complementary strand of a nucleotide sequence (the probe) with the DNA of interest (also called target). There are essentially five types of probe that can be used in performing *in situ* hybridization.

Oligonucleotide Probes

These are produced synthetically by an automated chemical synthesis. These probes have the advantages of being resistant to RNases and are small, generally around 40-50 base pairs (bp). Their small size allows for easy penetration into the cells or tissue of interest, and the single strand excludes the possibility of renaturation. This oligonucleotide probe allows the hybridization of short sequences, especially single-copy genes.

Single-stranded DNA Probes

These have similar advantages of the oligonucleotide probes except they are much larger, probably in the 200-500 by size range.

Double-stranded DNA Probes

These are the common commercial probes used. They are longer than the oligonucleotide probes and single-stranded probes allowing hybridization of repeated sequences or entire chromosomes. Because the probe is double stranded, it means that *denaturation* has to be carried out before *hybridization* in order for one strand to hybridize with the target.

RNA Probes (cRNA probes or riboprobes)

RNA probes have the advantage that RNA-DNA hybrids are very thermostable and are resistant to digestion by RNases. This allows the possibility of posthybridization digestion with RNase to remove non-hybridized RNA and therefore reduces the possibility of background staining.

PNA Probes

The peptide nucleic acid probes (PNAs) constitute a remarkable new class of synthetic nucleic acid analogues, based on their peptide-like backbone. This structure gives to PNAs the capacity to hybridize with high affinity and specificity to complementary RNA and DNA sequences and a great resistance to nucleases and proteinases.

The probes are synthesized using different existing protocols. Basically, one resynthesizes one strand of a double-stranded DNA molecule while incorporating modified nucleotides. A DNA probe could be labeled with a fluorescent dye (direct labeling) or a haptene (indirect labeling), usually in the form of fluor-dUTP or haptene-dUTP, that is incorporated into the DNA using *enzymatic* reactions, such as nick translation, random-primed or polymerase chain reaction (PCR).

If the probe has been labeled indirectly, an extra step is required for visualization of the *non-fluorescent* haptene that uses as enzymatic or immunological detection system. Whereas FISH is faster with directly labeled probes, indirect labeling offers the advantage of signal amplification by using several layers of antibodies, and might therefore produce a signal that is brighter compared with background levels.

A large variety of commercial probes are now available, so as to allow hybridization even in less optimal conditions. There are almost as many methods for carrying out *in situ* hybridization. The basic principles for *in situ* hybridization are the same. Here we report a brief outline of the common procedural *cytogenetic* steps.

Methodology

Proper technique for preparing slides is considered essential for both classical cytogenetics (banding) and molecular cytogenetic procedures (FISH). Different protocols were specifically designed for use with particular types of biological material or cells or for particular purposes such as detection of structural aberrations, numerical aberrations, or sequence amplifications. To preserve *morphology*, the biological material must be fixed. For metaphase chromosome spreads, methanol-acetic acid fixation is usually sufficient.

In the basic protocol, cells are resuspended in 3:1 methanol-acetic acid fixative(s) ,and dropped or smeared on commercially precleaned slides, which are dried at various temperatures. For *paraffin-embedded* tissue sections, use formalin fixation.

Cryostat-sliced sections fixed for 30 min with 4% formaldehyde have been used successfully. Unfortunately, a fixation protocol that can be used for all substrates has not yet been described. The fixation must be optimized for different applications.

Aging

Slides are subjected to dry heat and/or ethanol to remove water and fixative from the preparations, and to enhance the adherence of the material to the glass.

Enzymatic Treatment

RNase treatment serves to remove endogenous RNA and may improve the signal-to-noise ratio in DNA-DNA hybridization.

Pretreatment

The act of fixation results in crosslinking of proteins, which may present an obstacle to good accessing of the probe. *Pretreatment* serves to increase target accessibility by digesting the protein that surrounds the target *nucleic acid*. There are a number of different procedures for *permeabilization*.

Three common reagents used are HCl, detergents (Triton or sodium dodecyl sulfate [SDS]) and enzymes such as proteinase K, trypsin, or pepsin. The most common enzyme used is pepsin because it has the advantage that it can be easily inactivated by pH changes, and the reaction is easier to control.

Denaturation of DNA

Denaturation of DNA, which separates the two strands, is obtained by alkaline or heat treatment. Heat denaturation has become popular because of its experimental simplicity and greater effectiveness. Slide and probe denaturation can be performed together (called "simultaneous or direct denaturing") or separately (called "separate or indirect denaturing").

Numerous experiments with both commercial and custom-prepared probes showed that results were similar with both procedures. Variations in time and temperature should be evaluated to find the best conditions for denaturation.

Process of Hybridization

The hybridization step consists in simply mixing the single-strand probes with the denatured target DNA. The denaturation of DNA allows access of the single-strand probes to their respective complementary single-strand DNA. Provided that different probes are conjugated with fluorochromes with nonoverlapping emission wavelengths, multiple targets can be simultaneously hybridized in the same cells.

Washing

Labeled probe can hybridize nonspecifically to sequences that are partially but not entirely homologous to the probe sequence. Such hybrids are less stable than perfectly matched hybrids. They can be dissociated by performing washes of various stringencies.

The stringency of the washes can be manipulated by varying the formamide concentration, salt concentration, and temperature. Often a wash in 2× saline sodium citrate (SSC) containing 50% formamide suffices. For some applications the stringency of the washes should be higher.

Amplification of FISH Signal

The detection of the probes is the final step of FISH. When a DNA probe is labeled by a fluor-dUTP the probe-target hybrids can be visualized under a microscope immediately after the hybridization reaction. If the probe is labeled with haptene-dUTP, the fluorescent signal is usually provided by a fluor-labeled molecule (primary detection reagent) that binds the haptene with high affinity.

To amplify the signal, another fluorescent-labeled molecule (secondary detection reagent) that recognizes the primary reagent is used. Although these detection reactions sometimes are not based on the immunological principles of antigen-antibody recognition, but on protein binding with strong affinity (biotin/avidin), usually the term antibody is used to indicate the detection reagent. When a DNA probe is labeled by a fluor-dUTP, signals are usually weaker than the signal yielded by the same probe if it were detected with an antibody.

The reason is that an antibody molecule carries three or four fluors attached to it. The weak signal of a fluorescent-labeled probe can sometimes be amplified using antibodies raised against that fluor molecule. Antibodies against fluorescein, rhodamine, Texas Red, Cascade Blue, and other fluors are available commercially. These primary antibodies will bind to the respective fluor, and a secondary antibody labeled with the same fluor will be used to bind to the primary antibody and amplify the signal.

Counterstaining

Fluorescent DNA counterstaining is usually performed with red fluorescent propidium iodide (PI) or with blue fluorescent DAPI. An antifade agent should be mixed with the counterstaining dye solution to retard fading.

Optical Studies

Finally the signals are evaluated by fluorescence microscopy. A fluorescent microscope contains a lamp for excitation of the fluorescent dye and a special filter that transmits a high percentage of light emitted by the fluorescent dye. There has been considerable improvement in both the hardware and software that are used for the analysis of FISH images.

Cooled charge-coupled-device (CCD) cameras and fluorescence filter sets for microscopy that are more specific and efficient have improved the sensitivity and resolution of imaging, and sophisticated software facilitates the acquisition and processing of images. The general FISH procedure is illustrated in Figure 21.1.

In the following chapters, materials and methods are described that are common to the

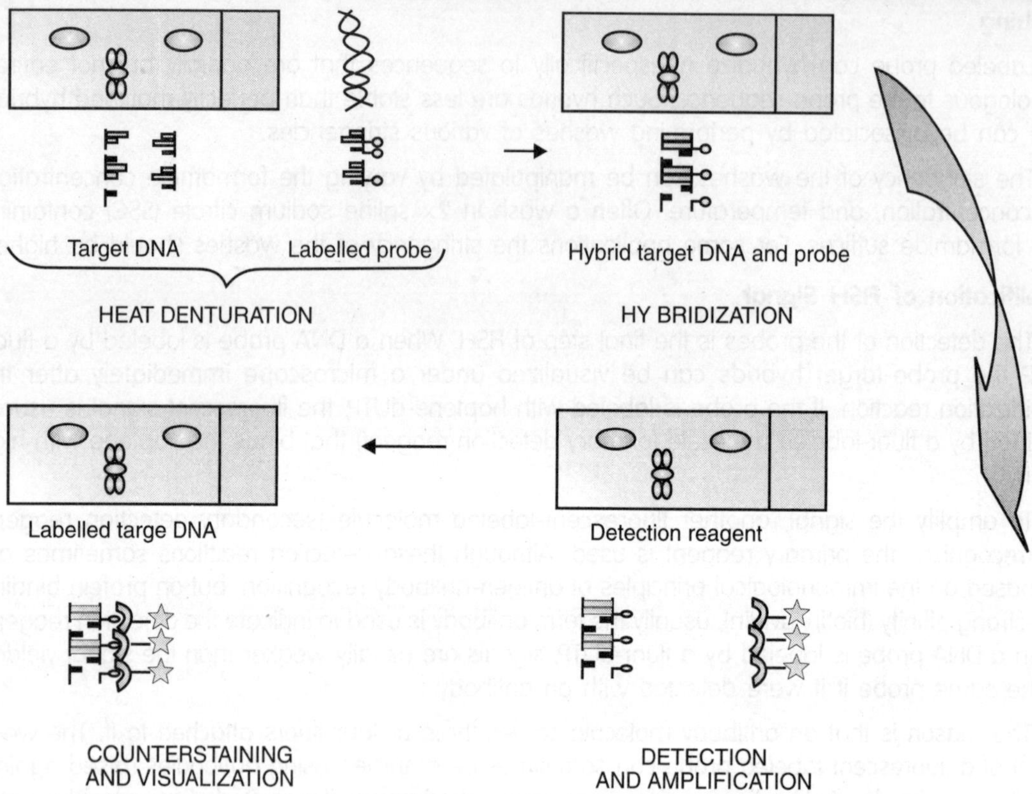

Figure 21.1: The basic principles of fluorescence in situ hybridization.

application of FISH to detection of chromosome aberrations in metaphase cells, detection of centromeric sequences in interphase nuclei and micronuclei, and estimation of telomeric sequence length. Whenever necessary specific aspects of these techniques are presented; in the end, criteria of slide analysis are separately described.

BASIC REQUIREMENTS

Fixation of Material

1. Cleaned slides.
2. Methanol-acetic acid (3:1).

Stepwise Methodology

1. 20× SSC: Dissolve 175.3 g of sodium chloride and 88.23 g of trisodium citrate in 900 mL of water. Adjust the pH to 7.0 with a few drops of concentrated HCl. Store at room temperature for up to 1 year. Dilute in water to make appropriate solutions for this protocol.
2. RNase A DNase-free (Roche, Mannheim, Germany) stock solution: 10 mg/mL. Store at-20°C. Prepare l mL of RNase working solution, 100μg/mL, by adding 10μL of RNase 10

mg/mL, 890 μL water, and 100 μL of 20× SSC. Prepare fresh each time.

3. Pepsin (Sigma-Aldrich) stock solution: 100 mg/mL in water. Store at -20°C.
4. 1× PBS + 50 MM $MgCl_2$ solution: Add 50 mL of $MgCl_2$ 1 M to 950 mL of 1 × phosphate-buffered saline (PBS).
5. 1% Formaldehyde: Add 2.7 mL of 37% formaldehyde to 100 mL of l x PBS + 50 MM $MgCl_2$ solution.
6. lx PBS solution: Dissolve 8 g of NaCl, 0.2 g of KCl, 1.44 g of Na_2HPO_4, and 0.24 g of KH_2PO_4 in 800 mL of water. Adjust the pH to 7.4 and adjust the volume to 1 L. Autoclave and store at room temperature.
7. Alcohol series: Prepare fresh 70%, 85%, and 100% ethanol and store at –20°C.
8. Plastic cover slips.

Denaturation

1. Denaturation solution (70% formamide-2 SSC): for 50 mL, add 35 mL of deionized formamide (Kodak, Rochester, NY), 10 mL of distilled water, 5 mL of 20× SSC. Adjust the pH to 7.0 with HCl.
2. Hot plate.

Hybridization

1. Probe.
2. Humidified chamber. Humidified chamber can be prepared using a clean plastic box with a lid. Soak several paper towels in water and placee them at the bottom of the tray. Put the lid on top and place chamber in the 37°C incubator.
3. Glass cover slips.
4. Glass cover slip sealant (rubber cement).

Washing

1. Wash solution: 50% formamide (Fluka, Buchs, SG, Switzerland) in 2× SSC. For 100 mL add 50 mL of formamide, 10 mL of 20× SSC, and 40 mL of distilled water. Adjust pH to 7.0.
2. 4× SSC + 0.1% Tween-20 solution: Add 100 mL of 20× SSC and 0.5 mL of Tween-20 to a total of 500 mL of distilled water, mix well on a magnetic stir plate, and store at room temperature (RT) for up to l month.

Amplification

1. Plastic cover slips.
2. Blocking solution: 3% bovine serum albumin (BSA) in 4× SSC + 0.1% Tween-20. Prepare 5 mL with 150 mg of BSA, l mL of 20× SSC, 3.95 mL of water, and 50 μL of 10% Tween-20.
3. Detection solution: 1 % BSA in 4x SSC + 0.1 % Tween-20. Prepare 5 mL with 50 mg of BSA, 1 mL 20× SSC, 3.95 mL of water, and 50 μL of 10% Tween-20.
4. Antibodies: To detect the haptene-dUTP labeled probe or to amplify the weak signal of

Table 21.1: Different Strategies to Label the Target and Amplify the Primary Fluorescent Signal.

Haptene	Primary detection reagent	Secondary detection reagent	Tertiary detection reagent
Digoxigenin	(Fluors) Mouse anti-DIG	Anti-mouse Ig-DIG	(Fluors) Anti-DIG
	} FITC, Cy3, Texas Red, Rhodamine		} FITC, Cy3, Texas Red, Rhodamine
Biotin	(Fluors) Avidin	Biotinylated anti-avidin	(Fluors) Avidin

fluorescent probe, primary, secondary, and sometimes tertiary detection reagents can be used (Roche, Mannheim, Germany).

Digoxigenin-labeled or biotin-labeled DNA probe can be detected with a variety of labeled antibodies (using different fluors) as shown in Table 21.1. Antibody combinatorial schemes allow, for example, triple colour detection and signal amplification for two or more probes labeled with different haptene-dUTPs.

Double Staining

With Propidium Iodide (for FITC-Labeled Probe)

1. Propidium iodide (Sigma-Aldrich) solution. Stock solution, 10 µg/mL in distilled water or PBS, can be stored at 4°C for several months. Working solution (PI/antifade): for each slide prepare in a microfuge tube 30 µL of solution with 6 µL of propidium iodide stock solution and 24 µL of mounting medium/antifade (Vector Laboratories, Burlingame, CA), mixing well (final concentration: 2 µg/mL).
2. Nail varnish.

With DAPI (for Cy3, Texas Red and Rhodamine-Labeled Probe)

1. DAPI (Sigma-Aldrich) solution. Stock solution, 100 µg/mL in PBS or 2× SSC, can be stored at 4°C for several months. Working solution: Prepare 50 mL of a solution with 100 L of DAPI stock solution in 2× SSC (final concentration: 0.2 µg/mL).

   Put the DAPI solution in a glass Coplin jar. The jar with DAPI solution is wrapped in aluminum foil or is taped with black tape (to protect it from light).
2. Mounting medium/antifade Vectashield (Vector Laboratories).
3. Nail varnish.

PROCEDURES

Fixation of Material

Prepare metaphase chromosome spreads or interphase nuclei on a glass microscope slide according to standard cytogenetic procedure currently in use in your laboratory. More details about metaphase and interphase cell preparation for subsequent FISH can be found elsewhere in this chapter.

Preservation

The purpose of aging is to fix the biological material to the glass surface and to make the chromosome structure resistant to the subsequent DNA denaturing steps. Very fresh slides, if not aged, will either lose most of the nuclei/chromosomes during denaturation or the shape of the chromosomes will become very distorted.

Too long aging will decrease the efficiency of hybridization. Slides are usually aged either by keeping them 2-3 days at' RT, or overnight at 65°C or by heating 30 min at 90°C.

Pretreatments of Material on Slides

1. Apply 100-200 µL of RNase stock solution to each slide and cover with plastic cover slip. Incubate for l h in humidified chamber at 37°C. Carefully peel back cover slip with forceps.
2. Rinse slides 3 times in 2× SSC at RT, 5 min each rinse.
3. Prewarm 0.01 M HCl in Coplin jars to 37°C in a water bath. Add 25 µL of pepsin 100 mg/mL per 50 mL of 0.01 M HCl (final concentration: 50 µg/mL). Place slides in Coplin jars for 10 min.
4. Wash slides in a glass Coplin jar 2 times in 50 mL of l x PBS at RT for 5 min.
5. Wash slides in 50 mL of l × PBS + 50 mM $MgCl_2$ at RT for 5 min.
6. Incubate slides for 10 min in 1% formaldehyde in PBS + $MgCl_2$ at RT.
7. Wash slides for 5 min in PBS at RT.
8. Dehydrate slides in a series of three cold ethanol washes: 70%, 85%, and 100%. All washes are done in Coplin jars for 2 min each wash, starting with the 70% wash.
9. Allow slides to air-dry or dry under an air jet.

Denaturation

1. Preset the hot plate to 70°C.
2. Prepare 100 µL of denaturation solution for each slide into a 1.5-mL microfuge tube. Denaturation solution must be prewarmed to 70°C.
3. Apply 100 µL of denaturation solution to each slide and cover with a glass cover *slip.*
4. Denature slides (no more than three slides at a time) for 2 min at 70°C on the prewarmed hot plate.
5. Immediately remove cover slip by soaking slides in a Coplin jar containing cold 70% ethanol

and rinse for 2 min. Remove the cover slips from the Coplin jar with forceps. Repeat rinse in cold 85% and 100% ethanol solutions.

6. Allow slides to air-dry or dry under an air jet.
7. Prewarm probe at 37°C for 5 min.
8. Vortex-mix gently and centrifuge 2-3 s to collect contents on the bottom of the tube.
9. Place 30 L of commercial probe for entire slide or 10 L of probe for a 22 × 22 mm area in a 0.5-mL microfuge tube and denature for 5 min in a 70°C water bath.
10. Immediately place the tube on ice until ready to use.

Hybridization

1. Apply 30 μL of probe on each slide and cover with 22 × 50 mm glass cover slip.
2. Seal the cover slip with glass cover slip sealant by applying sealant along the perimeter of the cover slip.
3. Incubate at 37°C for 12-16 h at 37°C in a humidified chamber.

Washing

1. Carefully remove the cover slip sealant with forceps. Do not remove the cover slip. This will fall off during the washing step. Take out the cover slips from the Coplin jar with forceps. Different probes require different stringency hybridization and postwashing conditions, according to the probe data sheet specifications. For this reason we can distinguish two separate post-hybridization washing protocols: low-stringency and high-stringency protocols.

   Low-stringency posthybridization washing (prewarm in water bath all the following solutions):

2. Wash slides 3 times in a glass Coplin jar containing 50 mL of wash solution at 37° C for 5 min with intermittent agitation.
3. Wash slides 3 times in 50 mL of 2× SSC, pH 7, at 42°C for 5 min with intermittent agitation.
4. Rinse slides in 4x SSC + 0.1% Tween-20 at RT in a Coplin jar. Do not allow slides to dry beyond this step.

   High-stringency posthybridization washing (prewarm in water bath all the following solutions):

2′. Wash slides 3 times in a glass Coplin jar containing 50 mL of wash solution at 42°C for 5 min with intermittent agitation.

3′. Wash slides 3 times in 50 mL of 0. l x SSC, pH 7, at 60°C for 5 min with intermittent agitation.

4′. Rinse slides in 4× SSC + 0.1 % Tween-20 at RT in a Coplin jar. Do not allow slides to dry beyond this step.

If the DNA probe is labeled by a fluor-dUTP proceed immediately to counterstaining. If the probe is labeled with haptene-dUTP, proceed to the following step (detection).

Identification

1. Remove slides from 4× SSC + 0.1 % Tween-20 and briefly blot excess fluid from the edge.
2. Apply 100 µL of blocking solution to each slide. Place a plastic cover slip over the solution. Incubate for 30 min in a humidified chamber at 37°C.
3. Carefully peel back the cover slip with forceps, tilt slide and allow fluid to drain briefly.

It is possible to detect the probe with individual antibodies or antibody combination. Antibody combination allows detection of two or more probes labeled with different haptene-dUTPs.

4. Apply 100 µL of primary antibody (fluors-labeled avidin and/or fluors-mouse anti-digoxigenin [anti-DIG]) to each slide and replace plastic cover slip. Incubate 30-60 min in humidified chamber at 37°C.
5. Wash slides 3 times in 50 mL of 4× SSC + 0.1% Tween-20 at 37°C for 5 min each wash with intermittent agitation.

Amplification

1. Apply 100 µL of secondary antibody (biotinylated anti-avidin and/or anti-mouse Ig DIG) to each slide and replace plastic cover slip over the solution. Incubate 30-60 min in a humidified chamber at 37°C.
2. Wash slides 3 times in 50 mL of 4× SSC + 0.1% Tween-20 at 37°C for 5 min each wash with intermittent agitation.
3. Apply 100 µL of tertiary antibody (fluors-labeled avidin and/or fluors-labeled anti-DIG) to each slide and replace plastic cover slip over the solution. Incubate 30-60 min in humidified chamber at 37°C.
4. Wash slides 3 times in 50 mL of 4× SSC + 0.1% Tween-20 at 37°C for 5 min each wash with intermittent agitation.

Double Staining

With Propidium Iodide

1. Apply 30 µL PI/Antifade (2 µg/mL) to each slide and cover with a glass cover slip.
2. Seal the edges of the cover slip with nail varnish. The signal keeps well for several weeks if slides are stored at 4°C.

With DAPI

1. Slides are incubated 10 min in the staining solution.
2. Rinse slides 10 min in 2× SSC to wash off excess dye.
3. Wash slides for a few seconds in distilled water to remove any salt traces.
4. Air-dry at room temperature and apply 30 µL of antifade solution.
5. Cover with a glass cover slip and seal the edges of the cover slip with nail varnish. The signal keeps well for several weeks if slides are stored at 4°C.

Studying Slides

Chromosome Painting

Chromosome painting was mainly developed to allow detection of stable exchange-type aberrations, such as reciprocal translocations, for biological dosimetry purposes. In fact, the analysis of dicentrics (unstable aberrations) in solid-stained chromosome preparations is very reliable to estimate recent and acute radiation exposures, but not for chronic or past exposures because the yield of dicentric chromosomes decreases with time after irradiation. For a correct analysis of aberrations in painted (as well as in solid-stained) metaphases, the detection of centromeres may be critical, especially in the case of mouse acrocentric chromosomes: it is obtained either by DAPI counterstaining that, by itself, after alkaline/heat denaturation, gives a bright signal to centromeric heterochromatin or by centromeric FISH staining.

In solid-stained metaphases, the conventional nomenclature classifies each simple interchange as a single event (e.g., one dicentric plus an acentric fragment), and distinguishes between complete (if no unrejoined breaks are found) and incomplete exchanges (e.g., a dicentric chromosome without any acentric fragment). The use of painting techniques has led to the development of new nomenclature systems.

These are mainly based on the recognition of colour switches between the fluor-conjugated probe and the counterstaining dye. According to the Protocol for Aberration Identification and Nomenclature Terminology (PAINT), each abnormal painted chromosome or fragment is described individually using the letters "A" and "a" to indicate counterstained chromosomal material, the letters "B" and "b" to indicate painted material, capital letters to designate centromere containing regions, lower case letters to indicate acentric regions.

Thus, typical PAINT-classified aberrations are t(Ab), dic(AB) or ace(ab), where t stands for translocation, dic for dicentric, and ace for acentric fragment. Another independently developed nomenclature system is that of Savage and Simpson (S&S), which is the method of choice for an interpretation of the mechanistic aspects of aberration formation.

When FISH painting is applied to the analysis of aberrations induced by chemical mutagens a further type of aberration, not encountered in radiation studies, must be considered for the correct assessment of induced effects: it is the presence of one or more supernumerary fully painted chromosomes in a metaphase with the euploid chromosome number for the specific species.

These supernumerary chromosomes derive from chromatid-type exchanges in the pericentromeric region, which are induced by radiation only when G_2 cells are targeted. Unless all chromosomes are hybridized with their specific painting probes conjugated with different fluorochromes and a computerized image analysis system is used to assign a different colour to each of them, only a portion of the whole karyotype is usually painted.

Aberrations involving painted chromosomes thus represent only a subset of all induced aberrations. It may be of importance to estimate the total number of aberrations induced to predict the fate of exposed cells or individuals. Assuming that aberrations are randomly distributed over different chromosomes, to convert the frequency of observed aberrations into an estimate of the frequency of total aberrations/genome, one has to know the proportion of the genome covered by each colour probe (or probe cocktail). Then, if p represents the proportion of genome painted in a given colour, *q* the proportion of genome painted in a second colour and *r* the unpainted fraction of the genome, the fraction of detectable exchanges is calculated as $S = 2pq + 2pr + 2qr$.

Multiplying the number of scored metaphases by S, one obtains the number of cell (genome) equivalents. The ratio between the number of observed aberrations and the number of cell equivalents gives the estimated frequency of aberrations/genome.

Also, assuming that n is the recommended number of metaphases to be scored when the whole genome can be scanned for aberrations, WS gives the number of painted metaphases to score to obtain the amount of information equivalent to that provided by n fully analyzed metaphases.

Detection and Aneugenic Agents

Centromeric FISH staining allows to detect potential aneugenic agents by measuring the frequency of chromosome loss and chromosome nondisjunction events. To estimate chromosome loss events, probes specific for the centromeric sequence of all chromosomes are used to detect the presence of centromere(s) in micronuclei.

Micronuclei labeled by the probe are classified as centromere positive (C+), assumed to contain one or more chromosomes, and ascribed to chromosome loss occurred at the preceding anaphase. To estimate nondisjunction events, probes that detect the (peri)centromeric region of specific chromosomes are used to assess the number of chromosomes present in each sister interphase nucleus of a binucleate cell.

Cytochalasin-B-blocked cells are used that retain sister nuclei of a single mitotic division within the same cytoplasm. Two spots for each probe are expected for any autosome in euploid nuclei. Nondisjunction events occurred at the preceding mitotic division are detected via deviations from the expected 2+2 pattern of signals in sister nuclei, e.g., 3+1 or 4+0 combinations; nondisjunction of the X or the Y chromosome in male cells gives rise to a 2+0 pattern of signals.

Only cells with the correct total number of hybridization signals with similar fluorescence intensity are analyzed. It has been demonstrated that nondisjunction events may be more sensitive indicators of aneugenic activity than chromosome losses. This emphasizes the importance of molecular cytogenetics in interphase cells as a fast and reliable approach to assess the frequency of spontaneous and induced nondisjunction events.

Elements of Telomere Length Estimation

Quantitative Image Analysis

Telomeres protect the end of chromosomes from end-to-end fusion. Shortening of telomeres beyond a critical length drives the cell into replicative senescence or is responsible of chromosome instability. Thus, the measurement of telomere length is a useful means for assessing the lifespan of a cell or for studying chromosome instability induced by physical or chemical agents.

Recently, a new *in situ* technique has been developed, the quantitative fluorescence *in situ* hybridization (Q-FISH). This technique, which can be applied to metaphase and interphase nuclei, is based on the use of a peptide nucleic acid (PNA) telomere oligonucleotide probe that generates stronger and more specific hybridization signals than standard DNA oligonucleotide probes.

Presently, Q-FISH is able to detect differences in individual telomere lengths below 1 kb. The assumption is made that the length of a telomere is directly related to its integrated fluorescence intensity (IFI) value, as the fluorescence probes used are assumed to hybridize quantitatively to telomere repeat sequences, allowing detection of short telomere length variations. Digital images are recorded with a camera on a fluorescence microscope equipped with a multiple filter wheel.

Image acquisition must be performed with dedicated software. Digital images are acquired from DAPI-stained chromosomes and from Cy-3-stained telomeres. A dedicated computer program was developed for image analysis (TFL-Telo program). In short, chromosomes and telomeres are identified through segmentation of the DAPI image and the Cy-3 image, respectively.

Both images are combined and corrected for pixel shifts. To prevent a possible selection bias, metaphases are chosen solely on the basis of a good chromosome spread. TFL-Telo program allows to make contours from individual chromosomes and individual telomere spots and the two images are successively combined. Finally, the fluorescence intensity of telomeres on all individual chromosomes is expressed in arbitrary units of fluorescence (a.u.f.).

Calibration

Two levels of calibration have to be used to ensure a reliable quantitative estimation of telomere length in various samples. First, to correct for daily variations in lamp intensity and alignment, images of fluorescent beads (orange beads, size 0.2 pm; Molecular Probes, Eugene, OR) are acquired and analyzed with the computer program in parallel to cell samples.

Second, to relate fluorescence intensity to number of $T2AG_3$ repeats, mammalian cells with a defined $(T_2AG_3)_n$ length of 10 and 80 kb (LY-S and LY-R lymphoma cell lines respectively) may be hybridized and analyzed. Based on the slope of the linear relationship between IFI values and telomere lengths of LY-S and LY-R cells, the telomere lengths of cell samples are estimated after their IFI values are acquired under the same conditions.

Alternatively to cell lines with defined telomere length, plasmids with telomere sequence inserts of known size can be used.

Optical Studies

A very important part of FISH is the visualization and recording of the results with a fluorescence microscope and camera system. A fluoreescencee microscope contains a lamp for excitation of

Table 21.2: Some of the Most Common Fluorochromes Used for the Detection of Chromosome Structural and Numerical Aberrations.

Fluorochrome	Color	Excitation (nm)	Emission (nm)
Propidium iodide	Red	535	605
DAPI	Blue	359	441
Rhodamine	Red	560	590
CY3	Red	552	565
Fluorescein	Green	494	523

the fluorescent dye and a special filter that transmits a high percentage of light emitted by fluorescent dye.

The filters allow certain wavelengths of light through, while blocking others. Therefore, the filters are individually designed for specific fluors and must be chosen properly. The propidium iodide or DAPI filter should be used to scan the slide for cells or metaphase spreads, while the filters specific for probe-conjugated fluors are used for visualization of the target(s) of interest. Table 21.2 gives general guidelines for choosing filters from your microscope manufacturer.

POINTS TO REMEMBER

1. Plastic cover slips can be prepared by cutting a piece of parafilm. They can be removed from the slide more easily than glass cover slips and should be preferred except in those steps in which high temperatures or sealing requirements are not compatible with plastics.
2. To prepare metaphase chromosome spreads, cultured cells are treated with colcemid (10 μg/mL-Invitrogen, Paisley, UK) for 3 h. This protocol can be adapted for many cell cultures. For cells with a short cell cycle, such as mouse cells, 2 h of colcemid treatment could be enough. For cells that grow more slowly, the colcemid time could be from 3 h to overnight. The colcemid concentration should be adequate to the treatment time: long treatment time requires lower colcemid concentration. To obtain metaphases from in vivo experiments, 0.3 mL per mouse of colchicine 10^{-3}M (Sigma-Aldrich) is used. To prepare binucleate cells cytochalasin B (3 μg/mL, Sigma-Aldrich) is added to the cultures for the duration of one cell cycle.
3. This protocol is recommended for cultured cells and cell suspensions obtained from experimental animals or humans, fixed in methanol-acetic acid (3:1).
4. The slide aging conditions are always related to the probe used. For this reason we suggest to age slides according to the probe data sheet specifications.
5. Many Incubations are performed in 50-mL glass Coplin jars. To obtain optimal results, verify the solution temperatures, placing a clean thermometer directly into the Coplin jar.
6. Ethanol should be stored at -20°C before use.
7. If multiple slides are processed simultaneously, each slide will cause the hot plate temperature to drop VC. Therefore, the temperature of the hot plate must be set up 1 °C higher for each slide to be denatured.
8. Care should be taken to avoid air bubbles.
9. Time and temperature are very important to maintain chromosome and nuclei morphology. Incomplete denaturation of target DNA can result in lack of signal. The suggested protocol has been verified for both metaphase and interphase human cells. Mouse cells may require higher denaturation temperatures up to 80°C.
10. Usually the probe is diluted in glycerol or formamide and the solution is very viscous. For this reason we suggest to prewarm the commercial probe at 37°C to decrease the solution viscosity.

11. For economy, only a target area of slides containing good quality and quantity of metaphase spreads or interphase nuclei could be hybridized. The selected area can be labeled under phase contrast with diamond pen.
12. We recommend wrapping in aluminum foil the tubes with the fluors labeled probe or fluors labeled antibodies, to protect them from light.
13. If two or more probes are to be used together, in the same hybridization, the final volume may be higher than the 10-12 μL accommodated under a standard 22 × 22 mm cover slip. To solve this problem, centrifuge the total volume of the mixed probes 10 min at 8400g and discard the supernatant. Allow the pellet to dry under air jet or by speed vacuum centrifuge, resuspend it in 10-12 μL of hybridization buffer (Hybrisol VI, Roche), and apply the mix to slides.
14. Temperature, time, and buffer concentration (stringency) of hybridization and postwashing solutions are important, as lower stringency can result in nonspecific binding of the probe to other sequences, and higher stringency can result in a lack of signal.
15. This step is used primarily to remove the nonspecific and/or repetitive DNA hybridized to the cells and chromosomes. It is impossible to present a standard procedure, but there are some guidelines that can help. The less concentrated the salt solution and the longer the duration of the wash and the higher the temperature, the higher will be the stringency and the more unbound or nonspecifically bound probe will be removed. For very short DNA probes (0.5-3 kb) or very complex probes (chromosome paint probes), the washing temperature should be lower (up to 45°C) and the stringency lower as well (1-2× SSC). When using single-locus, large probes, the temperature should be around 65°C and the stringency high (below 0.5x SSC). The temperature and stringency should be the highest for repetitive probes (such as a-satellite repeats).
16. If necessary, slides can be stored at 4°C in 50 mL of 4x SSC + 0.1 % Tween-20 up to 2 wk before detection.
17. Unless specified otherwise, the volumes of detection solutions are given for the whole slide area of 22 × 50. If only a 22 × 22 area has been hybridized, reduce the volumes to 1/3.
18. BSA blocks some of the unspecific binding of the antibodies to the glass and this is important to decrease background, but may also partially block antibodies from reaching the labeled DNA.
19. With commercially available, fluorescent-labeled antibodies, the manufacturer's instructions regarding antibody concentration for *in situ* procedures should be followed. As a general guideline, a 1: 100 dilution in detection solution of a 1 mg/mL antibody stock solution should work well, after 30 min of incubation at 37°C. Every time a new antibody is purchased, it should be tested at different dilutions, to find the optimal range for that antibody. In our experience, we have found the following antibodies at the following concentration to be good for *in situ* hybridization:
    a. *For biotin-labeled probes* Fluorescein avidin DCS (Vector Laboratories) and rhodamine avidin DCS (Vector Laboratories) 1:300 in detection solution. Biotinylated anti-avidin D

(Vector Laboratories) 1:100 in detection solution

b. *For DIG-labeled probes* Mouse anti-DIG (Roche) 1:100-1:200 in detection solution. Anti-mouse Ig-DIG (Roche) 1:100 in detection solution. Anti-DIG-rhodamine or anti-DIG-fluorescein (Roche) 1:50 in detection solution.

20. If using an antibody combination prepare a double concentration of antibody solutions and mix 50 μL of each solution on the slides.
21. This wash could be performed at 37°C or 42°C. The optimal temperature should be empirically determined to remove excess and unbound detection reagents.
22. The nail varnish may increase the shelf-life of FISH slides.

Chapter 22

Investing Fresh Virulents

In the past 15 yr, our knowledge of the extent of microbial diversity has been *revolutionized* thanks to two key discoveries: the development of the *polymerase* chain reaction (PCR) for rapid and specific amplification of DNA sequences, and the establishment of a *phylogenetic* classification system based on the sequence of genes encoding the RNA component of the *smallsubunit* (SSU) *ribosome*.

These breakthroughs have provided the means of identifying and classifying microorganisms in a variety of *ecological niches*, without the need for cultivation. The impact of these developments has been most apparent among environmental *microbiologists*, who had long been aware of what was termed "the great plate-count anomaly"; that is, the difference between numbers of *bacteria* estimated by plate counts of *cultivable* bacteria, and numbers estimated by direct microscopic counts.

All in all, it has been estimated that between 0.4 and 5% of all extant bacterial species have been cultured and characterized. The proportion of a bacterial community that cannot be cultured depends on the *ecosystem* studied, and may vary from >99% for saltwater to 85% in activated sludge.

Of all microbial ecosystems, the human body is undoubtedly the most extensively studied, yet our current knowledge of the presence and diversity of microorganisms associated with health and disease is almost entirely derived from cultivation-based methodologies.

Given the importance that is placed on microbial identification once humans develop signs of infection, it seems unlikely that the proportion of uncultivated bacteria present in the human microflora is as high as that for *environmental* niches.

Nevertheless, there are examples of pathological conditions associated with bacteria that have resisted isolation by existing culture methods, but which have been successfully characterized by molecular means. The aim of this review is to describe the *methodologies* employed in the *nonculture* analysis and *characterization* of bacterial *pathogens*.

CULTURE-RESISTANT STRAINS

The term *culture-resistant* was coined by Wilson to describe bacteria that could not be cultured by current methods and which were generally referred to as *unculturable* or *uncultured*.

Neither of the latter terms is satisfactory, as some bacteria originally so described have since been cultivated. *Culture-resistant* therefore seems to be a more accurate term. All bacteria are *intrinsically* capable of growing and multiplying. When we are unable to induce them to do so in the laboratory, it reflects our *inability* to fulfill the physiological needs of the microorganism in vitro.

In clinical bacteriology, it has long been recognized that many bacterial pathogens are fastidious in their requirements, but may be cultured once the appropriate conditions are provided. In diagnostic clinical bacteriology, the pressures of time, workload, and the necessity of identifying known microbial pathogens result in a somewhat narrow "window of cultivation."

Clinical samples with complex *microbial floras* (such as feces) are often cultured on selective media designed for the isolation of specific pathogens. Moreover, most samples are cultured *aerobically* at 37°C for 18 to 24 h, with only a subset of specimens being incubated for longer, or at different temperatures, or in different atmospheres.

While it is understandable that the function of a clinical bacteriology laboratory is to isolate known bacterial pathogens, this approach can lead to the misconception that the microorganisms isolated by routine laboratory culture methods represent the numerically dominant or significant species in that sample. An example of this narrow window of cultivation was provided by the discovery of *Helicobacter pylori*.

For many years, spiral or curved Gram-negative bacteria had been observed in gastric biopsies of patients suffering from gastritis and gastric ulcers. Efforts to culture these bacteria were unsuccessful or grew only contaminating bacteria. In the early 1980s, Marshall and Warren were attempting to isolate these spiral Gram-negative bacteria using plates incubated for 48 h in a *micro-aerobic* atmosphere.

Results were unsuccessful until a set of plates left by mistake for six days over the Easter holiday yielded growth of novel campylobacter-like bacteria. Subsequently, the bacteria, classified as *H. pylori*, became recognized as a cause of *gastritis* and *gastric ulceration*, and are associated with the development of gastric cancer.

It has been estimated that *H. pylori* infects 50% of the human population worldwide, ranking it as one of the most prevalent bacterial *pathogens*.

IDENTIFICATION OF CULTURE-RESISTANT BACTERIA

Evolutionary Approach

The seminal work of Woese and colleagues on the comparative analysis of SSU ribosomal RNA (rRNA; 16S in prokaryotes and 18S in eukaryotes) provided, for the first time, an objective framework for determining the evolutionary relationships between organisms, and gave a quantitative measure of *diversity*, expressed as sequence *divergence* on a *phylogenetic* tree.

Woese found that cellular organisms were not simply divided into eukaryote and prokaryote cell types, as had previously been supposed, but rather into three primary domains; the prokaryotic, termed *Bacteria* and *Archaea;* and a single eukaryotic domain, *Eucarya.*

Woese also defined I 1 major linkages, termed *phyla* or *divisions,* within the bacterial domain, on the basis of 16S rRNA sequences obtained from cultivated microorganisms. This 16S rRNA-based classification of bacteria has largely replaced existing schemes based on traditional *phenotypic* characterization.

16S rRNA

Genes encoding rRNA are ideal targets for phylogenetic analysis, as they are present in all organisms and, due to the functional constraints of the ribosome, contain highly conserved regions (being almost invariable for all bacteria).

Some variation is possible in other regions of the genes, and mutations accumulate at a slow but constant rate, allowing differences to evolve between the sequences found in different bacterial divisions, genera, and species.

Of the genes in the ribosomal operon, the gene encoding 16S rRNA (sometimes termed 16S rDNA) has become the most widely used in the study of bacterial phylogeny because it contains more information than the 5S rRNA, and is more easily sequenced and less prone to contain insertion sequences than 23S rRNA.

Eight variable regions have been identified in 16S rRNA, which are termed V1 to V3 and V5 to V9 (region V4 occurs only in eukaryotic 18S rRNA). Any *microorganism* can be characterized, provided that the 16S rDNA sequence can be determined. Ribosomal RNA sequence information can be retrieved from both cultured and culture-resistant bacteria present in mixed bacterial *communities* or infected tissue.

Broad-Range PCR

In the early years of rRNA-based analysis, rRNA sequences were obtained from RNase TI oligonucleotide catalogs of bulk rRNA, and later from reverse transcriptase sequencing of this RNA. Subsequently, PCR has been used to amplify rRNA genes for sequencing.

This has been achieved by using a broad-range PCR approach, using primers designed to target regions of the 16S rRNA gene that are universal to members of the bacterial domain, but distinct from eukaryotic 18S rRNA.

The design of the PCR assay requires only some knowledge of the nucleotide sequences that flank the target DNA. In broad-range PCR, the primers may be designed to target regions of the 16S rDNA that are universal (i.e., present in both Bacteria and Archaea) or specific for the bacterial domain or a specific division or genera within that domain.

The resulting amplified DNA contains internal regions of variable sequence that form the basis for specific phylogenetic analysis. To characterize a *culture-resistant, unpurified* bacterial pathogen from infected human tissue by PCR, total genomic DNA must first be extracted using a suitable protocol.

The nature of the extraction protocol used will depend on the specimen type; for example, feces and blood are known to contain substances that are inhibitory to PCR. The DNA sequences

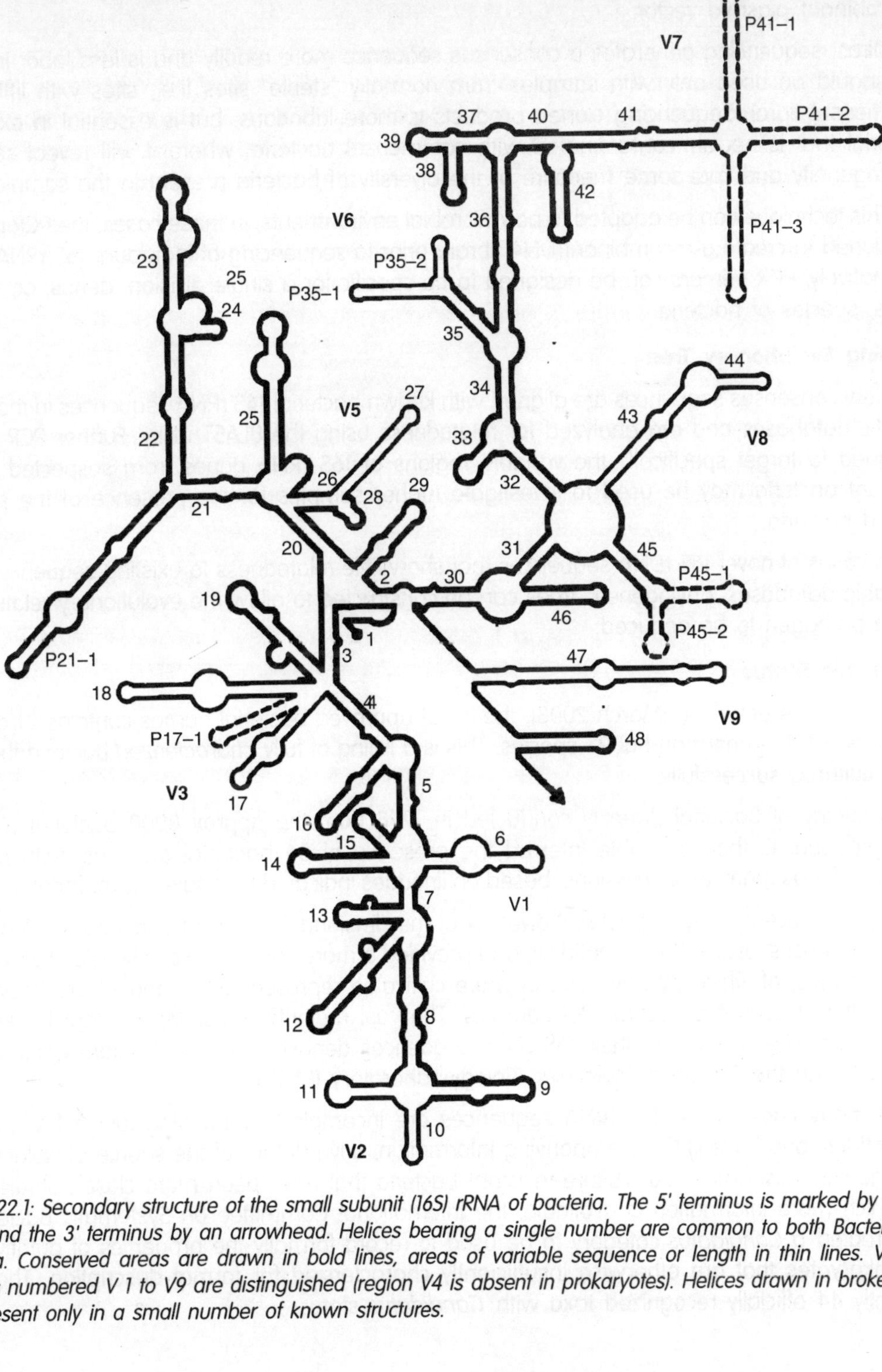

Figure 22.1: Secondary structure of the small subunit (16S) rRNA of bacteria. The 5' terminus is marked by a filled circle and the 3' terminus by an arrowhead. Helices bearing a single number are common to both Bacteria and Eucarya. Conserved areas are drawn in bold lines, areas of variable sequence or length in thin lines. Variable regions numbered V 1 to V9 are distinguished (region V4 is absent in prokaryotes). Helices drawn in broken lines are present only in a small number of known structures.

of the amplified products can be determined either directly or by first cloning the products in a recombinant plasmid vector.

Direct sequencing generates a consensus sequence more readily and is less labor intensive, but should be used only with samples from normally "sterile" sites (i.e., sites with little or no commensal flora). Sequencing cloned products is more laborious, but is essential in examining material that is heavily contaminated with commensal bacteria, where it will reveal sequence heterogeneity and give some measure of the diversity of bacteria present in the sample.

This technique can be adapted to polymicrobial environments; in these cases, the PCR products are cloned to create a recombinant DNA library prior to sequencing of individual 16S rRNA genes. Alternatively, PCR primers can be designed to be specific for a single division, genus, or, in some cases, species of bacteria.

Framing Evolutionary Tree

New consensus sequences are aligned with known bacterial 16S rRNA sequences in the current genetic databases and are analyzed for relatedness using the BLAST utility. Further PCR primers designed to target specifically the variable regions of 16S rRNA genes from suspected culture-resistant bacteria may be used to investigate further samples for the presence of the same or related bacteria.

In cases of novel 16S rDNA sequences that show little relatedness to existing sequences in the available databases, phylogenetic trees can be constructed to allow the evolutionary relationships of the pathogen to be deduced.

Taxonomic Status

At the time of writing (March 2003), the list of approved bacterial names contains 21 divisions comprising 1146 genera and 6093 species. This is a listing of fully *characterized* bacteria that have been cultured successfully.

A survey of bacterial diversity conducted in 1998 from the approx 8000 bacterial 16S rRNA gene sequences then available inferred the presence of 36 bacterial divisions, with possible evidence for as many as 40 divisions, based on lineages indicated by single environmental isolates.

A more recent study of bacterial diversity at the divisional level used only near-full length 16S rRNA sequences (over 1300 nucleotides), and provided a more conservative estimate of 35 division-level lineages, of which 22 have one or more cultivated representatives, and 13 are known only from culture-resistant environmental samples. The vast majority of culture-resistant bacteria are known from little more than their 16S rRNA sequences deposited in genetic databases such as GenBank and the European Molecular Biology Laboratory (EMBL).

In many cases these 16S rRNA sequences are incomplete, containing just a few hundred nucleotides, and have little accompanying information, save details of the source of isolation and phylogenetic association. For culture-resistant bacteria that have been more closely studied, and for which more information is available, the International Committee on Systematic Bacteriology has created *a Candidatus* category to be used to record formally the properties of putative taxa of prokaryotes that are otherwise insufficiently characterized for formal description. There are currently 44 officially recognized taxa with *Candidatus* status.

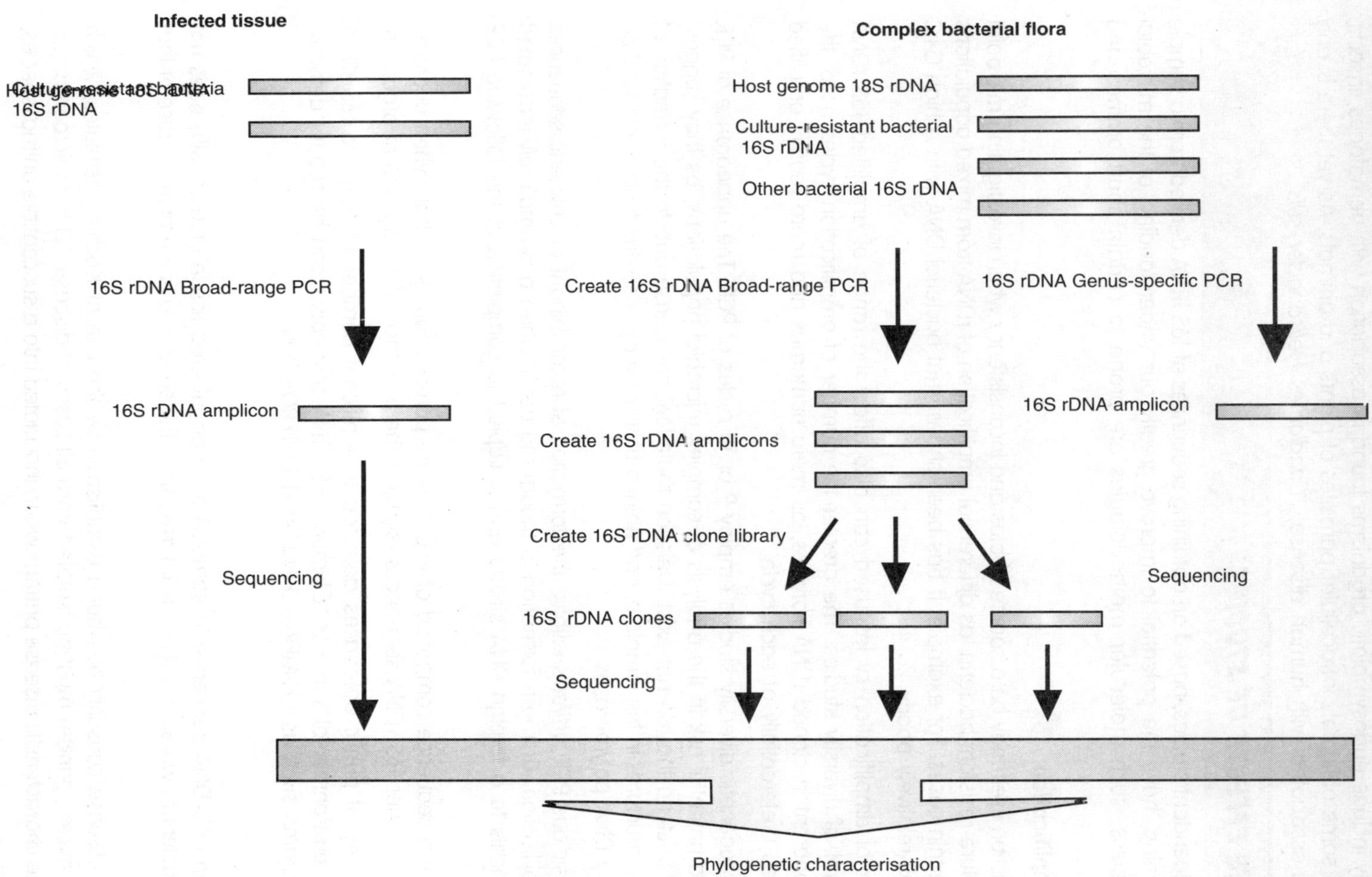

Figure 22.2: Summary of methods for PCR-based retrieval of bacterial 16S rDNA sequences from infected tissue and complex bacterial floras. In infected tissue, 16S rDNA-specific broad-range PCR is used to amplify sequences from culture-resistant bacteria from a background of host DNA. In samples from sites with complex bacterial flora, 16S rDNA-specific broad-range PCR is used to generate heterogeneous 16S rDNA amplicons, which are used to create a 16S rDNA clone library from which individual 16S rDNA clones can be sequenced. Alternatively, the 16S rDNA PCR can be designed to amplify only a subset of 16S rDNA from a mixed bacterial population. In this case the PCR assay has been designed to amplify I 6S rDNA from a specific bacterial genus.

These mostly include endosymbiotic organisms found in association with nematodes, insects, and plants, and some obligately intracellular parasites of plants and animals. As yet there is only one *Candidatus* associated with human disease, *"Candidates Helicobacter suis"*.

ADVERSE EFFECTS OF STUDIES

Culture-independent studies, based on obtaining sequences of 16S rRNA derived from a sample by PCR and cloning, have the potential to improve greatly our understanding of the microbial world. Nevertheless, such molecular methodologies are prone to certain shortcomings and disadvantages.

Differential Amplification

The PCR reaction itself may be a source of bias and inconsistency when investigating microbial diversity and culture-resistant bacteria, as differential amplification of rDNA from mixed populations may occur in certain cases. For example, it has been shown that bacterial DNA with a high G+C content amplifies relatively poorly.

The number of amplification cycles used can also affect the range of amplified 16S rDNA detected in microbial diversity studies. The greater the number of amplification cycles used, the less diversity apparent in cloned rDNA libraries, as more numerous and more readily amplified rDNAs are copied preferentially at each cycle.

Typically, phylogenetic diversity studies employ 9 or 10 cycles of PCR. The appearance of PCR artifacts is also a potential risk in the analysis of complex microbial populations, as they suggest the presence of bacteria that do not exist. Deletion mutations may arise due to the formation of stable secondary structures in the template, and point mutations may arise due to misincorporation of nucleotides by DNA polymerases.

The most insidious PCR *artifacts* involve the formation of recombinant or chimeric sequences during PCR amplification. Chimera formation is thought to occur when a *prematurely*-terminated amplicon re-anneals to a foreign DNA strand and is copied to completion in the following PCR cycles.

This results in a sequence composed of two or more parent sequences that, when analyzed comparatively with other 16S rDNA sequences, suggest the presence of a nonexistent organism.

A recent survey of public databases detected a significant number of chimeric 16S rDNA sequences from environmental samples. *Chimeric* 16S rDNA sequences can be readily detected by analysis programs such as CHIMERA-CHECK and BELLEROPHON.

Confirmation

The detection of a DNA sequence in samples from an infected tissue or body site does not prove that the bacterium whose presence is inferred from the sequence is the actual cause of that infection.

Traditionally, disease causality has been established by the use of Koch's postulates, which state that (1) the microorganism must be isolated from all cases of disease, (2) the microorganism must produce the characteristic disease pattern when inoculated into a susceptible animal species, and (3) the microorganism must be re-isolated from the experimentally induced disease. Obviously without a purified or cultivated microorganism, these criteria are difficult to fulfill.

The only alternative is to build a compelling body of circumstantial evidence to establish a causal relationship between the culture-resistant bacterium inferred from amplified 16S rDNA, and the disease under investigation. For example, DNA sequences belonging to the putative pathogen should be present in most cases of the infection and absent from hosts or tissues without disease.

The significance of an amplified 16S rDNA sequence can be increased by specific hybridization to visible microorganisms in infected tissue samples, or by *in situ* PCR.

EXPERIMENTALLY PROVED INSTANCES

"Helicobacter Heilmannii"

Following the successful isolation and characterization of *H. pylori* from human gastric biopsies in 1984, workers reported the microscopic observation of a large, Gram-negative, tightly-coiled spiral bacterium in the gastric mucosa of a small percentage of gastritis cases, that remained resistant to all attempts at culture.

This new bacterium was named *"Gastrospirillum hominis"*. Infected patients typically complained of upper-abdominal discomfort, and a diagnosis of chronic active gastritis was confirmed following endoscopic investigation. Treatment with bismuth salts resulted in eradication of the organisms, which coincided with resolution of the symptoms of chronic active gastritis, thus implicating the bacterium as the etiological agent of the infection. Subsequently, the 16S rDNA sequence of *"G. hominis"* was detected in specific pathogen-free mice fed homogenized gastric biopsy samples, and the presence of "G. *hominis"* in mouse stomach was confirmed by electron microscopy.

Comparison with other 16S rDNA sequences showed that "G. *hominis"* was a member of the genus *Helicobacter,* showing closest phylogenetic relatedness to *H. felts, H. bizzozeronii,* and *H. salomonis,* these last being readily cultivable species isolated from the gastric mucosa of cats and dogs. In light of this new phylogenetic information, the new name *"Helicobacter heilmannii"* was proposed (reports of similar organisms in animals seemed to suggest that the species designation *hominis* was inappropriate).

Although the isolation of *"H. heilmannii"* from human gastric mucosa using conventional culture techniques was reported in 1996, it was subsequently shown that the isolate was actually *H. bizzozeronii.* Other culture-resistant *Helicobacter-like* organisms from the gastrointestinal tract of animals have been described and accorded *Candidatus* status, namely *"Candidatus* Helicobacter suis" (formerly *"Gastrospirillum suis"* from pigs), and *"Candidatus* Helicobacter bovis" from the abomasum of cattle.

However, the 16S rRNA gene sequence of *"Candidatus* Helicobacter suis" shares over 99% homology with the corresponding sequence determined for *"H. heilmannii"* It has since been proposed that *"Candidatus* Helicobacter suis" and *"H. heilmannii"* should be referred to as *"Candidatus* Helicobacter heilmannii," to reflect the more frequent nomenclature usage and the nonspecific host range of this organism.

Tropheryma whippeli

Whipple's disease, or intestinal lipodystrophy, first described by the American pathologist George Hoyt Whipple in 1907, is an uncommon systemic infectious disease primarily affecting middle-aged *Caucasian* males. Patients typically present with weight loss, arthralgia, diarrhea, and *abdominal pain* and, less often, central nervous system or cardiac manifestations that may prove fatal if untreated.

The disease is commonly diagnosed by small bowel biopsy, where the appearance of the sample is characterized by inclusions in the lamina propria staining with *periodic-acid-Schiff,* which represent the causative bacteria. In his initial description of the disease, Whipple reported the presence of rod-shaped microorganisms in the vacuoles of macrophages, but did not link them to the causation of disease.

Thus, the bacterial etiology of this disease was not suspected until 1952, when a case was successfully treated with chloramphenicol. It was not until 196.1. that microscopic examination confirmed the presence of Gram-positive bacilli associated with areas of affected small-intestine mucosa; however, the organism resisted all attempts at conventional culture.

In 1991, Wilson et al. used broad-range primers to amplify bacterial 16S rDNA directly from the intestinal tissue of an infected individual, and suggested that the bacillus was most closely related to the actinomycetes. This work was subsequently confirmed with tissues from several other patients, and the provisional name *"Tropheryma whippeli"* was proposed.

The rDNA sequence of this organism forms the basis of a specific diagnostic PCR test, and further studies have demonstrated the presence of *"T. whippeli"* DNA in 25/38 sewage samples, suggesting that the organism, or a closely related species, was either a widely distributed commensal or an environmental microorganism. In 1997, the Whipple's disease bacterium was cultured from a cardiac valve, using human macrophages inactivated with interleukin-4, but the isolate could not be subcultured.

Subsequently, it was propagated on a human fibroblast cell line, employing a prolonged incubation period, with a reported generation time of 18 d, and the new genus, *Tropheryma,* and species, *T. whipplei,* were characterized and formally recognized. The genome of *T. whipplei* has recently been sequenced and lacks genes for key biosynthetic pathways and energy-generating metabolism, which may account for its host-adapted lifestyle.

However, the genome does encode a large family of surface proteins, which suggests that immune evasion and host interactions play an important part in its life cycle. Available evidence suggests that *T. whipplei* is a widely distributed human pathogen, and it may be that Whipple's disease is just one manifestation of its ability to infect humans.

Ehrlichiosis in Man

Ehrlichiosis is an arthropod-borne veterinary disease caused by several bacterial species of the genus *Ehrlichia,* which primarily affects dogs, cattle, sheep, goats, and horses. In 1987 an apparent case of human ehrlichiosis was reported, involving a middle-aged man who developed a febrile illness after being bitten by a tick during a trip to Arkansas.

Initially it was assumed that he was suffering from Rocky Mountain spotted fever, caused by the tick-borne bacterium *Rickettsia rickettsii,* which is known to infect humans. Serological tests for Rocky Mountain spotted fever were negative, but did provide evidence of infection with *Ehrlichia canis,* which causes ehrlichiosis in dogs.

Retrospective serological analyses demonstrated further unrecognized cases of human ehrlichiosis, and analysis of blood smears by electron microscopy showed the culture-resistant bacteria forming intracytoplasmic inclusions within leucocytes. In 1991, broad-range PCR with bacterium-specific primers was used to amplify 16S rDNA from the blood of a patient from Fort Chaffee, Arkansas.

Phylogenetic analysis of the retrieved 16S rRNA gene sequence showed that the organism from which the DNA was sequenced was closely related to, but not identical with, *E. canis.* This novel organism was provisionally named *"Ehrlichia chaffeensis,"* and has since been detected in other clinical samples and cultivated in a canine macrophage cell line.

In the case of *"E. chaffeensis,"* the nature of the clinical illness, the presence of leukocytic inclusions, and serological cross-reactivity with *E. can is are* all suggestive of a causal role for *"E. chaffeensis"* in disease.

Bartonella Henselae

Bacillary angiomatosis is an angioprollferative, AIDS-related disease involving the skin and numerous internal organs. The presence of clumps of pleomorphic bacilli in infected tissues, and a positive response to antibiotic therapy, strongly suggested a bacterial *etiology* for this condition, but the bacterium remained resistant to culture.

Broad-range PCR for 16S rRNA was used to amplify bacterial DNA from frozen tissue samples from an immunocompromised patient with disseminated visceral bacillary angiomatosis. The resulting phylogenetic analysis identified the pleomorphic bacilli as a proteobacterium.

Its closest relative was *"Rochalimaea quintana,"* which causes trench fever and is also associated with bacillary angiomatosis. This newly described bacterium was subsequently isolated and named "R. *henselae",* but was thereafter reclassified (along with *"R. quintana"*), as a member of the genus *Bartonella,* and renamed *Bartonella henselae.*

All *Bartonella* species will grow on blood agar, but are cultureresistant in that they typically take 12-14 d to produce visible colonies on primary isolation. On subculture, colonies usually appear after 3-5 d.

Bacterial Diversity

The endogenous microbial flora of the human body is extensive and varied, reflecting the wide variety of sites available for colonization, with distinct bacterial populations inhabiting the skin, the genitourinary tract, the oral cavity, and various niches within the gastrointestinal tract.

It has been estimated that the human body is colonized with 10 viable bacteria for every cell in the human body: 10^{14} bacterial cells compared with 10^{13} human cells. The endogenous bacterial flora is thought to play a beneficial role in nutrition and the physiological development of the host, in addition to stimulating the immune system and providing resistance to colonization by pathogens.

Nevertheless, the endogenous flora may also contain organisms capable of acting as opportunistic pathogens.

Periodontal Diseases

In excess of 300 bacterial species have been identified by culture from dental plaque, a polymicrobial matrix that is responsible for two of the world's most prevalent infectious diseases, dental caries and periodontal disease.

Certain species, such as *Porphyromonas gingivalis, Bacteroides forsythus,* and *Treponema denticola* have been identified as putative periodontal pathogens because they are detected in higher levels from diseased rather than healthy periodontal and gingival samples.

However, only 75% of the bacteria visible by microscopy in subgingival plaque can be cultured on nonselective media, and it is possible that the uncultured portion of the microflora may contain pathogens. Several studies have used broad-range PCR to amplify bacterial 16S rDNA directly from subgingival plaque in order to create clone libraries from the resulting heterogeneous population of 16S rRNA amplicons.

Analysis of the sequences of these clones has demonstrated that a significant proportion of the resident human bacterial flora remains poorly characterized, with between 40% and 50% of the directly amplified 16S rDNA sequences showing no close homology with 16S rDNA sequences in public databases.

Comparison of the bacterial flora of healthy subjects and that of subjects with periodontal disease showed that some 16S rDNA sequences from culture-resistant bacteria were associated with disease. Subsequently, Sakamoto and colleagues designed PCR assays specific for five distinct 16S rDNA sequences (called *phylotypes*) retrieved from cases of periodontal disease.

In a study of 45 patients with periodontitis and 18 healthy subjects, one particular phylotype, designated AP24, was significantly associated with plaque samples from subjects with periodontitis, suggesting that this phylotype represents the 16S rDNA sequence of a culture-resistant bacterium involved in periodontal disease.

Interestingly, phylotype AP24 showed a close phylogenetic relationship to the clone DA014 reported by Paster et al. in an earlier study of subgingival plaque.

Gastrointestinal Abnormalities

The majority of the estimated 10^{14} bacterial cells that comprise the endogenous microbial flora of humans are found in the lower gastrointestinal tract. Extensive efforts have been made to cultivate the bacteria found in human feces, making the microflora of the lower gastrointestinal tract one of the most widely studied bacterial ecosystems.

Despite a large degree of individual variation, the microflora of the lower gastrointestinal tract consists primarily of strictly anaerobic species coexisting with smaller populations of facultative organisms.

Nevertheless, the recent application of 16S rDNA PCR-based methodologies has demonstrated that between 50% and 90% of the lower gastrointestinal tract microflora is as yet uncultivated. Hence, the human lower gastrointestinal tract contains many undiscovered bacterial lineages.

Molecular studies of the bacterial diversity of the gastrointestinal tract have focused on the microflora of healthy individuals. A recent extensive study of infectious intestinal disease in England determined that microbial etiologies could be established in only approx 45% of cases of apparent gastrointestinal infection. While it is likely that some of the unattributed cases of gastroenteritis represent viral infection, it may well be that some are caused by culture- resistant bacteria. It is noteworthy that *Campylobacter jejuni,* now widely acknowledged as the single most important bacterial cause of human gastroenteritis worldwide, was itself regarded as culture-resistant prior to the 1970s.

It was not until the development of effective isolation techniques, such as filtration and antibiotic-containing selective media, that the true significance of this bacterium became apparent.

More recently, the application of a 16S rDNA PCR assay specific for genus *Campylobacter* detected sequences from a novel *Campylobacter* species in the feces of both healthy humans and those with diarrhea. This novel and initially culture-resistant organism was subsequently isolated and named *C. hominis.*

CONCLUDING REMARKS

The combination of PCR and molecular phylogeny in the form of broadrange 16S rDNA-based analyses has helped to elucidate the nature of several previously uncharacterized culture-resistant bacterial pathogens.

Using these methodologies, clinical bacteriologists now have the means to explore complex bacterial floras, and potentially to detect as yet undescribed pathogens.

The potential usefulness of this and related molecular phylogenetic approaches is considerable, as they offer a new perspective on the diversity of microbial life in human health and disease.

Index

A

B

C

D

E

F

G

H

I

J

K

L

M

N

O

P

Q

R

S

T

U

V

W

Figure 3.2: Model producing conserved synteny in descendant chromosomes B and C after independent inversion steps starting with ancestral chromosome A. Telomeres (/) divide the "genes" (numbered) into four chromosomes. Chromosomes are placed end-to-end before inversions (reversals) are performed to simulate translocation. Two inversion steps on the path leading to B and two inversion steps on the path leading to C are sufficient to generate syntenic relationships.